W9-AZR-841

Fundamentals of Radiation Protection

Fundamentals of

RADIATION
PROTECTION

HUGH F. HENRY

Department of Physics
DePauw University
Greencastle, Indiana

WILEY-INTERSCIENCE

A DIVISION OF JOHN WILEY & SONS

NEW YORK • LONDON • SIDNEY • TORONTO

10 9 8 7 6 5 4 3 2 1

Library of Congress Catalog Card Number: 70-82981
SBN 471 37294 3

Printed in the United States of America

Preface

An individual living during the second half of the twentieth century can hardly avoid being aware of the existence of something called radiation. This awareness, however, should not be confused with knowledge, since for most individuals radiation is something that, even though it may be capable of affecting their own futures, is understandable only to the "scientists." Thus, depending on his most recent source of information, a person may form the opinion that radiation promises an almost utopian tomorrow—or is a danger so great that it menaces any kind of tomorrow. As is almost invariably the case, the actuality for the future lies between these extremes— and understanding of this actuality is itself rather uncertain, clouded as it is by apparently contradictory data and suffering as it does from the voicing of the opinions and recommendations of the whole gamut of political, sociological, economic, and other self-interest groups.

Although the danger from radiation may not be so menacing as painted by some, even as its future may not be so bright as described by others, its hazards have been recognized almost from its introduction in 1895 and, most specifically, since its burgeoning use following the first nuclear fission some 50 years later. One result of this recognition, interest, and concern has been the mounting of a massive research program in the United States and elsewhere to determine the extent of the hazards associated with radiation and radioactive materials, to improve the means of recognizing them, and to evaluate ways by which they may be controlled. It is probable that no other potentially dangerous types of material, particularly those of a toxic nature, have received so much attention as have radiation and its associated problems.

An obvious result of this quite tremendous effort has been a correspondingly huge development of technical and scientific information, much of it applicable to rather highly specialized subfields of the overall topic. Concurrently, groups and organizations that deal routinely with radioactive materials have generally developed their own methods for coping with the potential hazards and problems of their own operations. All too frequently the rules and regulations thus adopted have become part of the general administrative requirements of these organizations, and they may thus reflect sparingly, if at all, the technical bases for their existence.

Since both conditions have resulted in a proliferation of material of particular interest to the specialist, who is thus assumed to have some appropriate background, it has become increasingly difficult for the beginner or someone otherwise interested in the basic premises of radiation protection to find a single source that treats essentially all aspects of the topic in an elementary fashion. Lack of ready access to such information at this level also tends to limit the ability of the public to evaluate properly the perhaps biased claims of the self-seeker and the partisan concerning matters associated with radiation that may be of wide general importance.

Hence this book is designed to introduce the reader not only to the technical aspects of radiation protection, including its physical and biological bases, but also to describe briefly the principles that will guide the implementation of a radiation-protection program and to help the novice understand some of the implications of actions taken with respect to matters of national and international concern as they are presented in the public press and elsewhere. References are provided for those who wish to study specific aspects of the subject more intensively; in fact, I hope that this introduction will so whet the appetite of those who read it that they will wish to delve more deeply into the topics that intrigue them.

Emphasis has been placed throughout the book on observational and experimental information and the conclusions developed therefrom. Hence, quantitative theory has been generally limited to those matters amenable to rather straightforward mathematical treatment at an appropriate level, which is no greater than that of elementary calculus. Even in these cases, an adequate understanding of the items concerned and the conclusions developed theoretically are not necessarily dependent upon a corresponding understanding of the mathematics employed.

Many people have contributed, directly and indirectly, to the development of this book. They include the DePauw University students who have used its earlier versions and have made helpful criticisms, vocally and otherwise; the health physicists whose professional competence, ingenuity, and common

sense have provided much of the intellectual stimulation for the book; and
the operating supervision and personnel of the Oak Ridge Gaseous Diffusion
Plant who, perhaps unwittingly but nonetheless surely, furnished the impetus
for my own professional development in the field. Special thanks are due
to my family who so kindly put up with my pangs of creativity and to the
members of the staff of John Wiley and Sons who have so patiently understood my delays and procrastinations.

If this book in any way helps to place the overall subject of radiation,
its hazards and control, into an appropriate perspective for the casual
reader while at the same time helping to clarify its basic concepts for those
with a more serious interest in the subject, my purpose in preparing it
will have been at least partly fulfilled.

Hugh F. Henry

Greencastle,
Indiana
April 1969

Contents

5 PHYSICAL ASPECTS OF RADIATION EXPOSURE

6 BACKGROUND AND OTHER RADIATION EXPOSURES

21 BENEFITS OF RADIATION

Fundamentals of Radiation Protection

CHAPTER ONE

Introduction

X-RAY DISCOVERY

Although man had been exposed to radiant energy as a part of his environment since before the dawn of history and had studied many of the properties of this radiation, principally that which is visible to the eye, his official acquaintance with what is rather generally classified today as nuclear radiation began with newspaper accounts in early January 1896 of a "sensational scientific discovery." A German physicist, Wilhelm Röntgen (1845–1923) had reported that in his researches with a Crooke's tube he had observed a new form of radiation that was capable of penetrating wood, human flesh, and other opaque objects; these rays he named X-rays. Röntgen first observed the new radiation on November 8, 1895, and by December 28, when he submitted his first preliminary report, he had apparently so thoroughly studied the properties of X-rays that no basically new ones were found for many years.

The new development quickly caught the popular imagination, and its practical aspects were promptly recognized in its worldwide use for medical purposes. In fact the first diagnostic radiograph in the United States was made by Pupin of Columbia University in January 1896 to locate gunshot pellets in a man's hand. The rapid and widespread medical uses of X-rays provide a shining example of the importance of basic research, since it is obvious that no one in his right mind would have been experimenting with a Crooke's tube if he expected to assist a physician in finding gunshot pellets or in setting a broken leg.

DISCOVERY OF NATURAL RADIOACTIVITY

On February 24, 1896, only a few months after Röntgen's announcement, Henri Becquerel (1852–1908), a distinguished French physicist with equally distinguished physicists as father and grandfather, first reported his observations of natural radioactivity from pitchblende, an ore that was considered valuable because it also was associated with gold and silver. His analysis and report were apparently triggered to a certain extent by Röntgen's discovery since he had already noted that, as in the case of X-rays, air would be ionized by "emanations" from this pitchblende ore; this was later found to be because of radiation from the uranium and the minute quantities of radium and polonium in the ore. In July, 1896, Pierre and Marie Curie (1859–1906 and 1867–1934), a husband and wife team in France, announced the concentration of the small fraction of radium in this ore, thus discovering a very highly radioactive material.

By 1899 the great New Zealand physicist, Ernest Rutherford (1871–1937), identified two types of radiation from pitchblende, which he labeled alpha and beta radiation, the latter having been identified at about that time by Becquerel as streams of electrons; these particles had been reported in 1897 by J. J. Thomson (1856–1940), a Scottish physicist. In 1903 the alpha rays were found by Rutherford to be helium nuclei and thus positively charged. Meanwhile Paul Villard (1860–1934), a French physicist, had identified a third type of radiation to be called gamma rays, which were later found to be very penetrating electromagnetic waves and thus similar to X-rays. With these discoveries at about the turn of the century the stage was set for essentially all of the problems and advantages of modern radiation protection concern except for the discovery of another particle, the neutron, which was identified in 1932 by James Chadwick (1891–), an English physicist working with Rutherford.

Although other nuclear particles are well known and their properties have been identified, only alpha, beta, gamma, and neutron radiations are of general importance today in the field of radiation protection, except for special laboratory conditions. Hence no attention is given to these other particles, although all of them are of considerable interest in nuclear structure and one of them, the proton, apparently has some interest from a radiation protection standpoint, primarily in high voltage accelerator applications. The positron, a particle discovered in 1932 by C. D. Anderson (1905–), an American physicist, and found to be very similar to the electron except for its positive electrical charge, is also of some slight significance.

NUCLEAR FISSION

X-Rays and naturally radioactive substances were the principal radiation sources available for general use until the discovery of nuclear fission in 1939 by the German chemists Otto Hahn (1879–1968) and Fritz Strassman (1902–), and the development of the first nuclear reactor prototype in 1942 under the leadership of Enrico Fermi (1905–1954), an Italian physicist. Subsequently the first harnessing of the atom for practical purposes culminated in the first nuclear weapon, which was exploded in 1945. These developments and their subsequent large scale use for both peace and war have not only introduced an exciting nuclear age but have, from the viewpoint of radiation protection, also produced very extensive sources of radiation, primarily from the accompanying large scale production of radioisotopes, both incidentally and by design. These include neutrons as well as alpha, beta, and gamma radiations from a bewildering variety of materials made artificially radioactive.

IDENTIFICATION OF INJURY FROM X-RAYS

As has been mentioned, X-rays were used medically almost immediately after their discovery for a multitude of purposes that involved some necessity for "seeing" the interior of the human body. During such enthusiastic and widespread use there was apparently little or no recognition of the possibility that they could be seriously injurious, although it was fairly quickly recognized that, if used long enough and at sufficiently high intensity, X-rays could produce a burn which looked very much like sunburn. This burn was called *erythema,* and the amount of radiation exposure that was sufficient to cause such skin reddening was called the *erythema dose,* which is now estimated to be about 1800 rad.* Investigators and practitioners quickly became careful to avoid the sufficiently high exposures necessary to produce this particular effect. As a matter of fact, the first actual experimental effort in the United States to determine whether X-rays were injurious was provided by Elihu Thomson, an American, in November 1896 when he exposed one of his fingers to X-rays for several days and noted injury over a week later.

However, within several years other and more unpleasant long-term effects of this new type of irradiation became apparent among those who had used it most extensively. These included cancerous sores of the skin as well as other more deep-seated cancers, principally of bone in the extremities. In fact it is a reasonably fair statement that most of the

* This unit is defined in Chapter 5.

doctors who worked extensively with X-rays in its early days eventually developed these cancers and in many cases suffered amputations of their fingers and hands in an attempt to prevent their spread. It thus became apparent that this scientific marvel was, as is the case of so many things, a good servant but a poor master.

There is today no way of estimating what exposure (doses) these early practitioners received with any degree of accuracy, although there seems little doubt from our present experience that they were in the thousands of roentgens (R), a unit that is used to measure exposure (dose). According to one estimate annual exposures prior to 1930 in the United States could have been greater than 100 R as compared with present averages of much less than 5 R; an estimate of a lifetime exposure of 2000 R was also stated, as was the uncertainty in these estimates. Among the reasons for this uncertainty was the technical fact that the early X-ray tubes were of the cold-cathode variety, similar to the Crooke's tube, which gave useful results only under rather limited ranges of air pressure and voltage that were interdependent. Since the pressure inside the sealed tubes changed with use and with time, wide variations also occurred in the voltages and exposure times and consequently in the total exposures necessary to produce a desired and useful result. It was not until the development in the 1920s of the hot-cathode X-ray tube invented in 1913 by W. D. Coolidge (1873–   ), an American physicist, that easily reproducible exposures could be produced.

Almost all of this early medical work was fluoroscopic, which, because of the comparatively long viewing time and the necessity for the viewer to be at least partly within the direct X-ray beam during exposures, recognizedly results in much higher exposures than does the photographic work that comprises almost all of current work. In addition, each user had his own manipulation techniques, and variations in the respective planned or unplanned shielding had unknown effects on their exposures.

By about the time of World War I enough injuries to those using X-rays had been recognized that there was increasing insistence on information concerning safe exposure limits and methods of determining them. The war itself had two effects of interest. First, the need for information obtainable only with X-rays in treating battle casualties pushed considerations of safety into the background, and, second, more physicians and other individuals became acquainted with the art. Thus the coming of peace in late 1918 permitted further attention to be given to the matter of limiting exposure.

RECOGNITION OF DANGER OF RADIOACTIVE MATERIALS

As was the case with X-rays, the possibility of injury from radioactivity, Becquerel's scientific discovery, apparently was not immediately recognized, although he reportedly received a fairly serious and slowly healing skin ulcer from exposure to the small bit of radioactive material he carried with him.

However, World War I also indirectly pointed to significant hazards from radioactive materials. Involved were the tragic and well-publicized cases of the radium dial painters. These cases involved several girls who were employed in a New Jersey plant to paint various dials with luminescent material containing radium salts. The war had brought on a great demand for such dials for night-time visibility. As was their usual custom with other paints, these girls sharpened their brushes with their tongues and thus inadvertently swallowed minute quantities of the radioactive radium and probably mesothorium. Within a few years some of these dial painters had developed bone cancer at the points of concentration of this radioactive material, and anemia had also been observed. The first fatality eventually attributed to this intake of radioactive material apparently occurred in 1922, although it was 1924 before an intractable osteomyelitis of the mandible of one of the girls was attributed to her previous occupation. By 1931 18 deaths had been attributed to internal deposits of radioactive materials, and the study has been continued until the present. In the past few years additional radium dial painters from this era and later periods have been located and are being closely followed clinically. However, other than those from the early dial painters of World War I described above, there have been few, if any, fatalities that are directly attributable to intake of radioactive material.

It was also at about this time that attention was directed to a previous observation that some of the miners of the Schneeburg and Joachimsthal mines in Czechoslovakia along the Saxony-Bohemia border had suffered an unusually high incidence of lung cancer. These mines had been worked for several centuries for a variety of minerals and were also the source of much of the world's pitchblende ore, this being the ore from which the Curies obtained their radium. A similar comparatively high incidence of lung cancer has also apparently recently been established among the uranium miners of the Colorado Plateau. Parenthetically it may be observed that the Curies were fortunate in that their funds were so limited that the "chemical separations plant" in which they isolated radium was a very open ex-cowshed; in an enclosed room they could easily have inhaled or ingested dangerously high amounts of material. However,

Marie Curie did die of aplastic anemia, a condition that has been connected with external radiation but not internally deposited radium. Pierre died in a traffic accident, being run down by a horse and carriage.

EXPOSURE LIMITS INITIATED

The growing number of serious injuries from X-rays, the fatalities of the radium dial painters, and the apparent high lung cancer incidence among miners of radioactive materials resulted in studies devoted to establishing limits on exposure to the new radioactive materials. It was in the early part of 1920 that the first limits, concerning X-ray exposure, were stated. These were apparently based on estimates, by some of those who had been prominent in the field for several years, of those exposures and exposure levels that would not cause apparent injury. The figures given were subsequently quoted and, appearing in enough places, became the unofficial and official values associated with the art. An increasing instrument technology that permitted a more accurate and consistent dose determination was also instrumental in providing reproducible criteria for measurements necessary to specify exposures.

The first formal national and international groups, the National Committee on Radiation Protection (NCRP) and the International Commission on Radiation Protection (ICRP) established to define safe exposure limits were organized in 1925, and their first official pronouncements were made in 1931, followed by revisions and amplifications in 1934 and 1936. Even at this time only a comparatively very few people were involved in radiation problems, these being principally in medicine. However, by 1942 and subsequently during World War II the number of people affected by radiation had been rapidly expanding, and it is a tribute to the careful work of these committees that there has been no evidence of injury among those whose exposures did not exceed the 1936 figures.

The explosion of the first atomic bomb not only ushered in a new era as far as sources of power and other applications were concerned but also introduced everyone to some of the problems of radiation and radiation protection and affected many people directly in their work as compared with the few individuals who were previously concerned. New problems resulted from the fact that no longer was radioactivity exhibited by only a few of the heavier and rather uncommon materials, such as uranium and radium, but neutron irradiation could also readily impart radioactivity to many of the more common elements, including those that make up an individual's body. Thus, in addition to the augmented problems of external radiation, considerations of internal exposure were significantly broadened and increased. Neutrons themselves, formerly only a laboratory curiosity, were produced in copious quantities and introduced some radiation problems of their own. However, even though the new science produced new sources, many more sources, and more intense sources than had been known previously, their basic hazards and problems were quickly recognized, as were proper protective procedures. Thus no unexpected injurious by-products marred this burgeoning developing adolescence of the nuclear era as was the case with the X-rays and the radium of its infancy.

The increased radiation problem introduced by war and the postwar exploitation of the atom has been accompanied by a concomitantly increased attention to the hazards of radiation. As a part of World War II activities, classified though they were, strenuous and very largely successful efforts as based on the limits prepared in 1936 were made to see that injuries did not occur as a result of the large sources being produced. After cessation of hostilities in 1945 the radiation protection committees again took up their work and in 1948 issued rather sweeping expansions of their previous efforts, as well as appointed subcommittees for more complete coverage of the various aspects of the problem. Subsequently the recommended limits and practices, both new and revisions of previous ones, have been published at irregular intervals.

CRITICALITY CONTROL PROBLEMS

Another and completely new consideration of radiation protection results from the fissionable properties of certain materials, primarily uranium-235 and plutonium-239. Thus "criticality control" or "nuclear safety," terms that are sometimes given to activities designed to prevent accidental nuclear reactions during processing and handling operations, can become serious and important programs. Although there will be radiation only if preventive efforts fail, such failure will result in dangerously high radiation levels, gammas and neutrons being especially significant, and other unfortunate consequences can be anticipated.

Contrary to experience with X-rays and naturally radioactive materials, which resulted in several significant injuries before the accompanying hazards were appreciated, the problems of nuclear safety were recognized even before the first nuclear reactor was built or the first weapon produced.

Accordingly many considerations for radiation safety have been incorporated into both the design of production facilities and all subsequent handling and development operations. These have subsequently been considerably refined so that there have been remarkably few accidents, despite the pell-mell development of the science. Such accidents do, however, present special considerations of radiation protection.

RADIATION PROTECTION ACTIVITIES TODAY

Radiation protection considerations continue today on the technical, political, and international fronts. Research efforts that can only be described as tremendous have been undertaken and are currently in progress to determine various aspects of radiation and its effects on the human body and its various organs. These have included studies of animals by the thousands, theoretical estimates of radiation effects, data from mock-up "human phantoms," and review of human data of all kinds (such as the results of radiation therapy to terminal cancer patients), to name only a few. In addition to observations of actual measurable injuries in people and animals have come estimates and theories of radiation injury. Much attention has also been given to considerations of the genetic effects of radiation on both an individual and his offspring many generations removed.

Almost as important as the generation of data has been its interpretation by various national and international scientific groups and others who have continued determinations and estimates of permissible exposure limits. These have subsequently been published and widely used as bases for action.

Professional organizations have not been omitted. Even in the early X-ray days the American Roentgen Ray Society and the Radiological Society of North America were organized with membership primarily among physicians who used X-rays regularly; similar organizations also sprang up in other nations. During World War II other disciplines, particularly physics, became very strongly involved in radiation and protection therefrom. Hence the term "health physics" has been used essentially as a synonym for radiation protection and those engaged in radiation protection activities identify themselves as health physicists. Their activities culminated in 1955 in the formation of the professional Health Physics Society in America and the establishment of the International Radiation Protection Association some ten years later.

Large research and development organizations have produced more useful and sensitive radiation determination and measuring devices. Methods have been developed for increasing the accuracy with which actual radiation exposures can be measured, and analytical procedures for bioassay determinations of potential exposure have been very carefully refined.

Governmental regulation has also kept pace with the expansion of the problem at the national, international, and, more recently, state and local levels.

EDUCATION AND UNDERSTANDING OF RADIATION PROTECTION

Education in the hazards of radiation has been broadened to the extent that there are probably few adults who are unaware of its existence, although this is where their understanding may cease. Accordingly, the development of a popular understanding of the problems of radiation control has been marked by what Marshall Brucer* calls three periods of hysteria. The first occurred at about the end of World War I and resulted in the specification of limits and the establishment of the various controls to prevent these limits from being exceeded. The second came immediately after World War II when all forms of radiation and reactors were unfortunately equated in the lay mind with the atomic bomb; conditions at this time were not helped by irresponsible predictions of doom which, if not malicious, were certainly uninformed. The third period of hysteria began developing in about 1959 and 1960 over the reported problem of fallout. This, too, had its genesis, to some extent at least, in some rather irresponsible interpretation of experimental data or unwarranted extrapolation of theory and subsequent wide dissemination to the public press. In all of these the possibility that the deep concern of many dedicated scientists to ensure that no one could possibly be harmed by this new development and their feeling of responsibility for the uninitiated may have resulted in unnecessarily conservative conclusions or have been misinterpreted.

TWO HYPOTHESES—ONE SET OF DATA

In the widespread discussion that has arisen in this field two basic concepts of the possibility of injury have been clearly enunciated. Both of these recognize that there are radiation exposure rates which are sufficiently low as not to produce any indication of an effect, much less one that is actually injurious, even though the exposure may be continued throughout life. One of these concepts is the so-called *linear hypothesis,* which in essence states that all radiation, no matter how small, is unqualifiedly harmful, even though the effect is unmeasurable. The other is the so-called *threshold hypothesis,* which takes the position that at levels of radiation exposure

* Marshall Brucer, *Jour. Amer. Medical Society,* **176**, p. 680, 1961.

below some threshold value, the effects of the radiation, if any, are not necessarily harmful.

The concept of a *permissible* exposure level, as based on the threshold hypothesis, is obvious. As based on the linear hypothesis, the concept assumes that the benefit to be derived from the radiation exposure at least compensates for the injury—recognized or unrecognized, real or imaginary—that has been received; for example, essentially all medical radiation exposures are considered justifiable, and no attention is given to the effects of background radiation despite the fact that exposure levels may vary by several factors from place to place on the earth's surface. In fact such background is frequently considered necessary to the evolutionary process and thus justifiable on that basis alone.

It is perhaps fair to conclude this brief analysis of conflicting opinions by pointing out that (a) the experimental evidence does not give a clear-cut justification for either position, (b) both are extrapolations of the same data, and (c) individual opinion will thus unavoidably be involved in any interpretation of protection requirements. Hence, those who are interested in the concepts of radiation protection for general information, as well as those who are interested in the subject professionally, are well warned to recognize the limitations of present knowledge and, insofar as possible, to recognize and avoid the substitution of opinion for fact.

REFERENCES and SUGGESTED READINGS

Most histories of science, particularly of physics, contain sections on this subject.

Shamos, Morris H., ed., *Great Experiments in Physics,* Holt, New York, 1959.

Taylor, L. W., *Physics, the Pioneer Science,* Houghton-Mifflin, Boston, 1941. (Reprinted by Dover, New York, 1963.)

Martland, Harrison S., *Collection of Reprints on Radium Poisoning,* AECD–2122, USAEC, Washington, D.C. 1943. Papers originally appeared in medical journals in the period 1920–1940.

Taylor, L. S., *Health Physics,* **1,** No. 1, 3 (June 1958).

Taylor, L. S., *Health Physics,* **1,** No. 2, 97 (September 1958).

Kathern, R. L., *Health Physics,* **8,** No. 5, 503 (October 1962).

QUESTIONS

1. What part did each of the following people play in the unfolding drama of the history of radiation physics: Becquerel, Pierre and Marie Curie, Rutherford, Röntgen, Thompson, Villard, Chadwick, Anderson, Hahn, Strassman, Fermi, Coolidge, radium dial painters?

2. What types of radiation are of importance in considering general radiation protection?

3. In 1939 what event occurred that added another dimension to radiation physics?
4. Both World War I and World War II were instrumental in drastically different ways in influencing investigations in radiation protection. Explain these outcomes of the wars and indicate their significance.
5. What is implied in the term "health physics"?
6. Is the understanding of radiation effects by most people indicative of the scientific data resulting from research done in the field of radiation physics?
7. Explain the difference between the linear and threshold hypotheses. The validity of which one impresses you most? Why?
8. In the tragic cases concerning the radium dial painters death was attributed to the intake of minute quantities of radium. Marie Curie also died as a direct result of her extensive contact with radiation. In what way was the cause of Marie Curie's death different from that of the radium dial painters?
9. According to Brucer there have been three major periods of hysteria. What do you think the fourth period, if any, will be?
10. How were the discoveries of X-rays and natural radioactivity related?

CHAPTER TWO

Atomic Structure

ATOMS AND MOLECULES

Unless one is to treat radiation protection as a blindly empirical art, some understanding of the origin of radiation and its effects is indicated. This then, implies some knowledge of the general structure of matter as it is now viewed by the physicist or chemist, that of living systems as proposed by the biologist, and the blurring of these disciplinary distinctions in the general study of biological systems. The next several chapters attempt to bring together and highlight some of the more significant concepts of the structure of matter and radiation effects thereon for review and reference.

In general present concepts of the structure of matter are rather loosely considered under the general title of "atomic theory," which assumes as a basic concept that matter is not infinitely divisible but is made up of small, discrete, and indivisible particles identified as atoms or, later, molecules. Democritus, an early Greek philosopher, is frequently given credit for enunciating this atomic concept; actually, his conclusion had nothing to do with measurement or observation beyond that of the existence of matter itself but was merely a philosophical approach to a paradox proposed by Zeno. An English chemist, James Dalton (1766–1844), is generally credited with the first scientific recognition of these atoms as necessary to an explanation of observed chemical phenomena.

By definition the molecule is the smallest unit of any chemically homogeneous substance or compound which has the chemical properties of that substance. A substance that cannot be chemically subdivided is called an element, and the atom is defined as the unit constituent of an element. Thus molecules generally consist of several different atoms tightly bound together; the molecule of an element may be a single atom or it may be several identical atoms. Obviously, a useful amount of a substance itself consists of a large number of these individual molecules more or less closely attached.

Most chemical and physical properties of a given substance which is not an element depend on the characteristics of its molecules, the characteristics of the individual atoms being submerged. However, early studies with naturally radioactive materials showed that their radioactive properties were independent of their chemical constitution or of the chemical compound of which they were a part. Hence it was concluded, and this was later verified, that radioactivity did not depend on any molecular factor but was some function of the atomic structure.

THE PLANETARY ATOM

For most accurate representation, the theory of atomic structure requires rather complicated mathematical symbolism and treatment. However, an adequate appreciation of the basic factors necessary for radiation study may be gained from a semiqualitative treatment based on a planetary model suggested by Niels Bohr (1885–1963), a Danish physicist; this in turn was a modification of an atom model earlier proposed by Rutherford. The Rutherford-Bohr atom still probably lends itself most easily to an elementary representation and interpretation.

According to this model the atom is considered to be a planetary system composed of three basic elementary particles—the proton, the neutron, and the electron. The protons and neutrons are contained in a very small nucleus about which the electrons travel in planetary orbits. A simple nuclear model of this type is indicated in Figure 2–1.

The dimensions of an atom are not particularly clear-cut entities, especially in the case of the nucleus, because the values obtained depend on the method of measurement. However, from "close-packing" estimates as well as other determinations, the diameters of the outer electron orbits of an atom, which thus represent the maximum atomic diameter, are on the order of 10^{-10} m. The nuclear diameter, primarily as based on measurements with electrically charged particles, is commonly accepted as being on the order of 10^{-15} m. It should be understood that for certain phenomena, important among them being the capture of neutrons as discussed later, there is no real meaning to the concept of a single, simple nuclear diameter since nuclear effects are apparently exerted over different volumes

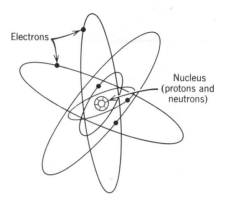

Figure 2–1 Schematic representation of atom.

and distances, which depend on the phenomenon studied, the incident-neutron energy, etc.

Each proton or neutron has a mass of about 1.7×10^{-27} kg, whereas the mass of the electron is about 0.05 percent as much, or about 9×10^{-31} kg. The neutron is electrically neutral, but each proton has a positive electrical charge of about 1.6×10^{-19} coulombs*; each electron bears the same quantity of charge, but it is negative. The atom in its normal state is electrically neutral; thus each atom has the same number of protons and electrons.

DETERMINANTS OF ATOM CHARACTERISTICS

Since the mass of the electron is negligible compared to that of either the proton or the neutron, essentially all of the mass of an atom is contained in its nucleus. The mass of a proton or neutron may be expressed by a convenient and frequently used unit called the atomic mass unit (a.m.u.); the atomic mass of an element, which is equal to the sum of the number of protons and neutrons per atom, is usually given in atomic mass units. Obviously, the total number of these elementary particles in the various nuclei determine the relative masses of the various atoms, and thus the relative atomic masses of the elements concerned.

Of perhaps more importance than the atomic mass of an element is its atomic number, which identifies the number of protons per atom and also the number of electrons in a neutral atom; this characteristic determines the chemical properties of the element concerned.

---

* The coulomb (C) is a unit of electrical charge.

In nuclear terminology the structure of a given element with the chemical symbol N is usually defined in terms of its atomic number Z and its atomic mass A by the relation $_Z^A N$. Thus carbon with an atomic number of 6 and an atomic mass of 12 is identified as $_6^{12}C$. Similarly, a specification of sodium as $_{11}^{23}Na$ means that the sodium atom is made up of a nucleus with 11 protons and 12 neutrons plus 11 planetary electrons. The atomic number Z is sometimes omitted in atomic designations, the chemical symbol being considered adequate to specify the atomic number, and a widely used notation scheme frequently found in the literature uses $_Z N^A$. Thus $_6C^{12}$, C^{12}, ^{12}C, or C–12 (or carbon-12) are considered equivalent to $_6^{12}C$; and $_{11}Na^{23}$, Na^{23}, ^{23}Na, or Na–23 (or sodium-23) are equivalent to $_{11}^{23}Na$.

With respect to nuclear radiation it has been observed that, with the exception of the production of X-rays, the source of radiation is the nucleus of the atom; hence the *nuclear* properties of various atoms are the ones of prime interest in the study of radiation.

ISOTOPES AND NUCLIDES

The isotopes of a given element are defined as materials with the same chemical properties but different atomic masses; the respective nuclei of such isotopes thus contain the same number of protons (and electrons in neutral atoms) but different numbers of neutrons. Uranium with an atomic number of 92 has two important isotopes with respective atomic masses of 235 and 238; nuclearly they would be identified as $_{92}^{235}U$ and $_{92}^{238}U$ (or U^{235} and U^{238}, or, also, U–235 and U–238). Radioactive isotopes are usually called radioisotopes. Most elements are actually mixtures of several isotopes, and their atomic masses as given in the periodic tables and elsewhere are weighted averages of the individual atomic masses of their respective components. Thus natural chlorine with an atomic mass of about 35.5 is composed of $_{17}^{35}Cl$ and $_{17}^{37}Cl$ in a constant proportion of about 3: 1.

The term "nuclide" identifies a species of atom that is characterized by the number of protons and neutrons in its nucleus, with the energy state also being implied. If the element is radioactive, the corresponding atom is called a radionuclide. Although the term "nuclide" is often used interchangeably with "isotope," it is a somewhat more inclusive expression.

ENERGY MANIFESTATIONS

At this point it is instructive as well as physically correct to consider an atom, its components, and changes involving them as different manifestations of energy;* work done, in this sense, is also considered as an energy manifestation. The Law of Conservation of Energy, which in essence states that energy can be neither created nor destroyed, is considered to be one of the most basic laws of the physical universe. Thus in the atomic and nuclear sphere electrons and other moving particles obviously have kinetic energy of translation, whereas electromagnetic waves, such as gammas or light, obviously have energy that is characteristic of wave motion. Electrical charge, and even mass as developed from relativity theory and confirmed experimentally, are also manifestations of energy. Under proper conditions these various forms of energy may be transformed from one into another. Thus the potential energy of water particles at the top of a dam becomes their mechanical kinetic energy at the bottom of their fall; this kinetic energy then becomes the kinetic energy of the rotating turbine, and this is then transformed into electrical energy which, in turn, can be transformed into heat, light, or other forms of energy. Certain forms of energy of value in subsequent sections are mentioned below.

Electrical energy E is usually simply expressed as the product of a charge Q and the potential difference V through which it is moved. Thus

$$E = VQ.$$

If V is given in volts** and Q in coulombs** (C), E is in joules (J). A convenient electrical unit is the electron volt (eV), which is the energy gained by an electron moved through a 1-volt potential difference. Since the electron charge is 1.6×10^{-19} C, an electron-volt is 1.6×10^{-19} J, or 1.6×10^{-12} erg. In many cases energies are expressed in million electron-volts (MeV). Thus 1 MeV $= 1.6 \times 10^{-13}$ J.

The kinetic energy E of a mass m traveling with a velocity v is usually given by the relation

$$E = \tfrac{1}{2}\ mv^2,$$

where E is in joules when m is in kilograms and v is in meters/sec. Relativistic determinations show that an object cannot attain the velocity

---

* An elementary definition of energy is the ability to do work. Its basic subdivisions are potential energy, which is energy because of position or state of being; and kinetic energy, which is that associated with motion. Certain other aspects of energy that are of importance nuclearly are described in subsequent sections. The erg and the joule (which is 10^7 ergs), as well as the foot-pound, are units of energy defined in terms of mechanical measurements, whereas the calorie and the British thermal unit (BTU) are energy units defined in terms of heat measurement.

** The volt and the coulomb are, respectively, units of electrical potential difference and of charge.

of light, c, which is 3×10^8 m/sec because of the relativistic increase in its mass, given by the expression

$$m = \frac{m_0}{\left(1 - \dfrac{v^2}{c^2}\right)^{\frac{1}{2}}},$$

where m_0 is the rest mass. The relativistic kinetic energy for a mass in motion with a velocity near that of light is given by

$$E = (m - m_0)c^2.$$

It may be shown that this expression for kinetic energy reduces to the value of $E = \frac{1}{2} m_0 v^2$ for velocities much lower than that of light. A very important conclusion developed from these relativistic determinations is the statement that the mass-energy E of a given rest mass m_0 is given by the famous expression

$$E = m_0 c^2,$$

where c is the velocity of light.

Obviously, if m_0 is in kilograms and c is given in meters/sec., E is also in joules. It may readily be shown that the rest mass of an electron, which is 9×10^{-31} kg, is equivalent to about 0.51 MeV, or about 8×10^{-14} J.

The energy E of a photon (see subsequent section on "Wave-Particle Duality") is given by the relation

$$E = h\nu,$$

where ν is the wave frequency and h is Planck's constant; E is in joules where ν is in vibrations/sec. and h is 6.62×10^{-34} J-sec. Thus, 1 eV, which is equivalent to 1.6×10^{-19} J, is also equivalent to a photon of a frequency of 2.4×10^{14} vibrations/sec and 1 MeV to a photon with a vibration frequency of 2.4×10^{20} vibrations/sec. The wavelength λ is related to the wave frequency by the relation

$$\nu \lambda = c,$$

where c is the velocity of light (3×10^8 m/sec, or 3×10^{10} cm/sec). Hence the respective wavelengths of the above photons are 1.25×10^{-6} and 1.25×10^{-12} m. Wavelengths are also expressed in millimicrons (1 mμ = 10^{-9} m) and angstrom units (1 Å = 10^{-10} m). Hence a wavelength of 1.25×10^{-6} m = 12,500 Å, or 1250 mμ.

WAVE-PARTICLE DUALITY

A major success of modern physical analysis, which has been implied above, is the observation of the wave-particle duality of the various forms of energy and matter. This may be anticipated since energy is transferred only by motion of particles or waves. The decision of whether light energy is transferred as a wave motion or as a motion of particles was a moot question for many years. Then in 1809 Thomas Young (1773–1829), an English physician and scientist, showed that the phenomenon of interference could be explained only by the concept that light was a wave motion. A few years later, in 1815, a French military engineer, Augustin Fresnel (1788–1827), reported his diffraction experiments as similarly being explicable only on the basis that light was a wave motion. Since particulate motion does not exhibit these characteristics, they have become identifying criteria for waves.

For about a century the fact that light was a wave motion was widely accepted in the scientific community. However, in 1905 the German physicist Albert Einstein (1879–1955), while an examiner in the Swiss patent office, produced a theory to explain the photoelectric effect which had previously been experimentally observed by Philipp Lenard (1862–1947), a German physicist. Although this phenomenon is inexplicable on the basis that light is a wave motion, it is readily explained by the concept that light also consists of particles, or photons, with an energy $E = h\nu$ as has been already described. It has also been indicated that under proper conditions other electromagnetic waves such as X-rays or gammas exhibit the properties of particles with specific amounts of energy; they are called quanta or photons. The wave-particle energy relationship, $E = h\nu$, is thus a basic quantitative link between a wave and its corresponding particle manifestation.

However, it was not until 1927 that a corresponding wave-particle duality of recognized particles was shown. In that year the American physicists C. J. Davisson (1881–1958) and L. H. Germer (1896–   ) showed that electrons also exhibit the characteristic of diffraction that is a criterion of wave motion. Hence it has been shown that electrons and other energy units which are ordinarily considered as being particles do similarly exhibit the characteristics of waves. It is because of this characteristic that electron microscopes are possible. It has been observed that the energy relationship, $E = h\nu$, is a fundamental quantitative connection between an electromagnetic wave and its corresponding particle manifestation. The other basic connecting link between particles and waves was the result of a suggestion in 1924 by Louis de Broglie, (1892–   ), the French physicist, that the momentum of the wave-particle, expressed as mv for a particle, was equivalent to h/λ, where λ is the wavelength of the associated wave and h is again Planck's constant. This wave-particle duality is at the heart of modern quantum mechanics, and recognition of this duality is useful in an understanding of some of the basic characteristics of radiation, even though it is not particularly necessary to an adequate practical understanding of the radioactive properties of materials.

QUANTIZATION LIMITATIONS

To return to the atom, the Bohr theory assumes that the planetary electrons apparently travel about the nucleus only in certain "permitted" orbits that are controlled by specific *quantization limitations*. In general, the angular momentum of an electron in one of these orbits is quantized (i.e., it is an integral multiple of a given constant), no two electrons can have exactly the same angular momentum and thus be in the same permitted orbit, and each electron takes the lowest energy orbit available to it.

Some of the more important basic principles of atomic structure may be illustrated by reference to the hydrogen atom, $_1^1H$, which, being composed of only one proton and one neutron, is the simplest available atom; in fact it was the success of the Bohr model in explaining some observations made with hydrogen that established the validity of his model. In this atom there are several permitted orbits for the single electron. It may be readily shown from electrostatics and mechanics that an electron's energy is determined by the position of its orbit—the further the electron is from the nuclear proton, the greater is its energy. Thus in the hydrogen atom's rest state, the electron is in its permitted orbit nearest the nucleus, where its energy is the least. Obviously, since only certain orbits are permitted, the electron can have only certain specific energies.

ATOMIC EXCITATION

If a comparatively small amount of energy is added to the atom (e.g., by heat or electrical excitation), it is apparently absorbed by the electron, which may then possess too much energy to stay in its lowest permitted orbit and will jump to a higher energy orbit, which in this model is an orbit further from the nucleus; the atom is then said to be excited. If sufficient energy is added, this electron may be removed completely from the remainder of the atom. This process is called ionization, the atom is said to be ionized, and two charged particles called ions are produced; one, the removed electron, is a negative ion, and the remainder of the atom is a positive ion. Subsequently, the electron in the excited atom may drop back into an orbit nearer the nucleus and emit the excess energy as radiation, the frequency of which is proportional to its energy change.

Obviously, a similar result is attained when the positive ion captures an electron.

The result of such emissions by a large number of excited atoms produces what is called an optical spectrum, which includes electromagnetic radiation in the visible, ultraviolet, and infrared ranges. This particular type of optical spectrum is called a line spectrum since only definite light frequencies are emitted as a result of the energy changes between the specific permitted orbits. The excited atoms of a given element thus emit a characteristic spectrum of which the red glow from neon lights is a typical example.

HEAVY ATOMS

The simplified model based on the hydrogen atom as described above may be extended qualitatively to elements that have many electrons. In this case the orbits of the individual electrons are arranged in *shells,* each of which contains several permissible electron-orbit positions. The number of electrons permitted in each shell is limited, and it is this shell structure which determines the periodicity of the elements, the first specification of which is attributed to the Russian chemist Dmitri Mendeleev (1834–1917). In general, each electron goes into the available position in which it has the least energy, the only deterrent to its gaining the shell nearest the nucleus or the lowest permitted orbit of a more distant shell being the fact that another electron already occupies the lower energy position. The energy differences between the different orbits of a single shell are very small compared with the average energy differences between the shells themselves. Similarly the energy difference between the two inner shells for some of the elements with low atomic numbers is much less than the corresponding energy difference between such shells for the metals and other materials of higher atomic numbers.

The electrons in the outermost shell are usually identified as *valence electrons,* and available evidence shows that atoms are held together to form molecules by interactions between their respective valence electrons; theoretical interpretations of these interactions form the basis of the chemical-bond approach to chemistry. For a given atom it is usually the electron in the position of highest energy that is excited to form optical spectra; however, such spectra may also result from excitation of more than one of the valence electrons.

If a comparatively large amount of energy is added to an atom, an electron may be removed from one of its inner shells. Its replacement by an electron falling from an outer shell results in the emission of a comparatively much greater amount of energy than is involved in the orbital changes of the valence electron, and the frequency of the resulting electromagnetic radiation (or photon emitted) is also correspondingly much higher. Such emission is in what is known as the X-ray range of frequency or wavelength.

Since X-rays are produced by bombarding metals with high-speed electrons, an accepted explanation for some of the frequencies produced is the impact removal of electrons from the inner shells and their consequent replacement by outer shell electrons. The energy lost by the electrons making this change appears as electromagnetic radiation with a frequency that is determined by the energy differences of the shells concerned. This emission results in the so-called X-ray *line spectra,* which refers to X-rays of certain specific frequencies that are determined by the target material and the accelerating voltage on the X-ray tube. Other X-rays in a range of frequencies are also observed, these forming the so-called continuous X-ray spectrum. These frequencies are produced merely by the slowing down of the incident electrons as a result of collisions with the atoms themselves (really deceleration as a result of electrostatic repulsion between the electron and the planetary electrons of the atom) and a consequent emission of electromagnetic energy by other mechanisms.

X-RAY SERIES

The highest X-ray frequency in a given line spectrum results from an electron coming to the vacated shell from a position at "infinity," which is a position at the limiting range of influence of the given atom. Electrons dropping from an outer shell to an inner one produce lower frequency radiation. Figure 2–2 shows a typical X-ray spectrum with two lines superimposed on a continuous component. Figure 2–3 indicates how several line-spectra frequencies may be produced. Even before the shell structure was proposed, the existence of these X-ray lines was recognized and the various series labeled. These have subsequently been identified with the atomic shells. An electron in the K-shell has the lowest energy, one in the L-shell next, and so on. If an electron is removed from the K-shell, its replacement results in lines in the K-series. Similarly, L-series lines are produced by replacement of an L-shell electron. Since the energy differences for electrons in the various orbits of a given shell are small compared with the energy difference between shells, neither the initial nor the final orbital position of the electron producing a given X-ray frequency is significantly important.

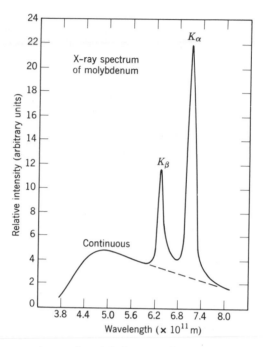

Figure 2–2 X-ray spectrum of molybolenum. (See Harnwell and Livingood, *Experimental Atomic Physics,* 1933, p. 360.)

X-RAY PRODUCTION

In practice X-rays are produced by accelerating electrons through a high potential and then allowing them to impinge on a target that is usually made of some of the heavier metals. Figure 2–4 is a diagram of a typical X-ray tube and associated electrical circuits. The cathode is heated so that electrons are "boiled" off and are accelerated to strike the anode target. If the potential difference between the cathode and anode is 50 kV, the electrons obviously have an energy of 50,000 eV (electron-volts) as they reach the target. It may thus be seen that the maximum X-ray frequency, which results from all of the energy of one electron being transformed into a photon, is 1.2×10^{19} vibrations/sec. Since the X-ray wavelength λ is related to the frequency ν and the velocity of light c by the relation

$$\nu\lambda = c,$$

it may be seen that the shortest wavelength that these X-rays can have is about 2.5×10^{-9} cm, or 0.25 Å. However, the actual wavelengths emitted in the line spectrum are normally greater than this value since they depend on the target (anode) characteristics as already described, and the continuous spectrum exhibits a wide range of frequencies down to very low ones. In practice some of these lower frequency waves are eliminated by shielding, but there still remains a rather broad spectrum of frequencies in the X-ray beam.

Other than X-ray production, all of the radiation phenomena of interest involve nuclear effects, and they are thus considered as resulting from changes in the energy of the nucleus.

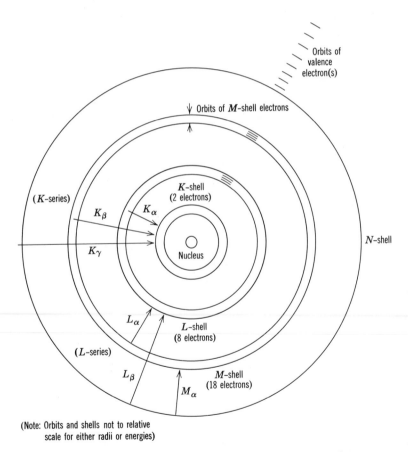

(Note: Orbits and shells not to relative scale for either radii or energies)

Figure 2–3 Electron orbital shells. Shell "thickenesses" represent the energy range of shell electrons. In this diagram it is assumed that the planetary electrons with the greatest energy (smallest negative energy) are in the N-shell.

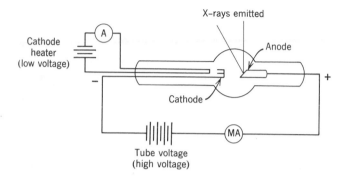

X-rays emitted

Cathode
heater
(low voltage)

Anode

Cathode

Tube voltage
(high voltage)

Figure 2–4 Schematic of typical X-ray tube and associated electrical circuits.

ELEMENTARY BOHR THEORY

Author's Note: This mathematical section may be omitted without disturbing the continuity of the text. However, for those with an elementary physics background, it may give a better appreciation of the way in which the various classical concepts of particle physics mesh with its earliest quantum aspects. However, this is still a far cry from formulations and postulates of the more recent quantum physics that have been developed from wave-mechanical considerations and which provide better mathematical interpretations of experimental data. Obviously, the wave-particle concepts of radiation and matter are indicated.

The elementary laws of electrostatics and mechanics may be used along with Bohr's hypotheses to express quantitatively the radiation frequencies emitted in optical and X-ray spectra. If Z protons make up the nucleus, then its charge is $+Ze$. Now if it is assumed that the valence electron is the atom's only planetary electron and it is circling the nucleus at a distance r, then the force of attraction between this electron and the nucleus when separated by that distance is

$$F = - \frac{Ze^2}{4\pi\epsilon_0 r^2},$$

which, assuming an attractive force is positive, takes account of the fact that electrostatic attraction exists only between unlike charges; ϵ_0 is a constant.

However, for an electron of mass m moving at a velocity v in an orbit with a radius r the electrostatic centripetal force above is balanced by a mechanical reaction or "centrifugal" force given by

$$F = \frac{mv^2}{r}.$$

Since the sum of these forces is zero for the electron to remain in orbit,

$$\left(\frac{mv^2}{r}\right) + \left(-\frac{Ze^2}{4\pi\epsilon_0 r^2}\right) = 0$$

or

$$\frac{mv^2}{r} = \frac{Ze^2}{4\pi\epsilon_0 r^2}.$$

Thus

$$mv^2 = \left(\frac{Ze^2}{4\pi\epsilon_0 r}\right).$$

Since the kinetic energy KE of a mass m is $\frac{1}{2} mv^2$, the kinetic energy of this electron is thus

$$KE = \frac{1}{2}mv^2 = \frac{1}{2}\frac{Ze^2}{4\pi\epsilon_0 r}.$$

The potential energy of the electron at a distance r from the nucleus is obtained from the simple relationship that the work done (and the potential energy PE gained) in moving the electron to this position from a point at a very great distance from the nucleus (assumed infinite so its potential energy is zero with respect to this particular nucleus) is given by the expression

$$PE = -\int_{\infty}^{r} F\, dr = -\int_{\infty}^{r}\left(-\frac{Ze^2}{4\pi\epsilon_0 r^2}\right) dr$$

because a force of attraction would be in the same direction as the movement, which is toward the nucleus and thus in the direction of decreasing distance r. The solution of this equation shows that

$$PE = -\frac{Ze^2}{4\pi\epsilon_0 r}.$$

Since the total energy E of the electron is the sum of its potential and kinetic energies,

$$E = PE + KE = -\frac{Ze^2}{4\pi\epsilon_0 r} + \frac{1}{2}\frac{Ze^2}{4\pi\epsilon_0 r} = -\frac{Ze^2}{8\pi\epsilon_0 r}.$$

Thus the energy of an electron in an outer orbit, E_2, with a radius r_2 is

$E_2 = -\dfrac{Ze^2}{8\pi\epsilon_0 r_2}$, and its energy E_1 in an inner orbit with radius r_1 is

$E_1 = -\dfrac{Ze^2}{8\pi\epsilon_0 r_1}$. Hence the energy difference

$$E_2 - E_1 = \frac{Ze^2}{8\pi\epsilon_0 r_1} - \frac{Ze^2}{8\pi\epsilon_0 r_2} = \frac{Ze^2}{8\pi\epsilon_0}\left(\frac{1}{r_1} - \frac{1}{r_2}\right).$$

Now Bohr's first hypothesis states that there are permissible electron orbits, and his second one declares that these permissible electron orbits are only those for which the orbital angular momentum of the electron, mvr, is equal to some integral multiple of $h/2\pi$, where h is Planck's constant. This constant had been determined by the German physicist Max Planck (1858–1947) in accounting for the radiant energy spectrum of a hot body. Thus

$$mvr = \frac{nh}{2\pi},$$

where n is an integer. From this relation and the previous determination that $mv^2 = \dfrac{Ze^2}{4\pi\epsilon_0 r}$ it may be shown that the radius of a permissible orbit is

$$r = \frac{\epsilon_0 n^2 h^2}{\pi m Z e^2},$$

or

$$\frac{1}{r} = \frac{\pi m Z e^2}{\epsilon_0 n^2 h^2}.$$

Similarly

$$E = -\frac{Ze^2}{8\pi\epsilon_0 r} = -\frac{mZ^2 e^4}{8\epsilon_0{}^2 n^2 h^2}.$$

Bohr's third hypothesis states that a quantum of energy, or a photon, is emitted as an electron drops from an outer orbit with an energy E_2 to an inner one with an energy E_1, and the electromagnetic wave frequency that is associated with this emitted photon may be obtained from

$$E = E_2 - E_1 = h\nu = \frac{mZ^2 e^4}{8\epsilon_0{}^2 h^2}\left(\frac{1}{n_1{}^2} - \frac{1}{n_2{}^2}\right).$$

For a given series of optical spectra n_1 has a given value and n_2 can have any integral value of $(n_1 + 1)$ or greater, the frequency corresponding to each value of n_2 thus representing a specific spectral line.

It was the success of this relation in explaining prior empirical observations of the relationships between the frequencies in various series of optical spectra, particularly hydrogen, that placed the Bohr atom on firm ground.

However, Bohr's first two hypotheses, introducing as they did the very important quantum concept, did require arbitrary assumptions. It was not long before this arbitrariness was justified by expressing these orbital concepts mathematically in terms of standing waves. Although such treatment is not susceptible to simple mechanical representation, it does account for the existence of permitted orbits and their quantum specification. In fact modern quantum mechanics is actually and successfully expressed almost entirely by wave equations.

For elements other than hydrogen the pictured structure rapidly becomes much more complex. In particular, although the nuclear charge is actually Ze, the inner shell and orbital electrons with their negative charges at least partially shield this positively charged nucleus from the valence electron, with the result that the equivalent nuclear charge for this valence electron is indeterminate but with a value not far from that of a single proton. This is usually expressed by saying that Z has a value not far from unity.

A relation similar to that of Bohr was very successfully applied to X-ray spectra. In this case the Z of Bohr's equation is replaced by an expression $(Z - \sigma)$, where σ is a so-called screening constant. The dependence of X-ray frequencies on Z for the respective spectral series had been established very early by the English physicist H. G. J. Moseley (1888–1915), who had observed, equally empirically, that σ was very nearly a constant for a wide range of the heavy elements of interest as X-ray-producing targets.

SUGGESTED READING

Almost any text in the field of modern physics covers the items reviewed in this chapter.

QUESTIONS

1. The eye can see light with wavelengths in the range of 4.2×10^{-5} to 7.1×10^{-5} cm (violet and red, respectively). What are the frequencies of the light of these two wavelengths?
2. What are the energies of the respective photons for light of the two wavelengths of Problem 1?
3. *a.* If an electron were to have the energy of a red light photon, through what voltage must it have been accelerated?
   b. What would have been the accelerating voltage for an electron gaining the energy of a violet light photon?
4. Determine the energy of a proton accelerated through 10 volts.
5. An electron is accelerated to an energy of 1 MeV. Determine its relativistic mass.
6. What is the velocity of a 1 MeV electron?
7. Assume you used the relation $E = \frac{1}{2} mv^2$ to obtain the velocity of a 1 MeV electron. What value do you obtain and why do you know it is incorrect?
8. Repeat Problems 5 and 6 but obtain values for the mass and velocity of a 1 MeV proton.

9. A thermal neutron has an energy of about 0.03 eV. What is its average velocity?
10. Express 50,000 eV in terms of joules.
11. In any chemical compound, what is the smallest unit that determines the characteristics of the compound?
12. In determining the mass of an atom (or molecule) why is the mass of the electron (9×10^{-31} kg) said to be negligible?
13. Given the information that $Z + A = 14$ and the total negative charge $= 9.6 \times 10^{-19}$ C, what element or isotope is thus characterized?
14. How many joules does the rest mass of an electron represent?
15. Define and differentiate between atoms, molecules, and elements.
16. What would be the relativistic mass of a body assuming its velocity could equal that of light?
17. What basic idea in radiation physics and quantum mechanics explains *line spectra?*
18. How is the wave-particle duality resolved?
19. How does the present theory of X-ray production account for the characteristic radiation pattern of the anode target?

CHAPTER THREE

Nuclear Transformations, Radioactivity, and Fission

NUCLEAR STABILITY

It is not at all clear how the transformations that result in the emission of radiation are caused in the nucleus or even how the various components of the stable nuclei manage to hold together as units, although such combinations obviously involve some form of energy. This so-called "binding energy" is thought to be closely connected with the fact that the actual atomic mass of a given nucleus is generally slightly less than the sum of the masses of the neutrons and protons making up the nucleus. Since mass is a form of energy, this mass defect is considered to be connected in some way with the binding energy. Certainly, such a condition implies that energy from some source must be supplied to the nucleus to permit separation of its constituent particles.

In addition to protons and neutrons, a rather large and bewildering variety of particles has recently been observed to be emitted from various nuclei as a result of bombardment by high energy particles. It is thus thought that various of these mesons and other nuclear particles, sometimes called *strange particles,* may have an important role to play in the "cement" which holds the nucleus together. Today nuclear exploration, particularly with high energy particles, is one of the most active fields in physics.

TRANSFORMATIONS OF NATURALLY RADIOACTIVE MATERIALS

In accord with some aspects of current theories the nucleus of an atom may be considered to have preferred positions of energy distribution in some of which it is in stable configuration, whereas in others the configurations are less stable or unstable. When the nucleus is in an unstable configuration it eventually goes to a more stable state with the release of some energy. This is identified as radioactive decay, or a radioactive transformation, and materials composed of atoms with nuclei in unstable configurations are called radioactive materials. The time required for this transformation to a more stable configuration by a given nucleus apparently involves statistical considerations; hence for a large number of atoms in a specific unstable configuration a given fraction will be transformed in some time unit. This is discussed in more detail in a subsequent section as is the fact that a given transformation may not result in a stable configuration but in a different unstable one and a succeeding transformation; this process, repeated through a succession of steps, results in a radioactive chain. In those radioactive materials that occur naturally, uranium and thorium being the most abundant thereof, each nucleus is thus originally in an unstable configuration, and transformations to more stable conditions occur spontaneously. Similar transformations occur in those elements which have been made radioactive artificially.

The first energy change in the transformation of the naturally radioactive uranium and thorium chains results in the release of an alpha particle, which is really a helium nucleus consisting of two neutrons and two protons. Subsequent transformations in these chains also involve beta particles which are electrons, and gamma rays which are electromagnetic waves or photons. Since all of these are forms of energy, the energy lost by a nucleus in the emission of a particle, such as an electron, includes not only that equivalent to the mass of the particle involved but also that appearing as the usually high kinetic energy of the particle itself. Similarly, energy emitted in the form of an electromagnetic wave, or photon, causes a mass decrease in the emitting nucleus. In fact, all energy lost from the nucleus results basically in a relativistic decrease in its mass, and this phenomenon is considered one of the strongest supports of the Theory of Relativity.

If an atom loses an alpha particle, it also loses two positive charges; it will thus no longer be neutral unless it also loses two of its orbital electrons. Actually this is exactly what happens; however, this is not energy lost by the nucleus and is thus not ordinarily considered a part of the reaction. Similarly, if a beta particle (or electron) is lost from the nucleus, the nucleus has an excessive positive charge. Apparently, this is quickly and simply rectified by the atom's capturing a free electron that is drifting by; except in rare cases it apparently does not capture as an orbital electron the beta particle emitted from the nucleus. The loss of a gamma ray, or photon, changes the mass of the nucleus only by its energy equivalent, and the charge of the nucleus is changed not at all.

TRANSMUTATIONS

As a result of changes in the nuclear charge some elements are actually transmuted into other elements, the dream of the early alchemists. Thus the transformation reaction for uranium-238 losing an alpha particle (helium nucleus) may be written as follows:

$$^{238}_{92}U \rightarrow \ ^{234}_{90}Th + \ ^{4}_{2}He.$$

Similarly the thorium-234 produced by this reaction loses a beta particle (or electron), and the resulting reaction is the following:

$$^{234}_{90}Pa \rightarrow \ ^{234}_{91}Th + \ ^{0}_{-1}e.$$

The protactinium product of this reaction then emits a beta particle; the transformation equation is the following:

$$^{234}_{91}Pa \rightarrow \ ^{234}_{92}U + \ ^{0}_{-1}e.$$

Of interest in this reaction is the fact that the product is another uranium, $^{234}_{92}U$. In the early days thorium-234 and protactinium-234 were respectively identified as uranium–X_1 (UX$_1$) and uranium–X_2 (UX$_2$).

NATURAL RADIOACTIVE CHAINS

The above equations actually represent the first members of a radioactive chain whereby the elements produced by radioactive decay themselves subsequently decay, producing new elements and eventually reaching a nonradioactive product. Some of the characteristics of the three naturally radioactive chains, radium (or uranium-238), thorium, and actinium (or uranium-235) chains, are given in Tables 3–1 through 3–3 (see end of chapter). In each case one of the isotopes of lead is the final nonradioactive product of each chain. It is conventional to refer to the members of such chains as parents and daughters. Thus thorium-234 is the daughter of uranium-238 but a parent of protactinium-234; similarly, protactinium-234 is a daughter of thorium-234 but a parent of uranium-234, and so on.

On the basis of the above hypothesis of an atomic mechanism, each particle emitted is the result of an energy change in a single atom, and this change does not affect the energy configuration in any of the other atoms. Similarly it should be reemphasized that all of these radioactive changes are considered to be nuclear effects, and thus the beta particles (or electrons) that are emitted also come from the nucleus and represent some energy change therein. It has already been noted that the new particles produced in the reactions of high energy physics also represent nuclear energy changes.

ARTIFICIAL RADIOACTIVITY

To make stable atoms artificially radioactive, energy must be added to the nucleus so that it goes into a less stable configuration. The resultant energy emission may be the result of a single transfer or it may be a series of such changes producing a radioactive chain; the energy emitted may be less than that added or it may be more; the energy emission may be in a different form than that by which it is added; and the energy release may take more than one path, as is shown in Tables 3–1 through 3–3. In any event the nucleus eventually reaches a stable configuration, as described above for naturally radioactive materials.

Most nuclear processes result from bombardment of nuclei by various particles; of these, alpha particles, neutrons, protons, and deuterons ($_{1}^{2}H$) are most frequently used. Nuclear reactions are also produced by gammas and X-rays. Since the electron has a small mass and its electrical charge has the same sign as the electron cloud surrounding the nucleus, it is not very useful in most nuclear studies—although X-rays are produced by electron bombardment of metals with a consequent disturbance of the shell electrons.

CHARGED PARTICLE REACTIONS

One of the earliest alpha reactions is represented by the following equation:

$$_{7}^{14}N + _{2}^{4}He \rightarrow (_{9}^{18}F) \rightarrow _{8}^{17}O + _{1}^{1}H.$$

The central part of this equation, $_{9}^{18}F$, represents an intermediate compound nucleus which is usually considered to be formed in the reaction; however, it is not often included as a part of the overall reaction. A particularly important alpha reaction wherein neutrons are produced is

$$_{4}^{9}Be + _{2}^{4}He \rightarrow (_{6}^{13}C) \rightarrow _{6}^{12}C + _{0}^{1}n.$$

A typical proton-bombardment reaction is that of

$$_{11}^{23}Na + _{1}^{1}H \rightarrow (_{12}^{24}Mg) \rightarrow _{10}^{20}Ne + _{2}^{4}He,$$

and a deuteron-induced reaction is

$$_{1}^{2}H + _{6}^{12}C \rightarrow (_{7}^{14}N) \rightarrow _{7}^{13}N + _{0}^{1}n.$$

In all cases a reaction energy Q should be added to the product side of the equation to represent the mass-energy differences of the reaction; similarly any kinetic energies of the initial or final particles should be appropriately included in the analysis, as should be any energy emitted as a gamma photon; this is particularly important since gamma rays are emitted in essentially all reactions. Such details are beyond the scope of this book but are readily available in the literature.

A frequently used shorthand method of indicating nuclear reactions such as those given above is of the general form

Target nucleus (incident particle; emitted particle) final nucleus.

With this model the above reactions would be expressed as follows:

$$^{14}_{7}\text{N} \; (\alpha, \; p) \; ^{17}_{8}\text{O},$$
$$^{9}_{4}\text{Be} \; (\alpha, \; n) \; ^{12}_{6}\text{C},$$
$$^{23}_{11}\text{Na} \; (p, \; \alpha) \; ^{20}_{10}\text{Ne},$$
$$^{12}_{6}\text{C} \; (d, \; n) \; ^{13}_{7}\text{N}.$$

An obvious difficulty in producing nuclear reactions by bombardment with positively charged particles, such as protons and alpha particles, is the comparatively tremendous amount of energy lost by the particle in overcoming the repelling force of the positively charged nucleus when it does approach the nucleus. Perhaps of greater importance is the comparatively high probability that such bombarding particles will not enter the nucleus because of reflection similar to that from elastic impact; it was the result of such reflection experiments with alpha particle bombardment that Rutherford concluded the radii of nuclei were on the order of 10^{-15} m.

NEUTRON BOMBARDMENT

The neutron, which is electrically neutral, does not suffer from the difficulties described above; it can easily pass through the electron cloud and enter the positively charged nucleus, where it may be captured, thus adding to the nucleus not only its mass-energy but also the part of its kinetic energy that does not appear as kinetic energy of the resultant compound nucleus. The adjustment process following neutron bombardment can result in the emission of one or more protons, neutrons with energies different from that of the invader, gamma rays (or other electromagnetic waves), beta particles, positrons (positively charged particles with the mass of an electron), and various other particles such as deuterons, or combinations of these particles. In fact, the energy emitted by the bombarded nucleus may be greater than that caused by the mass and kinetic energy of the entering neutron so that the final energy state of the nucleus is lower than that before the neutron entry.

Thus the neutron provides a simple tool for producing nuclear transmutations, many of which are radioactive. In fact neutron bombardment has produced isotopes of essentially every element; it is responsible for nuclear fission and it has also produced several transuranium (or transuranic) elements, which are elements whose atomic numbers are greater than the 92 of uranium. Radioisotopes are today predominantly produced by neutron irradiation.

It should not be surprising that very few radioisotopes occur naturally, since it would appear probable that, following the origin of the elements, whatever it may be, those which are stable would survive, even though they initially formed only a small fraction of possible nuclides or of those actually produced. In view of the short half-lives of most radioactive materials, as is discussed later, most of those produced would have long since decayed into stable isotopes.

ENERGY RELEASED IN NUCLEAR REACTIONS

In any nuclear reaction that results in the emission of radiation the energy change in the nucleus of the emitting atom involves the mass of the emitted particle; the energy represented by any difference of the initial masses of the interacting units and their final masses; the kinetic energy released in the reaction (which appears mostly as the motion of an emitted particle); and other energy release, which primarily occurs as gamma emission. It is observed that the alphas or neutrons which are emitted in a given reaction generally have one or more well-defined energies that are characteristic of the reaction concerned; similarly, specific characteristic gamma frequencies are also observed. Such observations have led to the conclusion that the nucleus may have a structure that is somewhat analogous to the shells and orbits of the extranuclear electrons. Although both gammas and X-rays are electromagnetic waves and may be expressed as having the same frequencies, a major difference between the two (which is important in protection considerations) is the fact that all the photons emitted in a given nuclear reaction have the same gamma frequency, whereas the photons emitted at a given X-ray frequency actually represent a broad and generally continuous range of frequencies below the given value.

However, no such relatively well-defined kinetic energy occurs in beta-emission, which has a spectrum of energies that is somewhat analogous to an X-ray spectrum although obviously not closely related. Such a spectrum is shown in Figure 3–1. This rather puzzling observation led to the conclusion that the maximum beta energy observed really represents the total non-mass energy change of the reaction. The difference between this energy and the actual kinetic energy of the beta particle emitted in a specific reaction is assumed to be the energy of another emitted particle, the neutrino—an uncharged particle with zero (or near zero) mass and several other interesting properties. Only in the past few years has the neutrino been comparatively directly observed. Thus the emission energies listed in an appropriate table of nuclear reactions involving betas (both electrons and positrons) are actually the maximum beta-energy values, and these may thus not be observed in a specific instance; on the other hand, the listed energies of other particles and of gammas are always observed. When both betas and gammas are emitted simultaneously the gamma frequency is a constant even though the beta energy varies. Where a given radioactive decay process can take more than one path, or where more than one energy is emitted in a given reaction, the relative probabilities of occurrence of the respective reactions depend on unidentified properties of the nucleus itself.

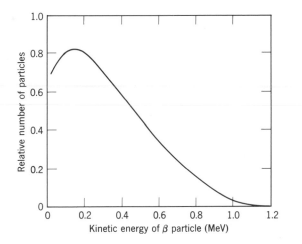

Figure 3–1 Beta spectrum of Radium-E ($^{210}_{83}$Bi).

TIME IN NUCLEAR TRANSFORMATIONS

To this point no mention has been made of any time factor involved in nuclear transformations, even though the fact that radioactive substances, whether natural or artificially produced, change only over a measurable period of time is probably the most important factor in identifying a radioactive substance. Nuclear transformations that occur essentially instantaneously (in times of about 10^{-9} sec, or nanoseconds) are not considered radioactive, primarily because the times may be immeasurably small.

It was only shortly after the discovery of natural radioactivity that it was also found that a radioisotope's activity, or rate of emission of the various "rays" observed, depended only on the amount of the radioactive material present and that this rate could not be altered by any of the available physical or chemical methods, such as chemical reaction, heat, pressure, etc. This observation is still true even though the methods of attack have been expanded. Thus, if it is accepted that the radiation emitted is an atomic (or, more specifically, a nuclear) phenomenon as described, then it must be further accepted that the length of time that a given atom remains in an unstable configuration before going to a more stable one with the emission of energy in the form of particles or photons has a statistical probability which depends only on unknown factors. For a large number of atoms this individual statistical probability results in a constant fraction of the unstable atoms present going to another configuration with the emission of energy in a given period of time. In particular the previous history, or time of actual existence, of a large number of these "excited," or unstable, atoms has no effect at all on this time fraction; neither does a change in one nucleus apparently affect another.

RADIOACTIVE DECAY

The observed facts that the total activity, or time rate of change, of a given amount of a radioisotope depends only on N, the number of atoms of the material present, and that radioactive changes cause the number of atoms in the original state to decrease may be expressed mathematically as

$$\frac{dN}{dt} = - kN,$$

where dN/dt represents the time rate of change in N and k is a constant of proportionality which is characteristic of the reaction concerned and is identified as the decay constant.

It may be readily shown that the solution of this simple equation is

$$N = N_0\ e^{-kt}\quad \text{or}\quad \frac{N}{N_0} = e^{-kt},$$

where N_0 is the number of atoms available at a time $t = 0$ and N is the number of unchanged atoms remaining after a time t. Obviously then

$$\frac{dN}{dt} = -\ kN_0e^{-kt}.$$

Since for a given element the mass is proportional to the number of atoms available, this phenomenon may also be expressed as

$$m = m_0e^{-kt},$$

where m_0 and m are the respective masses available initially and after a time t. As stated above, the equation holds regardless of when zero time on some larger time scale may be. Thus it is just as true where zero time is now as for zero time being a year or a million years ago. Obviously m_0 is the corresponding mass at whichever zero time is used.

HALF-LIVES

The probability of radioactive decay can also be readily expressed in terms of a radioactive half-life, which is defined as the length of time necessary for half of the atoms in a given amount of radioactive material to decay radioactively. Thus, if N is $0.5\ N_0$, which will be the case after a single half-life T,

$$0.5 = e^{-kT}.$$

From tables it is shown that $0.5 = e^{-0.693}$, and thus $kT = 0.693$. Therefore $k = 0.693/T$, and the basic equation becomes

$$N = N_0e^{-(0.693/T)t} = N_0e^{-0.693(t/T)}.$$

Since $e^{-0.693} = 0.5$, this may also be expressed as $N = N_0\ (0.5)^{(t/T)}$ or $N = N_0\ (0.5)^p$, where $p = t/T$ is the number of decay half-lives; for example, after three half-lives $N = N_0\ (0.5)^3 = 0.125\ N_0$.

To illustrate the effect of this half-life on the number of atoms available at a given time, assume that the half-life of a certain radioisotope is 30 days. Then after a period of 30 days the number of atoms of this material remaining will be 50 percent of those present originally. After a second similar period (or a total of 60 days) only 25 percent of the original material will remain. Similarly after a total period of 90 days (three half-lives) the remaining fraction will be 12.5 percent, after 120 days (or four half-lives) the fraction will be 6.25 percent and so on.

The activity, or rate of emission by decay, is given by the relation

$$\frac{dN}{dt} = -\frac{0.693}{T} N = -\frac{0.693}{T} N_0 e^{-(0.693/T)t}.$$

The fact that the time rate of change dN/dt is proportional to N, the number of atoms of the radioisotope remaining at any moment, is important in *counting,* which is discussed later. Briefly, the number of radiations, particles or photons, emitted in a given time is measured, and by use of proportionality factors that have been determined in calibration the mass of the radioactive material available may be determined.

Obviously, the number of atoms of the daughter of the radioisotope, n, produced by this decay during the time t is given by the relation

$$n = N_0 - N = N_0[1 - e^{-(0.693_T)t}]$$

Figure 3–2 shows graphically the relations $\dfrac{N}{N_0} = e^{-0.693t/T}$ and $\dfrac{n}{N_0} = (1 - e^{-0.693t/T})$ for a single radioisotope on Cartesian coordinate paper and Figure 3–3 shows the former on semilog graph paper. It will be noted that the decay curve is a straight line on the semilog graph paper; the half-life, or decay constant, may be determined from the slope of this line. This is actually the characteristic form of a true exponential relationship, of which radioactive decay is an example, and such graphs are frequently used to determine if a given set of data represents such a relation-

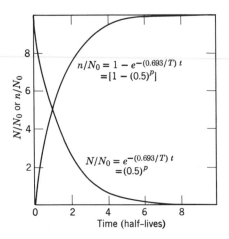

Figure 3–2   Radioactive decay and daughter buildup.

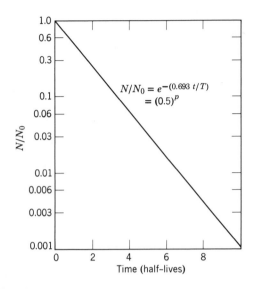

Figure 3–3   Radioactive decay.

ship. Since the decay rate dN/dt is proportional to N, a similar graph of dN/dt as a function of time will also be a straight line on semilog graph paper.

Practically, a graphical analysis made by plotting the count rate as a function of time is used to determine the half-life, or decay constant, of a given radioactive material.

MIXTURES OF RADIOISOTOPES

In many practical cases the actual measured activity of a given mass may be that from a mixture of emitters with different half-lives. This could be the case for mixtures of radioisotopes or even for the different reactions of a single radioisotope. In either case the emissions are always completely independent of each other; that is, one activity has no effect on the other, except for radioactive chains, which are mentioned later.

Thus, since the activities from the two components of a mixture are, proportional to N_1 and N_2 respectively, the combined activity is proportional to

$$N_1 + N_2 = N_{01}e^{-0.693t/T_1} + N_{02}e^{-0.693t/T_2}$$

where N_{01} and N_{02} are the respective initial numbers at time $t = 0$. Because of the different values of these initial numbers as well as their different half-lives, the total activity as a function of time is not a straight line on a semilog graph (as for decay with a single half-life) but a curve. In general, if the half-lives are very different, the initial slope of the line is closely related to the shorter half-life, whereas the final slope more nearly reflects the longer half-life. Figure 3–4 shows a curve for such a mixture of decay rates for two indium-115 reactions. Similarly radioactive decays of masses with more than two half-lives may be evaluated. Of particular interest in such mixtures is the apparent rate of decay of the fission products from a nuclear reaction since there is initially a great variety of radioisotopes with different half-lives. Figure 3–5 shows such a decay curve that has been determined empirically.

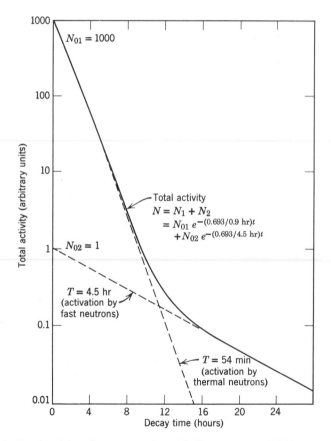

Figure 3–4   Total activity of neutron-activated indium. ($N_{01} = 1000 N_{02}$)

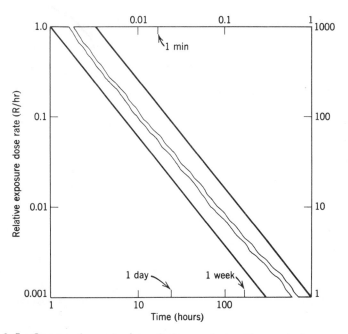

Figure 3–5   Gamma dose rate from fission products. Dose rate is assumed to be 1 R/hr at 1 hr after reaction. (*Note:* 1 hr after early "nominal" atomic bomb explosion total fission-product activity is 6 x 10^9 Ci. From *Effects of Atomic Weapons,* USAEC 1950, pp. 251–253.)

SOURCE OF RADIOACTIVE CHAINS

In much of the preceding discussion it has been assumed that when an excited atom loses energy it reaches a stable configuration. However, this is not necessarily true, and in practice it rarely occurs. Thus, although an excited atom in an unstable state loses energy in going to another configuration, this second configuration may also be an unstable one from which the atom will subsequently go by radioactive decay with a definite half-life to a third configuration, and so on. In fact, chemical procedures very early showed that the naturally occurring radioactive materials actually consisted of several elements in definite proportions, with radioactive changes from one to the other occurring in a constant order, and with definite half-lives which were independent of any preceding half-life. Such a series of radioactive changes were called radioactive chains of daughters and parents, and Tables 3–1 through 3–3 give current data on the products, half-lives, and energy emission types for the three naturally occurring radioactive series or chains.

For radioactive series or chains where the component half-lives differ the mathematical expressions for the number of atoms of each component available at a given time differ from the simple one given for a single radioactive change. In the case of a chain wherein material A decays into material B with decay constant k_A, or half-life of T_A, material B decays into material C with a decay constant k_B, or half-life of T_B, and material C is a stable isotope the basic equations are obviously

$$\frac{dN_A}{dt} = -k_A N_A,$$

$$\frac{dN_B}{dt} = k_A N_A - k_B N_B,$$

$$\frac{dN_C}{dt} = k_B N_B.$$

The above equations are not particularly difficult to solve but are somewhat tedious, and hints on their solution are readily available in the literature. In any event the numbers of atoms of each of the components after a time t are given by the following relations:

$$N_A = N_0 e^{-k_A t},$$

$$N_B = \frac{k_A}{k_B - k_A} N_0 (e^{-k_A t} - e^{-k_B t}),$$

$$N_C = N_0 \left(1 + \frac{k_A}{k_B - k_A} e^{-k_B t} - \frac{k_B}{k_B - k_A} e^{-k_A t}\right),$$

where N_0 is the original number of atoms of material A.

The symmetry of these equations indicates the way by which a much longer chain may be analyzed and the respective buildup of components determined. It should be particularly noted that mathematically and actually *none* of the radioisotopes completely disappears, even material A, although it can reach a very low level if given enough time.

RADIOACTIVE EQUILIBRIUM

Equilibrium for any one of the members of a chain is reached when its mass remains constant; that is, a daughter loses atoms by decay as rapidly as they are produced by the decay of its parent. Mathematically this is actually impossible. However, this condition may be approximately attained if the half-life of the parent is much greater than that of the daughter. Thus, if t is sufficiently long for $e^{-k_B t}$ not only to approach zero, which presupposes a period of several half-lives but also to approach zero more rapidly than does the factor $e^{-k_A t}$, the value of N_B becomes approximately

$$N_B = \frac{k_A}{k_B - k_A} N_0 e^{-k_A t}.$$

Since $N_A = N_0 e^{-k_A t}$, $N_B = \dfrac{k_A}{k_B - k_A} N_A$.

In terms of the respective half-lives this relation becomes

$$N_B = N_A \frac{T_B}{(T_A - T_B)}.$$

If T_A is somewhat greater than T_B, this is identified as transient equilibrium. If T_A is very much greater than T_B, which means that $T_A - T_B \simeq T_A$, the equation becomes simply

$$\frac{N_B}{N_A} = \frac{T_B}{T_A}.$$

This is called secular equilibrium and is particularly important because it occurs for several important decay chains. In terms of parent-daughter relationships this becomes

$$\frac{N_d}{N_p} = \frac{T_d}{T_p},$$

where the subscript p refers to the parent and the subscript d to the daughter.

Two conditions of secular equilibrium involving the uranium-238 chain are of particular interest. Thus, since uranium-238 has a half-life of 4.5×10^9 years and its daughter, thorium-234 (or UX_1), has a half-life of only about 24 days, the beta activity of the thorium will become essentially constant after several months—or rather it will change at the same rate as that of its uranium-238 parent. This uranium is called "old" uranium, and the thorium and other daughters are said to have "grown into" the uranium. If the uranium is chemically separated from its daughters it is said to be new uranium; however, after several more months with no further treatment the daughters will again grow into the uranium.

It will also be noted from Table 3–1 that uranium-234 with a half-life of about 2.5×10^5 years, is also a part of the uranium-238 chain. It also can be in secular equilibrium with its uranium-238 parent, several elements removed, and thus in natural uranium it should comprise only about 0.006 percent of the total. A different fraction indicates chemical actions between the environment and intervening elements. However, because of their relative half-lives, it may be seen that this fraction does not change markedly, even when the intervening parent-daughters are removed chemically, unless such intervening elements are kept removed for periods of time on the order of 10^5 years.

One important conclusion that may be drawn from this discussion of radioactive chains is the fact that it is not always possible to refer to the radiation problem of a given radioisotope in terms of its individual emission but it is also necessary to refer to the problem produced by any daughters. Thus, although uranium emits only alpha particles, uranium radiation problems also include those of its beta- and gamma-emitting daughters. It is this association of uranium with its gamma-emitting daughters that makes possible searches for underground uranium deposits by use of beta-gamma detectors, even by air.

A similar consideration also applies to artificially produced radioisotopes; for example, the yttrium-90 daughter plays an important part in the overall radiation problem of the strontium-90 isotope. In this case the short-lived yttrium-90 emits a much more energetic beta particle than does the strontium-90 itself, and, because of its short half-life, it forms a significant part of the radiation problem of strontium-90.

It has also been noted that in these radioactive chains there may be more than one path by which a stable configuration is reached; this is indicated in Tables 3–1 through 3–3 and also occurs for artificially produced radioisotopes. The relative abundance of the respective transformations and the radiation concerned must be included in any evaluation of the problems of these materials.

NUCLEAR CROSS SECTIONS FOR NEUTRONS

The fact that neutron approach to a nucleus is not affected by electrostatic forces has already been mentioned. However, the ability of a neutron actually to interact with a given nucleus depends on some as yet unexplained characteristics of the nucleus itself, the velocity of the neutron, and the type of interaction concerned; this last is elaborated on later. In general, a convenient measure of the relative probability that a given nuclear reaction will occur in a given element as a result of neutron bombardment is usually expressed in terms of the nuclear cross section, or the cross-sectional area σ of the nucleus concerned. Thus, if a neutron passes within a given distance r of the nucleus "center" (where $\sigma = \pi r^2$), the probability of a given nuclear event occurring is greater than 50 percent, whereas if its distance of passage is greater than r, the probability is less than 50 percent. On the average and for practical use the cross section may be conveniently considered as the area within which the event concerned always happens and outside of which it never happens. Thus, as far as the reaction under discussion is concerned, the nucleus may be considered to be a sphere with a radius r and a cross-sectional area σ.

Nuclear cross sections are usually measured in barns (1 barn $= 10^{-28}$ m²), a rather unglamorous unit that apparently received its designation by analogy with the faintly derisive comment on marksmanship from prior eras, to wit, "You couldn't hit the side of a barn with a shotgun." In fact a widely used technical reference giving neutron cross sections contains a cover picture of a barn and is popularly called *The Barn-Book*.

Nuclei have various cross sections which reflect the relative probabilities of their different neutron interactions. These include (a) elastic impact or scattering collisions for which the nucleus may be considered as a perfectly elastic sphere of cross section σ_e; (b) inelastic emission of a neutron at a different energy than that of the incident particle and with a cross section σ_i; (c) radiative capture, for which the incident-neutron capture results in the emission of any of the radiations described previously and with a cross section σ_r; (d) nonradiative capture, whereby no radiation is emitted after neutron capture and with a cross section σ_n; and (e) fission, which is important only for a few materials, with a cross section identified as σ_f. The nonradiative and radiative captures are usually combined to express a capture cross section σ_c, given by $\sigma_c = \sigma_r + \sigma_n$.

The total neutron cross section σ_T is the sum of all of the partial cross sections, the values of each of which are characteristic of the materials concerned and in each case also apparently depend independently on the incident-neutron energy, or velocity. Thus $\sigma_T = \sigma_e + \sigma_i + \sigma_c + \sigma_f$. Obviously, the fraction of the total interactions that result in a given reaction is equal to its proportional cross section. Thus σ_e/σ_T is the fraction of total reactions that give elastic scattering. Since the neutron actually is absorbed in all interactions except elastic scattering, an absorption cross section σ_a is defined as $\sigma_a = \sigma_i + \sigma_c + \sigma_f$. Obviously $\sigma_T = \sigma_a + \sigma_e$.

From the above it is obvious that the size of a given nucleus as far as its interaction with neutrons is concerned is a variable which depends on the neutron energy, the reaction concerned, etc. Although this may be a rather strange conclusion to reach with respect to a physical quantity that common sense dictates should be a constant, it must be remembered that a specific convenient model is specified and that this conclusion logically explains experimental results in terms of this model; it may not be an accurate description of the actual phenomena involved.

NUCLEAR FISSION

The most common source of neutrons today is the nuclear reactor, which depends on the fission process for its operation. In the first nuclear reactions only comparatively small units had been separated from the target nucleus, the largest of these being an alpha particle. However, in 1939 Hahn and Strassmann came to the conclusion that the results of certain of their experiments in which uranium was bombarded by neutrons could be explained only on the assumption that a neutron entering a uranium atomic nucleus (later determined to be that of uranium-235) caused this nucleus to split apart, or fission, and form two relatively equal masses.

In this process a certain fraction of the mass of the uranium-235 nucleus "disappeared" on conversion to energy, and additional neutrons were also released. An incredibly large amount of energy of about 200 MeV, which is about 3.2×10^{-11} J, was released in a single nuclear reaction. For comparison, the energy released in an ordinary nuclear reaction is no greater than about 20 MeV per atom and that in even the most violent chemical reaction only a few electron volts per atom.

CHAIN REACTIONS

Since addition neutrons are also emitted in the fission reaction, the possibility was also foreseen that a chain reaction could be produced whereby an initial neutron would enter a uranium-235 nucleus, releasing energy and other neutrons; these in turn would enter other uranium-235 nuclei, producing additional fissions with the release of additional large amounts of energy and other neutrons, and so on. Since the time between fissions would be measured in small fractions of microseconds, such a process in a sufficiently large mass of uranium-235 would release energy at explosive rates. The possibility of a tremendously powerful weapon was thus foreseen, and this became of particular interest with the outbreak of World War II at about that time.

SEPARATION OF URANIUM–235

Since uranium-235 comprises only about 0.7 percent of natural uranium with the remainder being essentially all uranium-238, it was quickly concluded that, except under special conditions, such a chain reaction would not be possible in natural uranium, because the neutrons emitted by one fission would be lost before they could produce additional fissions in other uranium-235 nuclei. Hence it was realized that the energy release desired depended on the concentration of the uranium-235 fraction, and this became a major wartime project, which, because uranium-235 and uranium-238 have identical chemical properties, was a formidable task. Suffice it to say that these efforts were successful, four proposed methods of concentration being taken to the pilot plant stage and two to the production stage. These were the gaseous diffusion process and the electromagnetic process, the latter of which has since been abandoned, although it was the method by which material for the first weapon was produced. Recently a third method that employs centrifuges and that did not originally get beyond the pilot plant stage has again become an important factor in the concentration process.

PLUTONIUM AND NUCLEAR REACTORS

Prior to Hahn's reports Fermi had also bombarded uranium with neutrons and had interpreted some of his results as indicating the production of transuranic elements with atomic numbers above uranium's 92, which is the highest found naturally. He concluded that the uranium-238 nucleus would capture a bombarding neutron and, by a process of beta emission, produce an atom of a transuranic element. The conclusion was also reached that one of these transuranic elements, with an atomic number of 94 and an atomic weight of 239, which was called plutonium, was relatively stable and also had fissionable properties similar to those of uranium-235 ($^{235}_{92}U$). Since plutonium is chemically different from uranium, it could be readily separated therefrom. Hence, a second method of producing fissionable materials by neutron irradiation of uranium was advanced. Various nuclear experiments then indicated that a uranium pile (later called a reactor) which was constructed with carefully calculated and maintained interspersal of natural uranium rods and very pure carbon could produce a chain reaction, or go "critical."

Again under wartime urgency Fermi's concept went to experimental determination, and the first nuclear reactor using natural uranium went critical late in 1942; in rapid succession the process went to pilot plant and then to production operation.

FISSION PRODUCTS

Fission produces a wide range of radioactive products, and a chain of reactions also usually results; for example, a typical result of uranium-235 ($^{235}_{92}U$) fission may give the following reactions:

$$^1_0n + {}^{235}_{92}U \rightarrow ({}^{236}_{92}U) \rightarrow {}^{141}_{56}Ba + {}^{92}_{36}Kr + 3{}^1_0n + Q$$

$$^{141}_{56}Ba \rightarrow {}^{141}_{57}La + {}_{-1}^0e \qquad\qquad {}^{92}_{36}Kr \rightarrow {}^{92}_{37}Rb + {}_{-1}^0e$$

$$^{141}_{57}La \rightarrow {}^{141}_{58}Ce + {}_{-1}^0e \qquad\qquad {}^{92}_{37}Rb \rightarrow {}^{92}_{38}Sr + {}_{-1}^0e$$

$$^{141}_{58}Ce \rightarrow {}^{141}_{59}Pr + {}_{-1}^0e \qquad\qquad {}^{92}_{38}Sr \rightarrow {}^{92}_{39}Y + {}_{-1}^0e$$

$$^{92}_{39}Y \rightarrow {}^{92}_{40}Zr + {}_{-1}^0e$$

The final result of this particular fission process, or chain, is the production of stable isotopes of zirconium and praseodymium plus the release of three neutrons, seven electrons (beta particles), and other energy in the form of gammas, heat (or kinetic energy), etc.

It should be reemphasized that not all fissions give the reactions shown, that this is only one of those which has been observed, and that it is entirely possible for one of the components of this chain to take a different path than the one indicated. Figure 3–6 shows the relative frequency with which various elements, radioactive or stable, are produced by fission. About 300 nuclides, both stable and radioactive, with atomic masses in the approximate range of 70 to 160 are produced as one result of a large number of fissions. In a few rare cases three nuclides are produced in fission, although the usual number is two. Three neutrons are not always released in fission, the number varying from one to five but with an average of about 2.43.

The reaction by which $^{239}_{94}Pu$ is produced is rather interesting; that is,

$$^1_0n + {}^{238}_{92}U \rightarrow ({}^{239}_{92}U) \rightarrow {}^{239}_{93}Np + {}_{-1}^0e,$$

$$^{239}_{93}Np \rightarrow {}^{239}_{94}Pu + {}_{-1}^0e.$$

If fission is not produced in the plutonium, which would lead to groups of fission-product series similar to that shown for uranium-235 ($^{235}_{92}U$) fission, the plutonium emits an alpha particle and goes to uranium-235 by the reaction

$$^{239}_{94}Pu \rightarrow {}^{235}_{92}U + {}^4_2He.$$

Actually the uranium-235 thus produced is an excited metastable atom that apparently has no daughter. Hence it is not considered to be the first

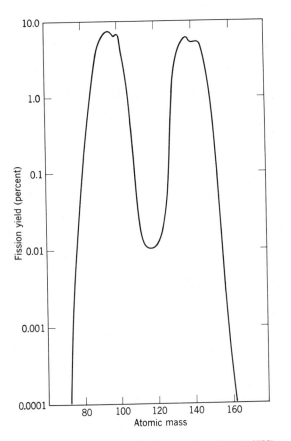

Figure 3–6 Thermal-fission-product yield for uranium-235 ($^{235}_{92}U$).

component of the actinium naturally radioactive series, which is shown in Table 3–3.

Uranium-233, the other fissionable material, is produced similarly by neutron bombardment of thorium-232 by the following reactions:

$$^{232}_{90}\text{Th} + {}^{1}_{0}n \rightarrow \quad (^{233}_{90}\text{Th}) \rightarrow {}^{233}_{91}\text{Pa} + {}_{-1}^{0}e,$$

$$^{233}_{91}\text{Pa} \rightarrow {}^{233}_{92}\text{U} + {}_{-1}^{0}e.$$

If fission does not occur in the uranium-233, it is the initial component of a radioactive series terminating in bismuth-209 ($^{209}_{83}\text{Bi}$), which, if radio-active, has a very long half-life.

From the above reactions it may be seen that the ordinarily nonfissionable uranium-238 and thorium-232 can be made to have fissionable products as a result of neutron irradiation. This is the basis of "breeding" reactors by which it is hoped to produce more fissionable material (uranium-233 or plutonium-239) than the uranium-235 burned up in the manufacturing process. To date such efforts have not been proven successful, certainly not on a commercial scale, and uranium-235 thus continues to be the basic source of fission for reactor use.

CRITICALITY CRITERIA

The condition of criticality that is necessary for reactor operation means merely that for every neutron which is absorbed by one nucleus of a fissionable material at least one of the emitted neutrons must enter another fissionable material nucleus and continue the process; that is, the fission reaction from any neutron absorption must result in the emission of more than one neutron, and the various characteristics of the reaction must be carefully balanced to maintain this one-for-one neutron economy. If the fraction of these emitted neutrons absorbed is so low that fewer fissions are produced in the second generation than in the preceding one, the system becomes subcritical and the chain reaction dies down; if the fraction absorbed to produce fission is so high that more fissions are produced in a given generation than in the preceding one, the system becomes supercritical and a condition of "runaway" results. It is beyond the scope of this book to discuss the methods used for providing these controls, although some illustrations of buildup of a supercritical reaction are given in a subsequent chapter. However, it may be mentioned that it is much simpler to make a reactor using uranium enriched in the uranium-235 isotope than to use only natural uranium, although some breeding does take place in the uranium-238 contained therein.

Obviously an initial neutron must be available to start the reaction. However, neutrons periodically appear in the air, primarily as a result of cosmic-ray activity, and these can start a reaction in a critical system. Accordingly a more controllable and convenient source, such as a mixture of radium and beryllium or plutonium and beryllium, is usually used for starting a reaction. Once started, the reaction will continue with the number of neutrons produced, or the reactor power level, depending on the neutron flux at which control is exerted; that is, a system can be critical when the flux is only 10 neutrons/cm^2-sec. or it can be critical when its internal flux is 10^6 neutrons/cm^2-sec, or 10^{10} neutrons/cm^2-sec.

RESULTS OF REACTOR OPERATION

The subject of nuclear reactions as opened by the plentiful supply of neutrons from fission is obviously a fascinating one, and its surface has been only barely scratched in this description. It may be of interest that the farthest transuranic element that has been produced and more or less identified is lawrencium, with an atomic number of 103; all of the transuranic elements are radioactive with rather short half-lives, and californium-254 ($^{254}_{98}$ Cf) undergoes spontaneous fission. Although fission can be produced in essentially all of the transuranic elements by neutron bombardment, the only other fissionable materials that are relatively stable are the uranium-233 isotope, which is produced as a result of the neutron bombardment of thorium-232, and plutonium-239 similarly produced from uranium-238.

The nuclear reactor and its plentiful neutron supply obviously makes possible the production of a wide variety of radioactive materials, or radioisotopes. In fact such activity now forms a rather large industry, the information concerning which, although beyond the scope of this book, is readily available in the literature.

NEUTRON-EMISSION HALF-LIVES

Although many of the same considerations of potential hazard apply to neutrons as to other radiations, the half-life, if any, for neutron emission from an excited nucleus is usually immeasurably small. There is some evidence that the emission of neutrons is not instantaneous, but since the time involved is on the order of fractions of nanoseconds, for radiation protection purposes the emission of neutrons can be considered to be instantaneous. An apparent exception to this rule is the emission of delayed neutrons from uranium-235 fission, which occurs in about 0.75 percent of the fissions; these are emitted in six distinct groups with mean delay times of as much as 80 sec. However, these are actually emitted as a secondary aspect of certain beta-decay processes.

Thus neutrons can be a hazard only as a result of (a) exposure to a neutron source such as radium-beryllium or plutonium-beryllium where the alphas emitted by the radium or plutonium produce neutrons from capture by the beryllium, (b) exposure to neutrons from a controlled reaction as in a specific reactor or an ion accelerator in which neutrons are produced, (c) exposure at the time of a criticality accident when neutrons are produced during the nuclear reaction, or (d) exposure at activation of an atomic or nuclear weapon. The last two are merely uncontrolled reactors.

OTHER RADIATION PROBLEMS

Some reactions produce protons and others produce positrons. In general the reactions that produce positrons are rather unusual, and the half-lives involved are short; thus they represent relatively small hazards. With respect to radiation problems the characteristics of positrons are very similar to those of electrons. Similarly, since protons have the low penetration capability of alpha particles and are produced only in certain types of experiments involving various accelerators, they usually present negligible problems.

Accordingly, for purposes of radiation protection as a continuing consideration except at reactor sites the principal attention is needed for the alpha, beta, and gamma radiations, and the primary emphasis will be concerned therewith.

It should be reemphasized that the biological significance of nuclear radiation depends on both its type and its energy. Thus a high energy, or fast-moving, beta particle is generally a greater hazard than is a lower energy one since it can penetrate greater thicknesses of the body. Similarly very low energy gamma rays may be of small overall hazard compared to those of higher energy. The injury-production problems of radioactive materials, however, are not tied directly and singly to their energies since a particle or gamma ray of sufficiently high energy may actually produce less injury than does a lower energy particle or photon. This is discussed subsequently.

REFERENCES and SUGGESTED READINGS

Texts on nuclear physics or nuclear reactor physics.
Kaplan, I., *Nuclear Physics,* Addison-Wesley, Reading, Mass., 1963.
Liverhant, S. E., *Elementary Introduction to Nuclear Reactor Physics,* Wiley, New York, 1960.
Glasstone, S., ed., *The Effects of Nuclear Weapons,* USAEC., Washington, D.C., 1962.
Stehn, J. R., M. D. Goldberg, B. A. Magurno, and R. Wiener-Chasman, *Neutron Cross Sections,* BNL-325, Second Edition, Supplement No. 2, Vols. 1, 2a-c, 3, Brookhaven National Laboratory, 1964-66.

TABLE 3–1

THE RADIUM SERIES (THE URANIUM SERIES)[a]

Parent Element	Nuclides		Particle Emitted	Particle Energy [b] (MeV)	Half-Life
	Parent	Daughter			
Uranium (U–I)	$^{238}_{92}U$ ⟶	$^{234}_{90}Th$	^4_2He (α)	4.20	4.51×10^9 years
Thorium (UX₁)	$^{234}_{90}Th$ (99+%) ⟶	$^{234m}_{91}Pa$ *	$^{\,0}_{-1}e$ (β)	0.19	⎫
	(0.63%) ⟶	$^{234}_{91}Pa$	$^{\,0}_{-1}e$ (β)	0.088	⎬ 24.1 days ⎭
Protactinium (UX₂)	$^{234m}_{91}Pa$ * ⟶		$^{\,0}_{-1}e$ (β)	2.29	1.2 min
		$^{234}_{92}U$			
Protactinium (UZ)	$^{234}_{91}Pa$ ⟶		$^{\,0}_{-1}e$ (β)	1.3	6.7 hr
Uranium (U–II)	$^{234}_{92}U$ ⟶	$^{230}_{90}Th$	^4_2He (α)	4.77	2.47×10^5 years
Thorium (ionium)	$^{230}_{90}Th$ ⟶	$^{226}_{88}Ra$	^4_2He (α)	4.68	8.0×10^4 years
Radium	$^{226}_{88}Ra$ ⟶	$^{222}_{86}Rn$	^4_2He (α)	4.78	1.6×10^3 years
Radon (Ra-emanation)	$^{222}_{86}Rn$ ⟶	$^{218}_{84}Po$	^4_2He (α)	5.49	3.8 days
Polonium (Ra–A)	$^{218}_{84}Po$ (99+%) ⟶	$^{214}_{82}Pb$	^4_2He (α)	6.00	⎫
	(0.02%) ⟶	$^{218}_{85}At$	$^{\,0}_{-1}e$ (β)	?	⎬ 3.0 min ⎭
Lead (Ra–B)	$^{214}_{82}Pb$ ⟶		$^{\,0}_{-1}e$ (β)	1.03	26.8 min
		$^{214}_{83}Bi$			
Astatine	$^{218}_{85}At$ ⟶		^4_2He (α)	6.70	2.0 sec

* Isometric state of Protactinum-234.

[a] Table adapted from Lederer, C. M., J. M. Hollander, and I. Perlman, *Table of Isotopes, Sixth Edition,* John Wiley and Sons, 1967.

[b] The particle energy listed for each reaction is the maximum given in the reference table; in many cases more than one energy is listed, and of course each beta reaction involves a spectrum of energies. Each reaction is also accompaned by the emission of a gamma photon; none of these is listed.

TABLE 3–1—Continued

Parent Element	Nuclides — Parent	Nuclides — Daughter	Particle Emitted	Particle Energy [b] (MeV)	Half-Life
Bismuth (Ra–C)	$^{214}_{83}Bi$ (99+%) →	$^{214}_{84}Po$	$_{-1}^{0}e$ (β)	3.26	⎫ 19.7 min
	(0.04%) →	$^{210}_{81}Tl$	$^{4}_{2}He$ (α)	5.51	⎭
Polonium (Ra–C′)	$^{214}_{84}Po$ →	$^{210}_{82}Pb$	$^{4}_{2}He$ (α)	7.68	1.6 × 10⁻⁴ sec
Thallium (Ra–C″)	$^{210}_{81}Tl$ →	$^{210}_{82}Pb$	$_{-1}^{0}e$ (β)	2.3	1.3 min
Lead (Ra–D)	$^{210}_{82}Pb$ →	$^{210}_{83}Bi$	$_{-1}^{0}e$ (β)	0.061	22 years
Bismuth (Ra–E)	$^{210}_{83}Bi$ (99+%) →	$^{210}_{84}Po$	$_{-1}^{0}e$ (β)	1.16	⎫ 5.0 days
	(2 × 10⁻⁴%) →	$^{206}_{81}Tl$	$^{4}_{2}He$ (α)	4.69	⎭
Polonium (Ra–F)	$^{210}_{84}Po$ →	$^{206}_{82}Pb$	$^{4}_{2}He$ (α)	5.30	138.4 min
Thallium	$^{206}_{81}Tl$ →	$^{206}_{82}Pb$	$_{-1}^{0}e$ (β)	1.52	4.2 min
Lead (Ra–G)	$^{206}_{82}Pb$	Stable	— —	—	Infinite

TABLE 3–2

THE THORIUM SERIES[a]

Parent Element	Nuclides — Parent	Nuclides — Daughter	Particle Emitted	Particle Energy [b] (MeV)	Half-Life
Thorium	$^{232}_{90}Th$ ⟶	$^{228}_{88}Ra$	$^{4}_{2}He$ (α)	4.01	1.41 × 10¹⁰ years
Radium (mesothorium I)	$^{228}_{88}Ra$ ⟶	$^{228}_{89}Ac$	$_{-1}^{0}e$ (β)	0.053	6.7 years

[a] Table adapted from Lederer, C. M., J. M. Hollander, and I. Perlman, *Table of Isotopes, Sixth Edition,* John Wiley and Sons (1967).

[b] See footnote b in Table 3–1.

TABLE 3–2—Continued

Parent Element	Nuclides		Particle Emitted		Particle Energy [b] (MeV)	Half-Life
	Parent	Daughter				
Actinium (mesothorium II)	$^{228}_{89}Ac$ ⟶	$^{228}_{90}Th$	$_{-1}^{0}e$	(β)	2.11	6.13 hr
Thorium (radiothorium)	$^{228}_{90}Th$ ⟶	$^{224}_{88}Ra$	$_{2}^{4}He$	(α)	5.43	1.91 years
Radium (Th–X)	$^{224}_{88}Ra$ ⟶	$^{220}_{86}Rn$	$_{2}^{4}He$	(α)	5.68	3.64 days
Thoron (Tn) (Th-emanation)	$^{220}_{86}Rn$ ⟶	$^{216}_{84}Po$	$_{2}^{4}He$	(α)	6.29	51.5 sec
Polonium (Th–A)	$^{216}_{84}Po$ ⟶	$^{212}_{82}Pb$	$_{2}^{4}He$	(α)	6.78	0.145 sec
Lead (Th–B)	$^{212}_{82}Pb$ ⟶	$^{212}_{83}Bi$	$_{-1}^{0}e$	(β)	0.58	10.64 hr
Bismuth (Th–C)	$^{212}_{83}Bi$ (64%) ⟶	$^{212}_{84}Po$	$_{-1}^{0}e$	(β)	2.25	⎫
	(36%) ↘	$^{208}_{81}Tl$	$_{2}^{4}He$	(α)	6.09	⎬ 60.5 min
Polonium (Th–C′)	$^{212}_{84}Po$ ↘		$_{2}^{4}He$	(α)	8.78	3.04×10^{-7} sec
	↘	$^{208}_{82}Pb$				
Thallium (Th–C″)	$^{208}_{81}Tl$ ↗		$_{-1}^{0}e$	(β)	1.80	3.10 min
Lead (Th–D)	$^{208}_{82}Pb$ ⟶	Stable	— —		——	Infinite

[b] See footnote b in Table 3–1.

TABLE 3–3

The Actinium Series[a]

Parent Element	Nuclides Parent	Nuclides Daughter	Particle Emitted	Particle Energy [b] (MeV)	Half-Life
Uranium (actinouranium)	$^{235}_{92}$U ⟶	$^{231}_{90}$Th	$^4_2$He (α)	4.58	7.1 × 10⁸ years
Thorium (U–Y)	$^{231}_{90}$Th ⟶	$^{231}_{91}$Pa	$^{\ 0}_{-1}e$ (β)	0.30	25.6 hr
Protactinium	$^{231}_{91}$Pa ⟶	$^{227}_{89}$Ac	$^4_2$He (α)	5.05	3.4 × 10⁴ years
Actinium	$^{227}_{89}$Ac (98.8%)	$^{227}_{90}$Th	$^{\ 0}_{-1}e$ (β)	0.046	⎫ 21.6 years
	(1.2%) ↘	$^{223}_{87}$Fr	$^4_2$He (α)	4.95	⎭
Thorium (radioactinium)	$^{227}_{90}$Th ↘	$^{223}_{88}$Ra	$^4_2$He (α)	6.04	18.2 days
Francium (actinium K)	$^{223}_{87}$Fr (99+%)		$^{\ 0}_{-1}e$ (β)	1.15	⎫ 22 min
	(4 × 10⁻⁸%) ↘	$^{219}_{85}$At	$^4_2$He (α)	?	⎭
Radium (Ac–X)	$^{223}_{88}$Ra ↘	$^{219}_{86}$Rn	$^4_2$He (α)	5.75	11.4 day
Astatine	$^{219}_{85}$At (97%)		$^{\ 0}_{-1}e$ (β)	?	⎫ 0.9 min
	(3%) ↘	$^{215}_{83}$Bi	$^4_2$He (α)	6.28	⎭
Radon (Ac-emanation)	$^{219}_{86}$Rn ↘	$^{215}_{84}$Po	$^4_2$He (α)	6.82	3.96 sec
Bismuth	$^{215}_{83}$Bi		$^{\ 0}_{-1}e$ (β)	?	8 min
Polonium (Ac–A)	$^{215}_{84}$Po (99+%)	$^{211}_{82}$Pb	$^4_2$He (α)	7.38	⎫ 1.8 × 10⁻³ sec
	(5 × 10⁻⁴%) ↘	$^{215}_{85}$At	$^{\ 0}_{-1}e$ (β)	?	⎭

[a] Table adapted from Lederer, C. M., J. M. Hollander, and I. Perlman, *Table of Isotopes, Sixth Edition,* John Wiley & Sons, (1967).

[b] See footnote b in Table 3–1.

TABLE 3–3—Continued

Parent Element	Nuclides		Particle Emitted	Particle Energy [b] (MeV)	Half-Life
	Parent	Daughter			
Astatine	$^{215}_{85}At$ ↘	$^{211}_{83}Bi$	^4_2He (α)	8.01	10^{-4} sec
Lead (Ac–B)	$^{211}_{82}Pb$ ↗		$^0_{-1}e$ (β)	1.36	36.1 min
Bismuth (Ac–C)	$^{211}_{83}Bi$ (99+%) →	$^{207}_{81}Tl$	^4_2He (α)	6.62	} 2.16 min
	(0.3%) ↘	$^{211}_{84}Po$	$^0_{-1}e$ (β)	?	
Polonium (Ac–C′)	$^{211}_{84}Po$ ↘	$^{207}_{82}Pb$	^4_2He (α)	7.45	0.52 sec
Thallium (Ac–C″)	$^{207}_{81}Tl$ ↗		$^0_{-1}e$ (β)	1.44	4.8 min
Lead (Ac–D)	$^{207}_{82}Pb$	Stable	———	——	Infinite

[b] See footnote b in Table 3–1.

QUESTIONS

1. If a certain element X emits an alpha (^4_2He) particle, it theoretically should then have a negative charge. This, however, is not found to be the case. What actually happens to keep the atom neutral?
2. When a stable atom is made artificially radioactive, energy is added to the nucleus. The resulting emission of energy may be less or greater than the energy input. Account for this fact.
3. Of all the known particles that are now used for nuclear bombardment to produce radioisotopes, which one is used the most and for what reasons?
4. Given a 100-g sample of pure uranium-238, how much of the sample is left after one year and how much energy has been emitted?
5. Given the same 100-g sample of uranium-238, how long ago would this sample have weighed 1 kg?
6. At what point can any one of the members of a radioactive chain be said to be in equilibrium?
7. How does emission of each of the following types of radiation change the nucleus of the atom: alpha particle, beta particle, gamma, and neutron?
8. Why is the neutron the most successful bombarding particle?
9. After two weeks, how much radon will remain if the initial mass was 2 kg?
10. Define criticality.

11. What are the instances during which neutrons can be hazardous?
12. Define radioactive equilibrium; discriminate between transient and secular equilibrium, using the equation

$$\left(\frac{k_A}{k_A - k_B}\right) N_0(e^{-k_A t} - e^{-k_B t}).$$

13. Complete

$$^{27}_{13}\text{Al} + ^{4}_{2}\text{He} \rightarrow ^{30}_{15}\text{P} + ?$$

$$^{30}_{15}\text{P} \rightarrow + ^{0}_{-1}e + ?$$

$$^{226}_{88}\text{Ra} \rightarrow ^{222}_{86}\text{Rn} + ?$$

14. On Cartesian Coordinate paper, draw a graph (very approximate) of the following curves concerned with radioactive decay: (a) activity versus time, (b) fraction of mass remaining after time t, and (c) graph (b) on semilog paper.
15. The half-life of Uranium-238 is 4.5×10^9 years. Its daughter has a half-life of 24 days (thorium-234). How long will it take each of these to reach three-fourths of its original mass? What will be the ratio of the masses at secular equilibrium?
16. Compare charged-particle reactions with neutron-initiated reactions.
17. From the following reaction, give a series of possible chain reactions:

$$^{1}_{0}n + ^{235}_{92}\text{U} \rightarrow (^{236}_{92}\text{U}) \rightarrow 2^{1}_{0}n + ^{145}_{57}\text{La} + ^{89}_{35}\text{Br}.$$

18. Given a mixture of radioisotopes such that $N_{01}/N_{02} = 10$ initially; if $T_1 = 0.5$ hr and $T_2 = 7$ hr, how long will it be until $N_1/N_2 = 1$?

CHAPTER FOUR

General Effects of Radiation on Cells

FUNDAMENTAL LIVING UNIT

Although a detailed description and analysis of biological systems is obviously beyond the scope of this book, some knowledge of the basic organization of living matter may be of assistance in understanding the effects of radiation on living systems, particularly the human organism. This is especially true since such effects are the only justification for radiation-protection programs.

The fundamental unit of living systems, whether plant or animal, is the cell. Basically the cell consists of a pliable, semipermeable cell membrane that encloses an aqueous suspension of a variety of organelles—the entirety sometimes identified as cytoplasm. The cytoplasm surrounds the nucleus, which consists of a fluid sometimes identified as the nucleoplasm which is also enclosed in a semipermeable nuclear membrane. Cells obtain necessary material and void wastes through the cell membrane.

Cell size obviously depends on its type, the living unit concerned, and so on. However, 10^{-5} m may be taken as a rough average equivalent cell diameter. This may be compared with the 10^{-10} and 10^{-15} m respective diameters of the atom and its nucleus. Molecular equivalent diameters depend not only on the number of atoms—which may vary from the one of monatomic materials or the three of water to many thousands for various organic molecules—but also on the arrangement of these atoms to produce the molecule. However, it is obvious that a biological cell must be composed of a very large number of molecules.

CELL CONSTITUTION

As much as 80 percent of the molecules of a given cell may be water, the remaining molecules making up the proteins, nucleic acids, fats and other lipids, enzymes, and so on. Practically all of the elements are found to some extent in body cells, although mostly in trace amounts. Table 4–1 gives an indication of the relative fraction of the more abundant elements and compounds in the body. Other than the hydrogen and oxygen in water, the most abundant elements in the body are carbon, nitrogen, phosphorus, sodium, and potassium.

Most of the principal cell components, or organelles, that have been identified are shown as a composite in Figure 4–1. The detailed functions of all of these various components have not been determined, nor are they all found in each cell; as an oft-quoted example, a red blood cell does not contain a nucleus. In addition to these identifiable cell components various cell chemicals also provide important functions, some of which are becoming better understood. Thus two of the nucleic acids found in cells, the ribonucleic acid (RNA) and the deoxyribonucleic acid (DNA), have become of increasing interest, the latter because of its importance in the reproductive process, and the former because it apparently controls the

TABLE 4–1

RELATIVE ABUNDANCES OF THE ELEMENTS IN THE BODY

Element	Percent Abundance	
	By Number of Atoms	By Weight
Hydrogen	63	10
Oxygen	30	76
Carbon	5.5	10.5
Nitrogen	1.14	2.5
Phosphorus	0.06	0.3
Potassium	0.05	0.3
Sulfur	0.04	0.2
Chlorine	0.02	0.1
Sodium	0.01	0.04
Magnesium	0.005	0.02
Calcium	0.003	0.02
Iron	0.001	0.01

Note: Copper, cobalt, zinc, manganese, and other elements are present in trace amounts.

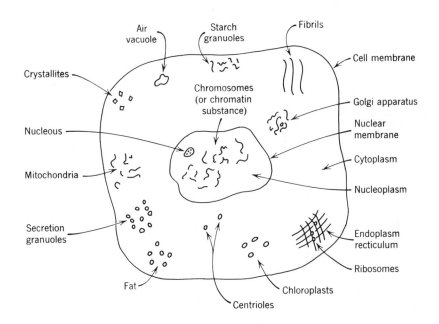

Figure 4–1   Idealized cell.

synthesis of proteins necessary for cellular life; this characteristic makes it of value in assessing radiation injury.

The cell nucleus apparently controls cell activities, and the chromosomes, which are microscopically visible as rodlike units in the nucleus during cell division, are of particular interest since they carry the genes that apparently determine the characteristics inherited by the cell and the organism. Each species has a definite number of chromosomes; thus man has 23 pairs of these homologous chromosomes, or a total of 46, with one of each pair being contributed by each parent. Each chromosome or chromosome segment carries an indeterminate number of genes, which may easily number in the thousands, and each gene of a chromosome has a counterpart on the other of the chromosome pairs; each gene, and its paired counterpart on the homologous chromosome, is related to some characteristic of the living system concerned, although more than one gene pair is usually involved in determining completely a given characteristic. A gene is generally considered to be a part of a DNA molecule, which has a molecular weight on the order of 10^6 to 10^7 amu. Cells with the complete number of chromosomes, 46 for man, are called diploid cells.

CELL MULTIPLICATION

Cells multiply by division. In all but the germ cells (egg and sperm) the chromosomes split in a process called mitosis so that each of the subsequent daughter cells is identical to the single parent. In this process the chromatin material first organizes itself into the filamentary rods (or threads) which are chromosomes, and the nuclear membrane begins to disappear. The chromosomes then split, or divide lengthwise, and replicate themselves, which means that each half then picks up the appropriate molecules and arranges them so as to reproduce two copies of the original chromosome. These two "identical" daughters of each of the original chromosomes then migrate to opposite ends of the cell, the original nuclear membrane having completely disappeared. At this point two new nuclear membranes are formed, each enclosing one of the sets of chromosomes collected at the opposite ends of the cell. This twin-nucleus cell then pinches itself into two units, each containing one of the nuclei, and two cells replace the original one; the rodlike characteristic of the chromosomes then disappears, thus forming the chromatin material that fills the nucleus, and the two cells are ready to take up their normal functions, which are those of the original parent cell.

Germ cells, on the other hand, go through a process called meiosis whereby the homologous chromosomes separate in cell division, and the resulting daughter cells, called haploid cells, each has only half of the original number of chromosomes possessed by other normal body cells; these cells thus have 23 chromosomes for man. However, every characteristic of the species is represented in these haploid cells, but by a single gene or gene combination on a single chromosome rather than the two genes or combinations on the paired chromosomes of the diploid cell. Since one of each chromosome pair originally came from each parent, this meiotic division means that each germ cell will contain some chromosomes and the accompanying characteristics as originally inherited from one parent and the remaining chromosomes and characteristics as inherited from the other parent. Since meiotic separation of the respective chromosome pairs with respect to original parental origin apparently occurs at random, a minimum of some 2^{23}, or 8.4×10^6, different chromosome combinations and corresponding combinations of characteristics are available to the germ cells produced.

Each of these haploid cells then divides by a process similar to mitosis, leaving a total of four germ cells, each with only half the proper number of chromosomes. For males all of these germ cells live and, as sperm, are capable of taking part in the reproductive process. However, in the female only one of these cells lives to become an egg cell, the others becoming polar bodies and disappearing. Union of these germ cells in fertilization of the egg by the sperm results in a diploid cell with the proper number of chromosomes, half from one parent and half from the other. In subsequent division the resulting cells divide mitotically. The sperm cells are developed throughout the reproductive life of the male from the immature spermatogonia. On the other hand the egg cells in the female (approximately 40,000 for humans) are apparently all available, though some are immature, at the beginning of the female's reproductive period if not at her birth.

TISSUES

Normally mitosis produces identical cells. However, during the development of an embryo following union of two germ cells, functionally differentiated cells appear "on schedule" as determined by a little understood mechanism. As a result the adult organism consists of a large number of specialized cells, as already indicated.

Groups of cells form tissues, which, in turn, may be specialized to form specific organs, the characteristic cells of which are said to be differentiated. Although there are many plans for specifying cell types, one of the more commonly accepted methods of identifying tissue types is to divide them into bone, muscle, blood, epithelial, connective, nervous, and germ (or reproductive) tissues. As noted above, not all cells have the composite structure indicated by Figure 4–1, and similarly not all of the respective tissues included in the above rough groupings behave alike; for example, with respect to their rate of division it is a general rule that most epithelial cells continue division through life, whereas muscle cells divide only occasionally, and mature bone cells divide little if at all in an adult.

RADIOSENSITIVITIES

Cells and tissues, or organs, vary widely in their response to radiation, and are thus said to have different radiosensitivities. Although the respective radiosensitivities of various types of cells differ markedly, a major difference for specialized cells depends on the readiness with which they normally undergo mitosis. In general, cells (such as those of the epidermis) in which mitosis regularly occurs are more radiosensitive than those (e.g., nerve cells) in which mitosis rarely occurs. Some aspects of these differences are presented subsequently.

It may be pointed out at this point and also later emphasized that, although much is known about living organisms and life, many observations do not have apparent explanations. Perhaps too many of these for comfort apply to radiation effects. Some of the uncertainties are mentioned in the following sections.

RADIATION INJURY TO CELL

It is frequently considered that the principal injurious effect of radiation at the lowest exposure levels is damage to the cell nucleus, particularly the chromosomes, which either results in chromosomal breaks or appears as a "stickiness" that affects the cell's ability to divide. This inability to undergo normal mitosis may indicate a functional abnormality of the cell that may eventually cause its death.

Another form of this type of cellular damage to the nucleus is the alteration of one or more of the genes constituting a chromosome so that, even though the chromosome itself is not broken and the cell itself may continue its relatively normal functioning, the change involved may ultimately affect a comparatively large number of cells produced by division of this damaged cell. In general such injury is usually minor unless a large number of cells are involved or it is the germ cells that are affected. In this latter case the abnormality may be transmitted to subsequent generations. In fact damage to a germ cell may eventually result in abnormalities in meiosis if the particular germ cell that is affected takes part in the reproductive cycle.

Recent experiments have emphasized the importance to the chromosome chemistry of DNA, the large molecules which apparently exist either as straight units or more generally as coiled structures similar to two intertwined helices. The helix, incidentally, has very interesting mathematical properties. As already noted, the DNA molecule is apparently intimately involved with the genes, or heredity carriers, of the cells and has the ability to replicate, or reproduce itself, by splitting with each half then "picking up" and properly arranging the molecules necessary to complete itself. Early experiments showed that radiation decreases the viscosity of DNA solutions, this apparently indicating breaks in the chromosomes themselves. In fact nuclear radiation is apparently capable of damaging systems of large molecules not associated with living systems by breaking individual units.

SENSITIVE VOLUME AND TARGET THEORY

Studies of the effect of radiation on various materials, including the large molecules and cells of living organisms, have resulted in a so-called target theory. This concept generally assumes that all portions of a given tissue are not equally susceptible to radiation damage but that there are *sensitive volumes* in which energy released by impinging radiation will cause more biological damage than in other locations. Hence the probability of injury to tissue depends on the probability that the incident radiation, considered as a particulate (which would mean consideration of gamma rays as photons), will enter one of these sensitive volumes. Such probability in turn depends on the relation of the total cross-sectional areas of these sensitive volumes to the total tissue area being irradiated. This of course introduces the concept of a *target*, which may be treated mathematically much as is the neutron impact probability given in Chapter 15.

Obviously the incident radiation must release energy in the sensitive volume to cause any effect (including damage), and the probability thereof depends on the pattern of such release. Therefore the linear energy transfer (LET), which is defined as the energy deposited in the attenuating material per unit path length of incident radiation, is also of importance. These concepts, as physical phenomena, are further described in Chapter 5.

The concept of a target and sensitive volume is particularly applicable to considerations of radiation effects on the cell nucleus. Thus it readily leads to the idea in genetic studies of chromosome breaks or point mutations, which are the result of localized changes in the genes without recognizable loss or alteration of the genetic material involved. As will be discussed later, some of the more easily caused point mutations (many of which are apparently well defined) have been studied in certain mammals as well as in lower forms of life. Such genetic changes also occur as a result of background radiation, heat, certain chemicals, and probably other unidentified causes.

On the other hand it appears that the cell cytoplasm is relatively resistant to radiation. However, the effect of radiation on the cell membrane is apparently not too well known, although it would appear that even comparatively slight damage or breaks in such membranes might lead to rather severe cell injury and perhaps destruction as a result of disturbances in such factors as the cellular water balance, permeability to various chemicals, and enzyme formation and action.

CELLULAR RADIOSENSITIVITY

In 1906 Bergonie and Tribondeau came to the conclusion that a given dose of radiation (X-rays at that time) causes the greatest damage in cells that reproduce rapidly, have long periods of mitosis, and are less morphologically and functionally differentiated. To this may be added the observation that the higher the order of the organism, the less radioresistant, on the average, are its cells.

It is generally accepted today that cells are most radiosensitive during division and that tissues whose cells normally divide readily are more radiosensitive than others, although this is not always so; for example, some types of rapidly growing tumors are relatively insensitive to radiation, whereas the opposite is true for lymphocytes, which divide only rarely. In agreement with the above general principle is the observation that the skin and the blood-forming processes in the bone marrow, the cells of both of which are continually dividing, are considered to be among the most radiosensitive tissues in the normal body, whereas muscles and the nervous system, the cells of which divide rarely, if at all, are considered among the more radioresistant types. However, some experiments in the Soviet Union indicate the possibility that the central nervous system may respond to very low radiation doses, though probably not in the sense of cell change or damage. The possibility has also been suggested that adjustment of the frequency of periodic radiation exposures to the rate of mitosis of specific cells or cell types may be extremely important in producing damage to the tissues concerned.

As a result of the relatively high radiosensitivity of dividing cells a given radiation exposure in a child is usually considered to be more significant than is the same exposure in an adult. Similarly, the success of radiation in checking some types of cancer is attributed to the high radiosensitivity of the rapidly proliferating cancer cells; this latter may be a two-edged sword, however, in that the comparatively huge doses of radiation necessary to injure enough cancerous cells to be useful to the body may also be sufficient to produce undesirable changes, including perhaps malignancies, in surrounding normal cells.

Although a great deal of research has been initiated in an attempt to find ways of reducing cellular radiosensitivity or damage from radiation, the results have not been too encouraging. It has been established that a decrease in the oxygen supplied to a cell tends to decrease radiation effects and that the presence of certain substances, such as the amino acid cystein, also tends to limit cellular damage.

REPARATIVE FUNCTIONS OF THE BODY

If a cell is actually killed, the body's normal reparative functions will endeavor to replace it as for similar destruction by any other means. Similarly, if a cell is injured, the body's reparative activity will behave normally. In this sense, then, radiation damage may be considered as being similar to that caused by other injuries, such as burns, which may destroy some cells and injure others. In all of these cases it is noted that the greater the total damage, or the greater the injury, the slower will be the repair and the greater the danger that the body cannot repair itself efficiently or completely.

The body's ability to repair various of its bodily parts also varies. Thus whereas skin is relatively easily replaced, the opposite is normally true of damage to eye or nerve fiber.

FACTORS THAT AFFECT REPAIR CAPABILITY

A major factor that affects the extent of damage to the body resulting from a given amount of radiation exposure is the period of time over which it is received. It is perhaps intuitively and qualitatively obvious that the longer this time, the greater is the opportunity of the body to repair itself and thus the less the apparent injury. This is generally true, but there is at least one exception, which will be mentioned later. This is no different from other and more commonly observed phenomena. Thus a person can obtain a tan by exposing himself gradually over a period of time to the sun, perhaps for a total of 10 hours during a 2-week period. However, if he exposed himself to the sun for a continuous 6-hour period, he well might not even get the desired tan but he would almost certainly get a very painful sunburn that could even be dangerous.

Similarly the body should be better able to repair a given amount of damage to a small part of itself than to repair the same unit cellular damage to a more extensive involvement of tissues. This means of course that a given exposure to any small part of the body is not as dangerous as the same exposure to the entire body. This is again similar to the observation that a person has a very good chance of surviving if he has received third-degree burns over only a small fraction of his body but that his survival chances become negligible if he receives such burns over his entire body. The implications of these general observations are more specifically noted subsequently.

If the radiation is intense enough to cause disruption and possible destruction not only of individual cells but of groups of cells that have a particular function in the body as does a single organ then the effect may be more serious than would be the case if the same total amount of damage were spread over all parts of the body. The possible effects of injury or damage to body cells observed may thus depend on the number of the cells damaged and their location. If a sufficient number of cells are injured or killed by irradiation then the effect of actual injury to the individual will appear; a lesser number might not be noted due to the body's reparative powers. If such injury involves internally functioning organs of the body the effect may take the appearance of the disease which would result from partial or total malfunction of those body parts.

OVERALL RADIOSENSITIVITY CONSIDERATIONS

One factor that complicates studies of radiation effects is the fact that different individuals apparently exhibit differences in overall sensitivity to radiation which cannot be detected without actual exposure. The reasons for these differences have not been identified, and their extent is very poorly known. However, statistical studies of lethal doses in animals suggest that radiosensitivity differences among members of a population are certainly no greater than threefold but more probably about twofold. Thus the lethal dose for the *least* radiosensitive individual is probably no more than twice (or three times) the lethal dose for the *most* radiosensitive person. Similar radiosensitivity ranges for other effects have not been stated but are probably of this same order of magnitude.

Two other possible radiation effects on the body should also be considered, even if only as interesting speculations. One may be identified as stimulation. Thus, although the immediate effect of a small dose of radiation may be injury to a comparatively few cells, such "injuries" may actually heighten the ability of the body to repair itself and remain "in shape." This may be likened to the effects of strenuous exercise, which not only increases the rate at which certain body cells are destroyed but also produces certain enzymes that cause "sore muscles" and thus appear to be immediately harmful; however, exercise is almost universally considered to be beneficial over the lifespan, especially if continued.

The other possible overall radiation effect may be likened to that of germicides. Radiation exposures at levels that have little, if any, effect on an individual may through direct or indirect effects be highly lethal to unidentified infections or other parasites with a net favorable result to the individual. An effect of this type has apparently been observed in agricultural studies involving poisons and is usually given as an explanation for the increased longevity noted in many chronic exposure studies, which are described later.

As an overall consideration it should be emphasized that, although radiation has peculiarities of its own, the injuries it produces are essentially the same as those that occur from other sources, and there is no particular reason to fear radiation injury more than injuries from these more "normal" substances or agents. All radiation effects involve the release of energy in the body of the individual concerned, and the results of this depend strongly on the extent and location of this release. Radiation injury and its significance is thus similar to the severity of injuries from other causes.

REFERENCES and SUGGESTED READINGS

Almost any text on biology and cell structure.
Ackerman, E., *Biophysical Science,* Prentice-Hall, Englewood Cliffs, N.J., 1962.
Andrews, H. L., *Radiation Biophysics,* Prentice-Hall, Englewood Cliffs, N.J., 1961.

QUESTIONS

1. Describe the processes of meiosis and mitosis.
2. What is a gene?
3. When are cells most radiosensitive?
4. Define target theory and linear energy transfer.
5. Comment on the validity of the following statement: "It makes no difference whether just my arm or my whole body is exposed, an injury I receive will be repaired by my body just as quickly in one case as in the other."
6. How wide is the apparent range of lethal doses for humans?
7. Radiation apparently has a genetic effect on the living organism. In what way does radiation thus affect the organism's ability to reproduce?
8. What is the reasoning behind the fact that different tissues of the body are affected to a different extent by radiation?
9. Which dose would have the greater effect on the body, a small dose over a long period of time or a large dose over a shorter period of time? Assume that the total dose is equal in both cases. Support your answer.
10. What are the major chemical elements in the cell?

CHAPTER FIVE

Physical Aspects of Radiation Exposure

IMPORTANCE OF ENERGY CHANGE

The fact that nuclear radiation, particulate or wave, possesses energy has been described, as has been the concurrent fact that some or all of this energy is lost or released if the wave intensity is reduced or if the particle is slowed down or stopped. In general this "lost" energy is imparted to, or released in, the environment or materials through which the radiation passes; if the radiation is incident on biological specimens, this release in individual cells may damage or even kill them. The detailed mechanism by which cell injury results from radiation exposure is very poorly known; particularly, this is true since the injury produced seems to be much greater than would be suspected from the amount of energy physically released or the number of cells or other units affected.

It has been shown that radiation essentially interacts with matter almost completely at the atomic or molecular level. Since each biological cell is an aggregation of a large number of molecules, it might well be expected that ionization or other radiation-induced effects would cause serious injury only if a relatively large number of atoms in a correspondingly large number of molecules were affected. However, this does not seem to be the case. Thus, it has even been estimated that the final effects of radiation exposure at lethal levels are somewhat analogous to assuming that a severely damaging action, or death, involving 20 people today will result in the death of the entire population of the United States 6 weeks from now. Assuming a population of 200 million people, the 20 people represent a population fraction of 10^{-7}. Even recognizing that the cellular mechanism and its interrelations are extremely complicated, that radiation can cause

70

chemical effects, and that there are unique biological molecules and vital cells, it is rather difficult to see how the serious overall effects that are observed can result from the energy released by interaction of electro-magnetic waves or particles with the atomic nuclei that make up the various components of the cells. Such is the case, however.

ENERGY RELEASE IN CELLS

Possible injury from nuclear radiation is caused by a particle or wave losing some of its energy to the body cells, primarily by ionization of a specific atom in the case of ionizing radiation; however, individual atoms may also absorb energy in being excited, as already described. Similarly, the incident radiation may lose energy by affecting the bonds between one or more of the atoms in a cell molecule. The molecule may then become excited, which is analogous to the atomic excitation already described, whereby the molecule itself absorbs energy and thus goes to a higher permitted energy level. However, if the input energy is sufficiently high, the molecule may be dissociated, with the bonds between one or more of its constituent atoms being actually broken even as a sufficiently high energy input to an atom results in ionization. Any energy lost by a particle or wave in producing ionization, excitation, or dissociation reduces its retained energy and correspondingly limits any further effects which it can produce. Of the various ways by which radiation may lose energy in traversing matter, that by ionization is among the more important and provides a relatively simple model for evaluating energy loss. Hence in succeeding sections of this chapter any energy release is treated as though it were ionization production except as specifically noted.

Without attempting to go into detail, it is perhaps obvious that the loss of energy by incident radiation has the overall effect of increasing the energy of the material through which the radiation passes. This may eventually appear as heat, which at the atomic or molecular level represents an increase in the kinetic or vibratory energy. Hence an actual effect of radiation absorption by a material may be a localized increase in its temperature. Although not currently considered a major factor in ac-counting for the injurious effects of radiation, the initial conclusion that localized overheating, or "burning," of the cells was a logical cause for injury was at least partly explained by the fact that skin burns, or erythema, were among the earliest recognized effects of radiation. An energy change, incidentally, is necessary for detecting any effect of nature!

DENSITY OF IONIZATION

The overall body injury caused by radiation depends not only on the total energy released or deposited therein, but also on the concentration of such energy release. If the radiation effect is that of ionization, this concentration is expressed in terms of the ionization density n, which is defined as the ionization, or number of ion pairs produced, per unit volume; n is thus expressed in the consistent units of ion pairs/cm.[3]

Obviously, the total energy E released by the ionizing radiation in this unit volume is the product of the number of ion pairs produced and the average amount of energy released per ion pair, W, or $E = Wn$. It should be noted that E, as thus defined, includes not only the energy actually deposited in the unit volume but also such secondary effects as Cerenkov radiation (which is emitted by an electron traveling in a transparent medium at a velocity greater than that of light in that medium), which may be deposited at a distance. In general, the energy released in a unit volume and that deposited therein are approximately the same. Hence, unless the difference in the two values is important and such is thus indicated, the two will be considered to be identical.

If W has the consistent units of ergs/ion pair, E is expressed in units of ergs/cm.[3] In material of density ρ this amount of energy is obviously released in a mass numerically equivalent to ρ. Thus E/ρ is the energy release per unit mass of the material concerned as expressed in the consistent units of ergs/gm. For indirectly ionizing particles, such as neutrons or the photons of electromagnetic radiation, a closely related factor is called the kerma, K, for which the energy concerned is the sum of the initial kinetic energies of released charged particles. Eventually this may include the energy radiated by the charged particles and the energies of any charged particles produced by secondary processes in the unit volume. The kerma has also been defined as $K = \Delta E_k/\Delta m$, where ΔE_k is the sum of the initial kinetic energies of all the charged particles liberated by indirectly ionizing particles in the mass Δm of an elementary volume element; a similar identification has not been given for directly ionizing particles, such as betas, alphas, or protons.

The ionization density n is also the product of the number of incident particles per unit area, called the particle fluence Φ, which may be expressed in the consistent units of particles/cm.[2], and the specific ionization S which is defined as the number of ion pairs produced per unit track length of an incident particle (in this case assumed to traverse completely the thickness of the unit volume) and may be expressed in the consistent units of ion pairs/cm.-particle. Thus $n = S\,\Phi$ and $E = S\,\Phi\,W$ for which the unit is ergs/cm.[3] The particle fluence Φ is also defined as $\Phi = \Delta N/\Delta a$, where ΔN is the number of particles entering an elementary spherical volume of cross section Δa; S may also be defined as $S = \Delta i/\Delta l$, where Δi is the number of ion pairs produced per elementary path length Δl per particle. Obviously, the energy released per unit path length traversed by an ionizing particle is $L = SW$, where L is identified as the linear energy transfer (LET), expressed in consistent units of ergs/cm-particle; L is also expressed as $L = \Delta E_l/\Delta l$, where ΔE_l is the energy released by a particle in traversing a distance Δl.

It may be noted that the value of L given above accounts for the total energy lost by the incident radiation, including energy released in other than the ionization event itself and thus not deposited at the primary interaction site. However, these secondary effects are usually comparatively minor, and the value of L as defined is thus approximately the same as the total energy deposited in an attenuating material per unit path of the incident particle. Actually the LET is also sometimes defined in this way, as in the equation above, although the volume concerned is also usually defined.

It is obvious that if incident particles are to traverse some specific thickness of material, they must have an overall energy no less than that released by ionization in the material concerned. Thus, if each incident particle has an energy A (exclusive of rest energy), the total energy incident per unit area is given by $F = \Phi A$; F is identified as the energy fluence and has the consistent units of ergs/cm.[2]. The energy fluence F is also defined as $F = \Delta E_F/\Delta a$, where ΔE_F is the sum of the energies, exclusive of rest energies, of all the particles that enter an elementary spherical volume of cross section Δa.

For continuous radiation over a period of time t the total energy released per unit volume, E, is given by the relation $E = SW\phi t$, where $\phi t = \Phi$ and ϕ is defined as the particle flux density, which is the number of incident particles per unit area per unit time expressed in the consistent units of particles/cm.[2]-sec. The particle flux density ϕ is more precisely defined as $\phi = \Delta \Phi/\Delta t$, where $\Delta \Phi$ is the particle fluence for the element of time, Δt. If $Q = E/t$, Q is the time rate of energy release per unit volume and is expressed in the consistent units of ergs/cm.[3]-sec. Obviously then, $Q = SW\phi$. It has been already shown that $E = Wn$.

The incident energy per unit area per sec, I, also identified as the energy flux density or the intensity, is defined as $I = F/t$ and is expressed in the consistent units of ergs/cm.[2]-sec. More precisely the intensity may be defined as $I = \Delta F/\Delta t$, where ΔF is the energy fluence for the element of time Δt. The above units are also those of the intensity of electromagnetic radiation incident on a surface. Thus, although E, Q, F, and I have been developed from a model involving particles, the energy and the energy time rates that they represent may be used not only for particulate radiation but also for gammas and other electromagnetic waves.

For gammas and other electromagnetic radiation the LET is not as simply demonstrable as in the case of ionizing particles for which the actual number of events per particle may be counted. However, an equivalent LET for a photon may be determined. It is experimentally shown that the change in intensity of a gamma beam, ΔI, in passing through a thickness Δl of some absorbing medium is proportional to that thickness and to the intensity I of the incident beam. Thus $\Delta I = - kI\Delta l$, where k is an experimentally determined constant of proportionality and its negative sign indicates an intensity decrease.

This intensity change in the incident gamma beam obviously represents a rate at which energy incident on unit area is released in a thickness Δl of the medium or absorbed therein. It is thus expressed in consistent units of ergs/cm²-sec, which are those for the incident intensity I. The product kI thus may be expressed in consistent units of ergs/cm³-sec. Since these are also the units of the product ϕL for particulate radiation, the two quantities are analogous and an equivalent LET for gammas may thus be determined.

Although the units in which the LET is expressed should obviously involve energy, it has been normal practice to express the LET in units that are appropriate to the specific ionization, one commonly used being ion pairs per micron (the micron is 10^{-6} m). Radiation that produces less than about 100 ion pairs/micron in tissue is normally considered to have a low LET, whereas radiation that produces more than about 1000 ion pairs/micron is considered to have a high LET. In general betas, gammas, and neutrons are considered to be of low LET, whereas alphas and other heavy charged particles are considered to have a high LET.

Since the energy that is expended in producing an ion pair in air is about 34 eV, (more precisely 33.7 eV), which is equivalent to about 5.4×10^{-11} erg, (34 eV $\times$ 1.6 $\times 10^{-12}$ ergs/eV) particulate radiation with an LET of one ion pair/micron is equivalent to about 5.4×10^{-7} erg/cm-particle, or 5.4×10^{-12} J/m-particle. Hence particulate radiation of a low average LET expends energy at a minimum rate of less than about 5.4×10^{-10} J/m-particle (or 5.4×10^{-5} erg/cm-particle), whereas the corresponding release rate for high LET radiation is greater than about 5.4×10^{-9} J/m-particle (or 5.4×10^{-4} erg/cm-particle).

As shown in Chapter 3, the energy of a gamma photon is $h\nu$, where h is Planck's constant (6.62×10^{-27} erg-sec, or 6.62×10^{-34} J-sec) and ν is the gamma frequency. It was similarly shown that the energy of a photon, usually expressed in electron volts, may also be stated in mechanical units by use of the transformation relation that 1 eV $= 1.6 \times 10^{-19}$ J, or 1.6×10^{-12} erg.

It is obvious that the intensity of an incident gamma beam, in ergs/cm²-sec, may also be expressed as the product of a photon flux density n in photons/cm²-sec and a photon energy $h\nu$ in ergs. Hence $I = nh\nu$ and $\Delta I = -knh\nu \ \Delta l$. Thus an equivalent energy loss per photon is

$$\frac{\Delta I}{n} = \frac{-kn(h\nu)}{n} \ \Delta l = -kh\nu \ \Delta l.$$

Obviously, $kh\nu$, the energy deposit in the medium per unit thickness per incident photon, is analogous to L, the LET coefficient for particles. It should be pointed out that generally attenuation, or energy loss, by an electromagnetic wave does not mean that each incident photon loses some part of its energy, as is generally true for betas and similar particles, but that the number of photons is reduced. This will be more fully discussed in Chapter 15.

To illustrate the evaluation of an equivalent LET for a gamma beam assume an attenuation constant k of 0.07 per centimeter in tissue for 1 MeV gammas. Hence the energy release per incident photon in a 1 micron tissue thickness is given by 1 MeV/photon $\times$ 0.07 per cm $\times$ 1.6×10^{-12} erg/eV $\times$ 10^6 eV/MeV $\times$ 10^{-4} cm/micron $= 11.2 \times 10^{-12}$ erg/photon-micron. Since the energy necessary to produce one ion pair is 5.4×10^{-11} erg, it is apparent that the fractional number of ion pairs produced per incident 1 MeV photon per micron in tissue is approximately 0.2. If a photon is considered to be a particle, an attenuation of 0.07 per centimeter for 1-MeV gamma rays is equivalent to such a photon's losing 7×10^4 eV/cm, or 7 eV/micron. This energy is obviously about 20 percent of that required to produce a single ion pair.

The above definitions and derivations have been developed on the basis that in a given situation all units of a given quantity have the same unique value; for example, a single value for L assumes that every particle produces the same number of ion pairs per unit path length and that the same energy is released at each ionizing event. Because of statistical fluctuations of various kinds, this is not actually true. However, if the energy released per ionization event is averaged over a large number of such events and if the number of such ionization events as actually produced by a large number of ionizing particles is averaged over the total distances traveled by these particles, a single useful value of L may be obtained. Useful values of the kerma, fluence, etc., are developed by similar averaging procedures, usually involving the techniques of finite or infinitesimal calculus as indicated above. In general the functional relationships are empirically determined, although theory has been developed for certain cases.

ATTENUATION AND PENETRATION

It is perhaps obvious that the greater the LET, or equivalent LET, of a given type of radiation in tissue, the more rapidly does the radiation lose its energy in traveling through matter and the less energy is transmitted through a given thickness of material. This transmission of energy, or radiation penetration, is also of a great deal of importance in radiation-protection considerations.

When a gamma beam passes through a given thickness of material, its fractional energy loss, or attenuation therein, is proportional only to the total energy, or intensity, of the incident beam. This was mentioned in the preceding section. Since this incident intensity steadily decreases with depth of penetration, the energy released per unit path length in general also decreases with such depth; this is discussed at greater length in Chapter 15, as is the fact that the attenuation per unit path length depends primarily on the mass density of the material. Accordingly a gamma beam theoretically does not lose all of its energy, regardless of the thickness of the material concerned, even though it may be so greatly attenuated as to have a negligible intensity. Although the attenuation factor for a given material depends on the gamma frequency and thus the photon energy, the velocity of incident gammas is unchanged by attenuation as is also the frequency of unscattered gammas. Thus the loss of energy by a gamma beam may be visualized as the complete loss of energy by some photons with others being unaffected. However, there may be a frequency change in gammas scattered out of the direct beam, as is described later.

On the other hand the loss of energy by neutrons, alphas, and betas may be considered to result from impacts of these incident particles with components, such as electrons, of the individual atoms in the material concerned. Since each impact causes a change in the velocity and thus in the energy of the colliding particles and since the impact probability as well as the average energy lost per impact depends on the velocity of the particle, it is obvious that the fractional energy loss per unit path length by a beam of such particles in tissue depends on both the initial energy of the particles and their depth of penetration. In any event, a series of such collisions can eventually result in the particles being essentially stopped. Thus such particles have a definite range, or maximum depth of penetration into a given substance. In addition to dependence on their own initial energies, the range of alphas and betas (as discussed in a subsequent chapter) depends primarily on the mass density of the material concerned, whereas the range of neutrons depends on the nuclear characteristics of the atoms composing the material, hydrogen being of particular interest.

However, the first few impacts of charged particles, particularly of betas, may result in comparatively small energy losses; thus for penetration distances that are small compared with their range, an attenuation factor similar to that for gammas may be developed. In addition, the impact probability for neutrons may be so small for many substances that an attenuation factor similar to that used for gammas may be developed for some useful conditions.

ENERGY LOSS BY ALPHAS AND BETAS

Atomic impacts by alphas and betas generally are either elastic, as in the case of billiard balls, wherein the incident particle loses energy in giving kinetic energy to the target, or inelastic, in which case the particle loses energy in producing ionization, dissociation, or excitation. In these latter cases low-energy electromagnetic radiation may be released as the atoms or molecules concerned return to "normal"; even though this photon energy may be too low to produce additional ionization, it may cause molecular or atomic motion and thus produce heat in the cells affected, perhaps at some distance from the site of the original ionization event.

NEUTRON ENERGY LOSS

With respect to neutrons, however, different mechanisms are assumed to be predominant in the slow and the fast neutron energy ranges. For low energy and thermal neutrons, it appears that the principal energy release occurs in the nuclear reaction of the neutron with the nitrogen and the hydrogen in the body, with the latter being predominant. These reactions are

$$_0^1 n + {}_1^1 H \rightarrow {}_1^2 H + \text{energy},$$

$$_0^1 n + {}_7^{14} N \rightarrow {}_6^{14} C + {}_1^1 H + \text{energy}.$$

A 2.2 MeV gamma is released in the hydrogen reaction, whereas the proton released in the nitrogen reaction has an energy of about 600 keV.

Thermal neutrons are those whose kinetic energies are on the order of 0.02 eV, which is equivalent to the average energy of molecules moving in air at a temperature of about 20°C.

For high energy neutrons, on the other hand, it is estimated that about 80 to 95 percent of the energy loss occurs in elastic scattering from the hydrogen nuclei in the body cells. A large fraction of the incident-neutron energy is normally absorbed by this hydrogen nucleus, which may thus be severely displaced or even removed from its molecules; these are called *recoil nuclei*. The neutrons are themselves eventually thermalized by these collisions and behave as thermal neutrons.

Thus from an elementary point of view it may be concluded that the damage from low energy neutrons is the result of actual chemical changes (or transmutations) in some of the molecules or atoms in the cell, whereas the damage from high energy neutrons results primarily from serious dislocations of hydrogen nuclei in the cell structure.

In the case of transmutation the affected nuclei may emit gammas or charged particles, and these may be sufficiently energetic to cause ionization. Thus, whereas ionization is not generally a primary effect of neutron irradiation of tissue, it may be an important secondary one. The resultant ionization density thus depends on the incident-neutron energies, the neutron-capture characteristics of the irradiated material, and the type and energy of the radiation emitted. Both the form and the quantity of the energy emitted by neutron-induced nuclear transmutation is, in general, independent of the incident energy of the captured neutron. On the other hand, the range of the neutrons in a substance depends strongly on the neutron energy and the nuclear characteristics of the atoms involved, principally their elastic collision properties. This is discussed in Chapter 15.

Generally neutron effects are developed from a so-called first collision dose concept that takes account of the energy released at the first collision of a neutron with an atomic nucleus and ignores secondary effects. It is commonly accepted that this accounts for practically all of the energy actually released by neutrons in slowing down. Figure 5–1 indicates the relative absorbed tissue dose per neutron for neutrons of various energies. Further implications of this figure are discussed in Chapter 10.

PARTICLE RANGES IN TISSUE

The range of most betas of interest is a few centimeters in flesh, about 2 mm in aluminum, and about 10 m in air; betas with energies lower than about 70 keV do not penetrate the dead outer skin layer. Although thermal neutrons have a range of about 1 ft in tissue, fast neutrons have such a

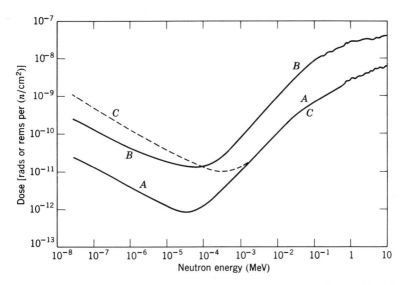

Figure 5–1   Relative absorbed tissue dose for neutrons. Curve *A:* first-collision-dose curve (rads/*n*-cm²); curve *B:* first-collision RBE dose curve (rems/*n*-cm²) (curve *A* x RBE); curve *C:* total energy deposited (multiple collisions) (rads/*n*-cm²). (From NBS Handbook 63, p. 7, 1957.)

great range that their attenuation by the human body may be considered to be essentially of the same type as that of gammas. Alphas of concern have a definite range of a few centimeters in air and a few fractions of a millimeter in tissue, with those whose energy is lower than about 7 MeV being completely stopped by the dead outer layer of the skin. Hence, as long as an alpha emitter remains outside an individual's body, even though such material may coat the skin, there is little radiation concern since essentially no radiation energy will be released in living tissue. Accordingly, alphas and other heavy particles which are electrically charged are considered to be nonpenetrating, but neutrons, gammas, and most betas readily penetrate the outer skin layer and are thus frequently identified as penetrating radiation.

EXTERNAL AND INTERNAL EXPOSURE

External exposure is produced by a radiation source outside the body and thus depends on the capability of the radiation concerned to penetrate the part of the body that is irradiated. The external radiation source could be either highly localized, as in the case of a single sealed radium-beryllium or cobalt-60 source; it could be diffused, as would result from radioactive materials spread over surfaces or other extended locations; or it could be an intermediate case, as in a large reactor in operation. As may be surmised only beta, gamma, and neutron emitters are external exposure hazards.

Internal exposure results from radioactive materials having been introduced into the body. This generally results from inhaling airborne radioactive materials, from ingesting such materials in food or drink, from their entry through broken skin (as may occur from a puncture wound by a splinter or nail contaminated with radioactive material), or from absorption through the unbroken skin. These factors are detailed in subsequent chapters. Obviously, alpha, beta, and gamma emitters may all be taken into the body and are thus internal exposure hazards. However, except for the comparatively few current neutron sources (principally mixtures of radium-beryllium or plutonium-beryllium), essentially all neutrons are produced only during a nuclear reaction and are thus negligible as internal hazards since, so far, it has been impossible to make a reactor small enough to be inhaled or swallowed!

Qualitatively, it may be noted that because of their relative penetration capability gamma beams from an external source should cause more injury to deep-seated internal organs than should corresponding beta beams of the same energy, but the opposite should be true for the skin and body organs that are near the surface. On the other hand, a gamma-emitting substance taken into the body and deposited in one of the body organs should cause less injury to that organ than a similarly deposited beta emitter of the same energy. Alphas expend their entire energy in traversing only a fraction of a centimeter in tissue, and thus internal deposits of alpha emitters should cause much more damage to nearby cells than either betas or gammas of equivalent energy from similar deposits.

There may also be combinations of radiations. Although there are pure gamma emitters, most beta emitters also emit gammas; these are frequently grouped as beta-gamma emitters and are the principal external radiation sources of general concern. Mixtures of radioisotopes may exhibit radiation-type combinations, as is the case for most small neutron sources that are mixtures of beryllium and such strong alpha emitters as plutonium or radium; they are thus of concern for both alpha and neutron radiation. The fact that beryllium is itself a dangerously toxic material points up the possibility of nonradioactive hazards from materials associated with those of radiation concern. Although such problems are beyond the scope of this book, they do require consideration in an overall attack on a given situation.

RADIATION DOSE

In principle, it should be comparatively simple to compute the ionization produced by incident radiation and thus the energy lost to the various tissues in its path. However, for reasons such as those indicated previously the theoretical interpretations of the energy transfer relations become rather complicated and are beyond the scope of this book; they are available in the literature, as are also the empirically determined formulas by which such energy release may be approximated. On the other hand, for practical purposes extremely detailed analyses are usually unnecessary since "averaged" responses for various groups of tissue types and for appropriate ranges of radiation energies generally provide adequate information.

The energy released in tissue is called a dose, although the terms "exposure" and "radiation exposure" are sometimes used with the same meaning. Two types of "dose," the absorbed dose and the exposure (or exposure dose), are usually considered for reasons that will become apparent subsequently.

The absorbed dose D is currently defined as the energy ΔE that is imparted to the mass Δm of a given volume element, or $D = \Delta E/\Delta m$. However, an initial definition which may actually give a better physical interpretation of this important quantity describes the absorbed dose of any ionizing radiation as the energy imparted to matter by ionizing particles per unit mass of irradiated material at the place of interest. The unit of absorbed dose is the rad. 1 rad is 100 ergs/gm. Obviously also 1 rad = 0.01 J/kg.

The exposure* X is currently defined as $X = \Delta Q/\Delta m$, where ΔQ is the sum of the electrical charges on all the ions of one sign produced in air when all the electrons and positrons liberated by photons in a volume element with a mass of Δm are completely stopped in air.

An early definition states that the exposure (dose) of X- or gamma-radiation at a certain place is a measure of the radiation that is based upon its ability to produce ionization. The unit of exposure (dose) of X- or gamma-radiation is the roentgen (R). One roentgen is an exposure (dose) of X- or gamma-radiation such that the associated corpuscular emission per 0.001293 gm. of air produces, in air, ions carrying one electrostatic unit of quantity of electricity of either sign. Since 1 esu quantity of electricity

* See Author's Note on page 93.

(or charge) is about 3.33×10^{-10} C, 1 R $= 2.58 \times 10^{-4}$ C/kg. Although not included in the definition, the air considered is dry air, earlier definitions specifying dry air under normal conditions of temperature and pressure. The upper limit of energy for which the roentgen should be used is 3 MeV, and for practical purposes a lower limit of 20 keV is also usually specified. The earlier roentgen specification referred to ionization produced by 200 to 300 keV (or 0.2- to 0.3-MeV) X-rays.

COMPARISON OF ABSORBED DOSE AND EXPOSURE

Since the absorbed dose actually represents the energy that is released in the tissue concerned, it should give a truer indication of an actual radiation effect than does the exposure dose. Obviously, what is considered to be the total-body dose for an individual exposed to radiation should be related to some average of the doses absorbed in the various tissues of his body. However, absorbed dose measurements are difficult, and, as is discussed in a subsequent chapter, ionization of a gas is one of the simpler effects that can be detected; many monitoring instruments reflect such ionization. Thus they measure the exposure (dose), which may actually be considered an air-absorbed dose. In most cases the absorbed dose can be calculated with fair precision from the measured value of the exposure (dose). Thus, although the roentgen is a unit of X- and gamma-ray exposure only, the term has also been more widely used in practical dosimetry to indicate absorbed dose from X-rays or gammas; the earlier literature reflects this usage. Similarly the rad (or the earlier rep), an absorbed dose unit, has also been frequently used to indicate an exposure (dose) of nonelectromagnetic radiation.

The absorbed dose in rads, D, may thus be expressed in terms of the exposure in roentgens, R, by the expression

$$D = fR,$$

where f is a factor that depends on the type of irradiation, its energy, and the characteristics of the irradiated material. Table 5–1 gives values of f for certain materials of interest for various energies of gammas. However, it is instructive to indicate the types of analyses used in determining values of f for specific conditions.

For these calculations, ionizing particulate radiation is assumed, with the measurements made in small volumes or thicknesses of material. Then the absorbed dose, which implies irradiation by particles, depends on a material's mass stopping power, which is defined as the loss of energy per unit mass per unit area by an ionizing particle traversing a material

TABLE 5–1

RATIO OF ABSORBED DOSE TO EXPOSURE FOR GAMMA RAYS

Photon Energy (MeV)	f (rad/roentgen)		
	Water	Muscle	Compact Bone
0.01	0.92	0.93	3.58
0.05	0.90	0.93	3.61
0.10	0.95	0.95	1.47
0.20	0.98	0.97	0.98
0.50	0.97	0.96	0.93
1.00	0.97	0.96	0.92
2.00	0.97	0.96	0.92
3.00	0.97	0.96	0.93

medium. The mass stopping power s/ρ of a material is also defined as

$$\frac{s}{\rho} = \frac{\Delta E_s}{\rho \Delta l},$$

where ρ is the material density and ΔE_s is the energy lost by a particle of a specified energy in traversing a path length Δl in that medium. It may thus be expressed in terms of ergs-cm^2/g.

If energy is lost by ionization, the mass stopping power depends on the energy lost at each ionizing event and the probability of ionization being produced in the mass or volume of interest. Thus, E_m, the energy absorbed per gram of material, is given by the simple relation

$$E_m = J_m W_m,$$

where J_m is the ionization produced (or number of ion pairs produced) per gram of material and W_m is the average energy expended per ionization event (or ion pair produced).

It has been experimentally determined that W_m for air is in the range of 33 to 35.5 eV/ion pair formed, and a value of 33.7 eV/ion pair is recommended for calculations involving gammas above 20 keV, whereas a value of 35 eV/ion pair is recommended for fast-neutron calculations. Since 1 eV $= 1.6 \times 10^{-12}$ erg, $W_m = 1.6 \times 10^{-12} \times 33.7 = 5.4 \times 10^{-11}$ erg/ion pair formed.

The value of E_m varies somewhat from material to material, this variation being reflected in the respective mass stopping powers of the substances concerned. For materials of principal interest, such as body tissues, the variation is generally rather small, even with respect to air, and the assumption has been made that the variation primarily reflects differences in the respective values of W_m. Hence, the above relation may be expressed by

$$E_m = J_g W s_m,$$

where W is the value of W_m for air (33.7 eV), J_g is the number of ion pairs produced per gram of air, and s_m is the ratio of the mass stopping power of the material to that of air; there are thus no units for s_m, and E_m may be expressed in consistent units of ergs/g. Table 5–2 gives values of s_m for water, carbon, and "tissue" for some electron energies; s_m for air is obviously unity.

In this form the above energy relation is generally identified as the Bragg-Gray theorem or principle which is usually applied to cavity ionization chambers where the cavity contains an ionizable gas (e.g., air) and the cavity wall is the material of interest; ionization chambers are discussed in a subsequent chapter. In essence E_m thus represents the energy released in the chamber wall by particles produced by an ionizing event in the gas, and s_m is the ratio of the stopping power of the wall material to that of the cavity gas for these particles. It is assumed that the energy lost by the ionizing radiation at one event in the cavity gas is measured by

TABLE 5–2

RATIO OF MASS-STOPPING POWERS[a]

Electron Energy (MeV)	s_m		
	Carbon	Water	"Tissue"[b]
0.01	1.05	1.20	1.18
0.05	1.04	1.18	1.16
0.10	1.03	1.17	1.15
0.50	1.02	1.16	1.14
1.00	1.01	1.15	1.13
2.00	1.00	1.14	1.12
3.00	0.98	1.12	1.10
10.00	0.94	1.07	1.05

[a] As defined, $s_m = \dfrac{\text{mass-stopping power of material}}{\text{mass-stopping power of air}}$

[b] Mixture of 13 percent carbon and 87 percent water.

absorption of the energy of the ionization products in the cavity wall. Thus, the cavity must be small compared to the range of ionized particles in the gas contained therein; however, the wall thickness must be much greater than the range of ionized particles therein but small enough to attenuate only slightly the gamma beam that is assumed to produce the initial ionization in the cavity.

If E_m is expressed in ergs/g, the absorbed dose in rads, D, is given by the relation

$$D = 0.01 \ E_m = 0.01 \ J_m W s_m,$$

since an absorbed dose of 1 rad represents an energy release of 100 ergs/g.

Since ionization means the separation of an electron from the remainder of the atom, a single ionizing event produces a charge of each sign equal to e, the electron charge, which is 1.6×10^{-19} C. Hence n_0, the number of ion pairs formed or ionizing events produced, in releasing a total charge q is given by the simple relation

$$n_0 = q/e = q/(1.6 \times 10^{-19} \ \text{C}).$$

The relations given above may be readily used to determine the value of f_a, the ratio of absorbed dose in rads to exposure dose in roentgens, for air. Since the roentgen is defined as the release of 1 esu (or 3.33×10^{-10} C) of electrical charge in 0.00129 g of air, the number of ionizing events (or ion pairs formed) in 0.00129 g (or 1 cm³) of air per roentgen is

$$n_0 = 3.33 \times 10^{-10}/1.6 \times 10^{-19} = 2.08 \times 10^9 \ \text{ion}$$
$$\text{pairs/roentgen-cm}^3 \ \text{of air.}$$

Since this represents only the number released in 0.00129 g of air, the number of ion pairs released/g of air per roentgen, J_{ar}, is

$$J_{ar} = (2.08 \times 10^9)/0.00129 = 1.6 \times 10^{12} \ \text{ion pairs/g of air}$$
$$\text{per roentgen.}$$

Since 5.4×10^{-11} ergs are expended in producing one ion pair, $E_{ar} = (1.6 \times 10^{12}) \times (5.4 \times 10^{-11}) = 86$ ergs/g of air per roentgen. Thus, it is obvious that the relation $D = fr$ becomes, for air, numerically $D = 0.86r$, and f_a is thus 0.86 for air.

Since air under normal conditions of temperature and pressure contains $(6.06 \times 10^{23}/2.24 \times 10^4) = 2.7 \times 10^{19}$ molecules/cm³, it is apparent that a 1-R exposure (dose) ionizes only a molecular fraction of $(2.7 \times 10^{19})/(2.08 \times 10^9)$ or one of 1.3×10^{10} molecules, which corresponds to an ionization fraction of only 7.6×10^{-11}.

For materials other than air relative energy absorptions must also be considered. Hence, the more general relation for f is

$$f = 0.86 \ (\mu_m/\mu_a),$$

where μ_m is the energy-absorption coefficient for the material concerned and μ_a that of air. This ratio is analogous to s_m for particles and is actually probably conceptually more nearly correct even for particles than is that of the relative stopping power.

Related to this energy-absorption coefficient is the mass-attenuation coefficient μ/ρ of a material for indirectly ionizing particles; this is given by

$$\frac{\mu}{\rho} = \frac{\Delta N}{\rho N \Delta l},$$

where N is the number of particles normally incident on a layer of material of thickness Δl and density ρ and ΔN is the number of particles that experience interactions in this layer.

For most substances of interest biologically the ratio μ_m/μ_a is near unity for gammas with a wide range of energies. Specifically, as may be seen from the values of f in Table 5–1, it is in the range 1.0 to 1.2 for water and muscle tissue for radiation energies above 20 keV and the same for bone for energies above about 200 keV. Similarly, as shown in Table 5–2, the value of s_m is in the same approximate 1.0 to 1.2 range for monoenergetic electrons in water and "tissue" for the same energy range. Thus for most cases of gamma radiation f is in the range of $0.86 \times (1 \text{ to } 1.2)$ or 0.86 to 1.03; this also represents the range of the ratio of the absorbed dose in rads to the measured exposure (dose) in roentgens. A commonly used ratio is 0.93, which implies an average energy release of 93 ergs/g tissue by 1 R of gamma- or X-rays.

The fraction of tissue molecules ionized per rad of absorbed dose may also be approximately found from an experimentally determined value W_m for tissue of 36.4 eV per ion pair. On the basis of the 0.93 average value of f for tissue a corresponding μ_m/μ_a ratio of about 1.08 is required in the relation

$$W_m = W_a \ (\mu_m/\mu_a),$$

where W_a has the value of 33.7 eV/ion pair formed in air. Hence in tissue the average ionization energy is $(36.4 \times 1.6 \times 10^{-12}) = 5.82 \times 10^{-11}$ erg/ion pair. Since 1 rad represents an energy release of 100 ergs/g of tissue, an absorbed dose represents $(100/5.82 \times 10^{-11}) = 1.72 \times 10^{12}$ ion pairs/g of tissue-rad. The number of molecules/g of tissue depends on the average molecular mass of these molecules. For water with a molecular mass of 18 the number of molecules of water/g $=$ $(6.06 \times 10^{23}/18) = 3.37 \times 10^{22}$. Thus, for water the fraction of molecules ionized per rad is some $(1.72 \times 10^{12}/3.37 \times 10^{22}) =$ 5.1×10^{-11}. This means that only one of some 1.96×10^{10} molecules is ionized per rad dose. Obviously the ionization fraction will be larger if the average molecular mass of tissue is greater than that of water; for example, it would be about $(\frac{120}{18} \times 5.1 \times 10^{-11}) = 3.4 \times 10^{-10}$ for molecules of molecular mass 120, which is an average value attributed to some amino acids; similarly, the fraction would be on the order of 1.85×10^{-7} for hemoglobin with an approximate molecular mass of 63,000.*

Although Table 5–2 shows that the value of s_m is in the 1.0 to 1.2 range for "tissue" and water for most electron energies of interest, there is a significant practical difference in the exposure of tissue to beta particles and to gammas. This results from the fact that as betas lose energy their velocities change, and the absorbed dose in a given thickness of tissue depends on the energy of the beta particles incident thereon. In addition the range for many beta energies is less than various tissue thicknesses; for example, 70 keV betas will not penetrate the dead outer skin layer. Most gammas of concern readily penetrate tissue, and since there is generally little or no change in the energy of a photon with penetration, the exposure at any depth depends only on the gamma beam intensity and thus the number of photons of the initial energy available at that depth.

The energy deposited in tissue as a result of a decrease in incident gamma beam intensity with tissue depth penetration produces what is usually identified as a first-collision dose. However, energetic gammas may also produce high-energy ionization electrons as well as other secondary energy emission in tissue (this is more fully reviewed in a subsequent chapter), and these may then add to the absorption dose in those tissues. Since the secondary ionization electrons are preferentially scattered in a forward direction, the absorbed dose at the tissue surface may be less than that at a small depth, and the actual absorbed dose at greater depths continues above that predicted by tissue attenuation alone. The increase due to secondary effects is sometimes identified as buildup and various

* As noted in Chapter 7, a total body dose of about 600 rads of gamma radiation is lethal for essentially all people. A 600 rad dose to water molecules would result in an ionization fraction of about 3×10^{-8}, or one molecule ionized in 3.6×10^7 molecules available. This molecular fraction for "destroyed" molecules is obviously approximately the same as the population fraction assumed "killed" or severely damaged with a resultant death of the population itself as noted in the hypothetical illustration at the beginning of this chapter.

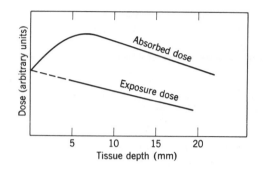

Figure 5–2 Relative tissue depth-dose pattern for cobalt-60.

buildup factors (the constants by which the first-collision dose must be multiplied to give a total dose) have been given in the literature. A typical curve for tissue depth versus dose is indicated in Figure 5–2 for cobalt-60 radiation. A curve of similar shape will frequently indicate the correction necessary for wall thickness in measurements made with a cavity detector.

Beta-radiation depth-dose curves for small distances in tissue have a shape similar to that of Figure 5–2. In this case the increase is attributed to the fact that particle impact can result in the beta's being reflected, or back-scattered, with the reflected particles themselves producing subsequent ionization in the region initially traversed.

DOSE MEASUREMENT

Although the exposure in roentgens is specifically defined only for electromagnetic radiation, a similar comparison of exposure and absorbed dose may also be applied in principle to betas. This requires (a) that the exposure "dose" measuring device use air or other appropriate gas in a volume which is small enough for the average beta energy to be only slightly changed during gas ionization and (b) that the gas be encased in material which could be penetrated by betas with little attenuation. Many present-day monitoring devices use ionizable gases as the detecting unit; however, the detector case thickness is usually not only as great as the range of many betas but it also strongly and differentially attenuates the high-energy ones. Similarly, the comparatively high attenuation of betas in tissue and the fact that the range of many betas is less than the body thickness would indicate that any external exposure measurement could be generally applicable only to the absorbed dose in the skin or organs that are not deep seated. Hence, exposure beta measurements are not generally applicable to absorbed dose determinations. This is discussed at greater length in Chapter 13.

There is obviously little meaning to an exposure (dose) of alpha particles or similar heavy and relatively nonpenetrating radiation. In the case of neutrons, there is also little meaning to an exposure (dose) since the absorbed dose is so strongly dependent on neutron energies; however, what might be identified as a neutron exposure is usually expressed as neutron flux density, or neutron particle flux density, in units of neutrons/cm^2-sec. This is further considered in a subsequent chapter.

THE ROENTGEN-EQUIVALENT-PHYSICAL (REP) UNIT

It may be of historical interest that the exposure (dose) and its unit of the roentgen were really defined before the absorption dose. What is now called an absorption dose was then called the rep (roentgen equivalent physical) and defined in terms of relative energy release. Thus a rep of any radiation was the quantity of that radiation which would release the same energy in a given tissue or other material as would 1 R of X- or gamma radiation of specified photon energy. The amount of energy released by 1 R in soft tissue was originally specified as about 83 ergs/g; this changed over the years with better information on the value of W, and the final specification before introduction of the rad was about 93 ergs/g. Obviously, the rad has not only replaced the rep but it is now considered to be the basic unit for dose measurements in radiation protection.

THE DOSE EQUIVALENT AND THE REM

In addition to their obvious dependence on the purely physical effect of ionization, the overall effects of radiation on biological systems also depend on frequently unidentified characteristics of the radiation type itself; the relative LET has already been mentioned as one such characteristic. However, the observed relative effects of neutrons and gammas, for example, do not lend themselves to such a comparatively simple evaluation. Hence still another term, that of dose equivalent, has been recently suggested to relate the absorbed dose in rads to its overall biological effect. The unit of this dose equivalent is called the rem, from the term "roentgen equivalent man" (or "roentgen equivalent mammal").

The dose equivalent in rems is the product of the absorbed dose in rads, and any other modifying factors that express the dependence of biological effects on other than the total ionization, or energy release, itself. Hence, the dose equivalent A may be expressed in terms of the absorbed dose D and various modifying factors $f_1, f_2 \cdots f_n$ by the product

$$A = D(f_1 \times f_2 \times \cdots \times f_n).$$

The modifier that rather empirically expresses the biological differences caused by radiation types and certain other factors is now called the quality factor; another modifier, called the dose distribution factor (or relative-damage factor) may be used to express the biological differences caused by nonuniform internal deposits of radioisotopes.

By use of the dose equivalent, the effects of various types of radiation on biological systems may be considered additive, even though the sum of the absorbed doses in rads may not be so treated. This permits the total dose equivalent to an organ because of 10 rems of gamma rays and 10 rems of neutrons to be considered a 20 rem total dose, although the sum of the respective absorbed doses in rads may not be so added.

What is now called the quality factor is very nearly the same as what has long been known as the relative biological effectiveness (RBE) of various radiations. Specifically, the RBE has been defined as the ratio of the amount of ionization produced by X- or gamma rays in causing a given biological effect to the ionization produced by some other type of radiation in causing the same biological effect. Thus, if a given number of rads of a certain type of radiation causes greater effects to man than does the same number of rads of X- or gamma rays, this different radiation is considered to be more biologically effective, or to have a higher quality factor, than X- or gamma rays. Obviously, then, a greater number of rads of X- or gamma rays would be necessary to cause the same effect as that of the other radiation, and the RBE for this second type of radiation is greater than unity. It has been recommended that the term "relative biological effectiveness" should refer only to radiobiological studies and not to dose evaluation, even though what is now called the dose equivalent is very closely analogous to what has long been known as the RBE dose. However, the latter term is still widely used and understood, and current RBE values probably give the best method for intercomparisons of the biological effects of various types of radiation. Accordingly, it will be used subsequently with its earlier meaning which is very near that of the quality factor.

Tables 5–3 and 5–4 show some quality factors that are rather generally accepted today. It is particularly noteworthy that the quality factor for betas is unity, whereas those for neutrons and alphas are greater. Because of their inability to penetrate the dead outer skin layers, alpha

TABLE 5-3

RELATIVE BIOLOGICAL EFFECTIVENESS OF VARIOUS RADIATIONS[a]

Radiation Type	Quality Factor (or RBE)	Radiation Type	Quality Factor (or RBE)
Gamma rays	1	Neutrons:	
X-rays	1	Thermal	3
Beta particles:		0.005 MeV	2.5
< 0.03 MeV	1.7	0.02 MeV	5
> 0.03 MeV	1	0.10 MeV	8
Conversion electrons	1	0.50 MeV	10
Alpha particles	10	1.00 MeV	10.5
Recoil nuclei	20	5.0 Mev	7
		10 MeV	6.5

[a] Sources: National Bureau of Standards Handbooks 59, 63, 69.

TABLE 5-4

RELATIVE BIOLOGICAL EFFECTIVENESS OF HEAVY IONIZING PARTICLES[a]

Specific ionization[b] (ion pairs per micron of water)	Quality Factor (or RBE)
100	1
100–200	1–2
200–650	2–5
650–1500	5–10
1500–5000	10–20

[a] Source: National Bureau of Standards Handbooks 59, 63, and 69.
[b] An ionization of 100 ion pairs per micron of water is equivalent to a linear energy transfer of 3.5 keV per micron of water.

particles are radiation problems only as internal emitters, and their high quality factor reflects the opinion that energy release in the body by particles of high LET is more biologically damaging than is a similar release by particles of lower LET. Other than this consideration, no attention is given to penetration effects in expressing quality factors. Similarly, it will be observed that the factors for neutrons depends on their energy, the body organs concerned, and strongly on the duration of the exposure. Thus neutron quality factors vary all the way from about 1.5 for short-term lethal effects to almost 20 for long-term eye cataract formation. The qualitative observation that radiation of a high LET is more injurious than that with a low LET is indicated by the quality factors (or RBE values) assigned in Table 5–3 for heavy ionizing particles such as alphas and recoil nuclei.

Although all the units given above for the radiation dose are in use, a general feeling of uncertainty by the medical profession about the quality factors (RBE values) for the various types of radiations and the conditions of exposure has led to a fairly general practice of expressing high radiation doses or exposures in terms of rad, with the type of radiation also specified. In assessing actual potential injury from radiation, then, the responsible individual, normally the physician, establishes his own RBE value, or quality factor, for the specific condition concerned.

ACTIVITY AND THE CURIE UNIT

One other unit should be defined. This is the curie (Ci), which is the unit used to express the activity, or time rate of disintegration, of a given amount of a radioisotope. The unit itself was originally defined as the disintegration rate of 1 g of radium, which is 3.7×10^{10} disintegrations per second; thus 1 Ci $= 3.7 \times 10^{10}$ disintegrations per second. A microcurie is 3.7×10^4 disintegrations per second, and a picocurie (or micromicrocurie) is 3.7×10^{-2} disintegrations per second.

It is thus obvious that the disintegration rate of a 2-Ci cobalt-60 source is 7.4×10^{10} disintegrations per second, regardless of the actual mass of cobalt used or that of the source itself. One exception to this use of the curie unit is in an original specification of the disintegration rate of recently extracted natural uranium for which 1 Ci means 3.7×10^{10} disintegrations per second of uranium-238, plus 3.7×10^{10} disintegrations per second of uranium-234, plus 1.7×10^9 disintegrations per second of uranium-235; 1 Ci of uranium would thus have about 7.6×10^{10} disintegrations per second in total. Similarly, 1 Ci of recently extracted thorium is considered as 3.7×10^{10} disintegrations per second from thorium-232 and the same from thorium-228. Such definitions are a result of long term practical usage of the unit before codifying efforts became important. However, such usage by these definitions is not completely universal today, so it may be advisable to check carefully the meaning of the activity, in curies, as applied to natural uranium or thorium.

The term "curie" has also been frequently used, although incorrectly, to indicate a unit of quantity, or mass, of a radioisotope. Although this practice may not be unduly ambiguous for the naturally radioactive materials for which the custom really originated, there are significant difficulties in such use for artificially produced radioisotopes; for example, the activity of a given mass of some material which has been made radioactive by irradiation in a reactor depends on the irradiation rate (which is the production rate) and the time of irradiation, or activation time.

Author's Note

What it is now recommended should be identified as *exposure* is really an air-absorbed dose from X- or gamma-rays. Until recently, it has been treated as a radiation dose, generally called an *exposure dose*. In this form, the term has also been used in connection with nonelectromagnetic radiation. Hence I have employed the term *exposure* for matters of current applicability, but have modified it with the word *dose,* generally in the form *exposure (dose),* for historical reference, which includes application to nonelectromagnetic radiation, recognizing the incorrectness of such usage in current terminology.

REFERENCES and SUGGESTED READINGS

National Bureau of Standards Handbooks 59, 63, 69, 78, 84, 85. (See Appendix I for complete list.)

QUESTIONS

1. What is linear energy transfer and in what units is it expressed?
2. Convert 5.4×10^{-12} J into electron volts.
3. What is the difference between exposure and absorbed dose and in what units are each of these measured?
4. How do neutrons lose their energy? Does this depend on their energies?
5. Explain the seemingly paradoxical situation in which alphas are least harmful from an external source whereas gammas are very harmful, but from internal sources the results are the opposite—alphas are very harmful and gammas are less harmful.
6. Assuming a photon energy of 1 MeV, what would be the absorbed dose in muscle and bone if the exposure dose was 10 R?
7. Define equivalent dose and its relationship to the rem unit.
8. Define relative biological effectiveness.
9. Define absorbed dose and exposure dose; differentiate between them.
10. Relate density of ionization, energy fluence, flux density, and linear energy transfer.
11. Compare the penetration and attenuation of gamma radiation with the penetration and attenuation of other forms of radiation.
12. Define the rad and the roentgen. These are units of what?

CHAPTER SIX

Background and Other Radiation Exposures

GENERAL COMMENTS

Important to an evaluation of radiation effects is an analysis of available experimental data, which includes their applicability to situations of interest. Thus, prior to discussing various radiation effects, the sources of data on which present evaluations are based are reviewed with particular attention to the omnipresent background.

Information on radiation effects is primarily based on observations of actual exposures in humans and other biological entities, the data developed from exposures of mammals being of particular interest. In fact animals of many species have been given almost every conceivable type of exposure, both external and internal, and under almost all conditions imaginable. Obviously a wide variety of results have been obtained, much too extensive to be discussed herein, although the data are available in a rather vast literature to which the reader is referred for specific information. Although human data are probably of the greatest value, they are rather limited, and thus the summaries of radiation effects given in subsequent chapters are also based on experimental animal data, extrapolations therefrom, and application of theories developed thereby.

Many of the exposures of human beings have resulted from accidents, which have actually been surprisingly few considering the extensive potential for them. Some of the accidental conditions that have resulted in significant human exposures are briefly summarized with respect to the radiation concerned, the types of data thus available, and the applicability of such data to the general subject of radiation protection. The patterns of effects have generally been consistent with the extent of the exposure, as described in subsequent chapters. General background radiation, both natural and man-made, is discussed in somewhat more detail.

ACCIDENTS

Criticality Accidents

These are all short-term exposures (probably not more than a few minutes duration) to high levels of gamma-neutron radiation produced by systems of fissionable materials that accidentally attained criticality. To date (early 1968) the only five fatalities in the history of radiation to occur within a short time after a single exposure have been caused by such accidents. Two of these occurred in experimental facilities, two during production-type operations, and one in preoperation reactor testing. In addition, some 30 to 35 other individuals have received sufficiently high exposures in the various incidents to necessitate hospitalization or other special treatment of various duration, although none has subsequently succumbed. There is some evidence that similar incidents have occurred in Soviet operations but details are unknown.

Accidental Short-Term Gamma Exposures

Included in this category are short-term (less than a day) exposures to gamma radiation in which the doses received were more than 25 rems. Recent incidents of this type have occurred principally when individuals approach unrecognized sources too closely or worked an excessive time in a high field. In at least one instance, such exposures were incurred during attempts to rescue individuals injured as a result of a steam explosion in a reactor malfunction. Some survivors of nuclear-weapon blasts probably received exposures in this range, although, as is shown later, those who were close enough to the explosion to have received fatal radiation exposures were probably more severely injured by other and more conventional effects of explosion than by nuclear radiation.

Accidental Beta Exposures

In some instances persons have worked with beta-active materials without appropriate shielding and have principally received beta-burns of their skins. However, such injuries have also occurred when radioactive material released into the atmosphere settled on various people and caused such burns. The fallout from nuclear tests that descended on the Marshall Islanders and the 23 fishermen in the Japanese fishing boat *Lucky Dragon* (Fukuru Maru) also comes within this category since, although the betas were accompanied by gammas, the principal injuries observed were as typical of high beta exposures as of gammas. In these cases, the exposure continued for an extended period, 13 days in the case of the fishermen who were estimated to have received total-body doses in the 200 to 250 rem range. It is of interest that one of the fishermen who had infectious hepatitis subsequently died and his death was widely attributed to his radiation exposure, although he could readily have succumbed to the hepatitis alone. Otherwise those who received these exposures have apparently completely recovered.

Exposures to Accelerator Beams

A few experimenters have received varying exposures from various types of radioactive particles, including neutrons, while working with cyclotrons, van de Graaff generators, and other ion accelerators. These were generally received more or less periodically over a long time, probably several years. Some eye cataracts have apparently resulted from such exposures, although no other types of injury have been specifically noted.

ROUTINE AND EXTENDED EXPOSURES

High Exposures to X-Rays

High exposures to X-rays today are usually received only by individuals receiving therapy; they are given periodically over an extended period of time and are partial body only. Certain diseases and disabilities have apparently been relieved by such exposures. However, it is known that in the early days of X-ray practice many experimenters, technicians, physicians, and perhaps patients regularly received high exposures, frequently on a day-to-day basis, for many years. The total doses involved are difficult to estimate but must have been in the thousands of rads; similarly, the exposure rates were probably on the order of several rads per day. Some persons who had been so exposed developed aplastic anemia, whereas others developed skin and bone cancer—particularly of the extremities, which, in some cases, necessitated amputation; these diseases generally resulted in the early death of the individuals concerned.

On the other hand, there were many technicians, physicians, and perhaps others who received X-ray exposures for a year or so at a comparatively high rate but then ceased such work; in many cases, there has been no evidence of any effects from such limited exposures.

Exposure to Radium (and Mesothorium) Radiation

Two types of exposures are involved. One is the external exposure to beta-gamma radiation incurred from radium used in early experimental work as well as that used for therapy in much the same manner as X-rays. Some skin burns, as well as other effects of beta-gamma exposure, have been noted. However, the most important effect, as already noted, is the result of internal exposure of radium-dial painters and others using radium during World War I and later. In these instances bone sarcomas developed, pernicious and other types of anemia (but not aplastic) were observed, and death came to several individuals some 4 to 6 years and longer after radium intake. The total deposits of radium in the bones of these individuals are not too well known but were recognized as being well above 1 μCi. However, many persons who received the same exposure have apparently developed no symptoms of the type that caused such severe effects in others.

Exposure to Radon Contamination in the Air

Miners of the Schneeberg-Joachimstal mines in Czechoslovakia and, more recently, uranium miners on the Colorado Plateau have apparently developed a higher statistical incidence of lung cancer than normal. This is attributed to their inhalation of radon and its daughters, particularly polonium-218, lead-214, bismuth-214, and polonium-214. As would be expected the radon and daughter concentrations in these mines have varied widely. A 1952 survey of some 157 mines of the Colorado Plateau showed a median daughter concentration of 1200 $\mu\mu$Ci/l of air and a mean radon concentration of 4200 $\mu\mu$Ci/l of air; an average of the medians for groups of these mines gave a value for daughter concentrations of about 2100 $\mu\mu$Ci/l of air, and a corresponding figure for the radon air concentration itself was about 4650 $\mu\mu$Ci/l of air. The measured radon and daughter concentrations varied widely, the ratio of the maximum to minimum sampled values varying by factors of 20 to 200. At that time it was estimated that some 64 percent of the miners worked in average daughter concentrations of more than 1000 $\mu\mu$Ci/l air. This was also the estimated average concentration in the European mines.

Thorium Exposures

Patients who were given injections of thorotrast (thorium oxide) to assist in X-ray work some 20 or more years ago have only recently begun to develop malignancies of the reticuloendothelial system in which the thorotrast was apparently deposited and was not removed by normal body processes.

Medical Data

In addition to radiation used in diagnosis and therapy (which is discussed at greater length subsequently), patients who were terminal cases for a variety of reasons have been given various treatments and tests involving radioisotopes as well as external radiation. Radiation effects have been observed in these patients.

MISCELLANEOUS EXPOSURES AND ANALYSES

A variety of miscellaneous exposures have been observed. In general they have produced little or no measurable effects, particularly deleterious ones. More details on some of these types of exposure and the results of studies made are given subsequently. Unless otherwise indicated in the following notes, no effects have been observed.

1. During the middle 1940s comparatively large numbers of people worked in locations where comparatively high levels of uranium air concentrations probably existed.

2. People have been regularly employed in the 100-year-old gas mantle manufacturing industry where thorium is a specific ingredient.

3. A few people have incurred normal industrial-type injuries as a result of which small amounts of plutonium have become imbedded in their tissues. Some of these particles are estimated to be large enough to give "dangerously high" internal exposures, and the material itself is apparently slowly going to permanent deposit locations in their bodies.

4. Some internal exposure is unavoidable as a result of radioactive fallout of nuclear weapon tests; the Marshall Islanders have probably been the most highly exposed group.

5. In addition to the groups that have been discussed earlier, some individuals have received rather limited but comparatively high exposures periodically or irregularly over an extended period. These include shoe salesmen using X-ray fitting machines, dentists, and a wide variety of industrial workers.

6. Populations of similar character have lived for many, many generations in locations having widely different background radiation levels. This subject is discussed at greater length in the subsequent sections.

7. Various statistical studies have been made of rather limited groups which have been thought to receive greater radiation exposure than similar groups in a common environment. Such studies have principally involved longevity, abnormalities among their offspring, and abnormalities in the incidence of such diseases as leukemia. A favorite study population has been the radiologists. Other groups have included persons who have received single high but nonlethal exposures, such as the survivors of the weapons released in 1945; those receiving X-ray and radioisotope therapy; and those apparently receiving essentially chronic exposures, such as the populations living under different conditions of background radiation. These studies are discussed at greater length in the sections that follow.

NATURAL BACKGROUND RADIATION

External Penetration Radiation

Everyone is inevitably exposed to background radiation, which varies from place to place and from time to time in both amount and type. Some of this exposure is caused by external penetrating radiation that comes from both cosmic rays and radioactive materials in the ground. Cosmic radiation of concern is principally a result of protons and the products of their interaction with various nuclei. Exposure resulting from such radiation depends on the latitude and the altitude. The dose rate of radiation from cosmic rays ordinarily measures 28 mrad/yr when tested at latitude 70°, sea level; appropriate corrections may be made for other latitudes and altitudes. Thus at latitude 50° and sea level the dose rate is probably on the order of 20 to 22 mrad/yr.

Environmental gamma activity results primarily from uranium, thorium, and their daughters in various rocks and soils; estimates of such activity at a height of about 1 m over granite areas are typically on the order of 150 mrad/yr and over limestone on the order of 20 mrad/yr. Obviously the actual level of radiation caused by the radioisotope content of rocks and soil varies widely from place to place, and the actual background at a given location can be determined only by measurement. In many places in the United States average levels of 120 to 200 mrad/yr are indicated by simple Geiger-Müller-tube survey-instrument measurements; much higher levels are observed in some locations—and lower ones in others.

Because of their materials of construction, buildings can also be responsible for background radiation—or their shielding can reduce its effect.

Five areas show notably high background levels. The two in which the largest populations dwell are probably on the monazite sands in the Travancore region on the east coast of India and in the monazite areas in the states of Rio de Janeiro and Espirito Santo in Brazil. About 100,000 people in India thus receive average annual doses of the order of 1500 mrad/year, whereas 30,000 Brazilians average 500 mrad/yr, with some receiving about 1000 mrad/yr. One Brazilian village of 350 inhabitants averages about 1600 mrad/year, with a peak possibility of about 12,000 mrad/yr.

Radioisotopes in the Air and Water

Radioisotopes in the air, in water, and in foodstuffs add to the natural background radiation. The principal air component is the radon ($^{222}_{86}$Rn) of the uranium-238 decay chain and the thoron ($^{220}_{86}$Rn) of the thorium-232 decay chain, with the average concentration of the former being on the order of 0.1 $\mu\mu$Ci per liter of air and that of the latter being about 2 percent as much. The actual concentration varies widely with location, and there are also marked diurnal and atmospheric effects as a result of which the activities measured at any location can vary by factors of 100 or more. In many mines the levels are of course much higher.

The principal source of background radiation in water is radium ($^{226}_{88}$Ra) and its radon daughter, together with uranium-238. In the United States the average concentration of the former in the tap water of several cities is on the order of 0.04 $\mu\mu$Ci/l, whereas one deep well tested has a concentration of 37 $\mu\mu$Ci/l. After normal purification treatment the activity of tap water in most cities is probably 10 to 50 percent that of untreated water. Several streams in parts of the west show much higher concentrations, but these are not used as sources for drinking water.

Surveys of various foodstuffs have shown average radium-226 concentrations of about 1 $\mu\mu$Ci/kg in many locations.

Body Deposit

In addition to radiation from environmental sources exposure also occurs from body deposits of radioactive materials. All told it is estimated that radiation from this source averages about 25 mrem/yr, with potassium-40 ($^{40}_{19}K$) accounting for about 90 percent of the total. This, combined with the average of external radiation in "normal" areas of the order of 100 mrem/yr, results in an estimated overall dose rate of about 125 mrem/yr per person.

ARTIFICIAL BACKGROUND RADIATION

Man-made radiation also adds to the total background exposure of individuals and populations. Sources of such exposure are primarily those of medical diagnosis and therapy; fallout from nuclear weapon testing; industrial exposure, which includes that of medical technicians; and those from a variety of sources and activities such as luminous dial watches (much less pronounced today than formerly), X-ray shoe-fitting machines (now very rare), improperly confined polonium electricity suppressors, and so on. In general diagnostic and therapeutic medical exposures, although recognized as existent, are normally always considered as being overall advantageous.

Medical Exposures

Although natural background radiation results in total body exposures, both diagnostic and therapeutic medical exposures normally involve only some part of the body; obviously the organs or other parts affected depend on the purpose of the exposure itself. Hence, in evaluating the overall effect of such exposures it has been customary to consider the principal problem as that of the genetic effects of gonadal exposure.

Genetically Significant Dose

As discussed in Chapter 9, genetic effects are of concern to the populations that include the individuals actually receiving some given exposures. Hence the overall effect of these exposures is expressed in terms of the *genetically significant dose,* which is defined as "the dose, which, if received by every member of the population, would be expected to produce the same total genetic injury to the population as do the actual doses received by the various individuals".* In essence this means that total gonadal doses received by the various members of the population are weighted by the age-considered individual child-producing probability of

* *Report of the United Nations Scientific Committee on the Effects of Atomic Radiation,* General Assembly Official Records, Seventeenth Session, Supplement No. 16 (A/5216) (1962), p. 416

those actually irradiated and this is then averaged over the entire population. In this sense all total body exposures have genetic significance.

The genetically significant dose may thus be considered to be represented by an expression of the type

$$G = \frac{\Sigma nDkA}{N},$$

where n = the number of individuals of a specific age and sex group annually receiving a given type of radiation dose such as that from a chest X-ray,

D = the average dose received in such exposure,

k = the fraction of the dose received by the gonads,

A = the fraction of those in the exposed group who will bear or sire children subsequent to the exposure, and the number of such children expected,

N = the total population concerned.

Thus

$$G = \frac{n_1 D_1 k_1 A_1 + n_2 D_2 k_2 A_2 + \cdots + n_q D_q k_q A_q}{N}.$$

There are assumed to be q groups, and it will be noted that the various groups may overlap. Thus, for example, a 30 year old man may receive both a chest X-ray (thus being included in a group, n_1) and may also receive a dental X-ray (thus being included in another group, n_2). Obviously the values of D and k for the respective groups would differ, and the sum, Σn, given by $n_1 + n_2 + \cdots + n_q$, will not equal N. Even if background radiation received by everyone is excluded, Σn could be greater than N, as might be the case in a highly developed culture, or it could be less than N, as would be the case in a more primitive population. This illustration is an application of the conclusion, further discussed in Chapter 9, that it is the total genetically weighted dose to all members of a population that has genetic significance. Thus, an individual 10 rem dose of a certain type has the same genetic significance as 1 rem doses of the same type received by each of 10 individuals in the same group (age, sex, etc.) of that population.

In countries for which such estimates have been made, the total genetically significant dose from X-ray diagnostic exposures appears to be in the range of 5 to 60 mrem/yr; similarly that for external radiotherapy for both malignant and nonmalignant conditions is on the order of 5 mrem/yr, and that from the administration of radioisotopes is in the 0.20 to 0.40 mrem/yr range. By contrast occupational exposures of atomic-energy workers contribute some 0.1 to 0.2 mrem/yr to the genetically significant dose, and all other occupational workers contribute about the same amount.

The actual gonad doses received from various exposures, Dk in the above equation, obviously depend on the exposures required as well as the techniques used. In diagnoses mass chest surveys are uniformly low and are on the order of 1 to 10 mrem/exposure. As would be expected, exposures in the pelvic region would be the highest, yielding exposures of 500 to 5000 mrem. Similarly gonad doses due to external radiotherapy can range from 100 mrem or less for skin treatments to 100,000 mrem for hip exposures; it is of interest that the genetically significant dose for this latter case is on the order of 1 mrem/yr.

In contrast to gonadal doses are the actual doses received by the individual during diagnostic and therapeutic X-ray exposures. These are expressed in terms of skin dose, which approximates the exposure dose normally measured, as already described. For mass chest X-ray examinations and most other diagnostic X-rays these exposures are on the order of 50 to 500 mR per film for roentgenography. The highest X-ray exposures are normally those incurred in fluoroscopic examinations, those for chest examinations being approximately 50,000 mR; however, exposures of 100,000 to 200,000 mR may be given in fluoroscopic examinations of the stomach and gastrointestinal tract.

It should be emphasized that the values given above are merely apparent averages of a large amount of data given in the literature and that United States data are not included in the genetically significant, or gonadal, doses given. It should also be pointed out that these data refer only to specific nations which, in general, are the more advanced technologically, with the result that the corresponding values for less developed nations or parts of the world would probably be lower.

Fallout

One product of nuclear weapon explosions is radioactive debris which, in the form of dust, may be widely distributed by winds and other meteorological phenomena. This is ordinarily identified as *fallout* and is further considered in Chapter 20. Although much of this debris is in the form of large particles that descend fairly rapidly, some particles are so small that they may remain in the air for long periods. Since radioactive decay continues, the long-lived radioisotopes are of principal concern in these cases. The principal long-lived radioisotopes of concern in fallout are carbon-14, strontium-90, and cesium-137; other elements are problems on a short-term basis.

Such debris, principally from weapon tests, adds another component to artificial background radiation. This is ordinarily expressed as the *dose commitment,* which is identified as the total doses that will be received by mankind in all generations from given periods of weapon testing;* it is thus the dose received during the given period of testing plus that received subsequently (to infinite time) from the testing of that period.

As an example of their effects on background, it is estimated that the dose commitment to the gonads from tests of the period 1954–1961 was about 76 mrad, with a corresponding figure for bone marrow of 163 mrad and of cells lining bone surfaces of 257 mrad. It should be noted that these values were, respectively, 40, 75, and 116 mrads prior to the Soviet Union tests of 1960–1962.

Inhalation of fallout materials during periods of testing may result in lung doses of the order of 5 mrem/yr.

Other Sources

Background radiation also comes from sources other than those reviewed above; from present data it appears that these contribute about 2 mrem/yr to the genetically significant dose, this being principally from luminous dials of clocks and other instruments.

EFFECT OF DIFFERENCES IN BACKGROUND RADIATION

The general effects of radiation exposures are discussed in Chapters 8 and 9. However, before leaving the subject of background radiation it should be pointed out that there is literally no even reasonably conclusive evidence that differences in background radiation have caused any effects, much less deleterious ones. Several attempts have been made to correlate birth abnormalities with these differences as determined not from actual measurements of penetrating radiation in the respective locations but from estimates of potential background differences; this has sometimes been based on corresponding estimates of the differing potentialities of uranium content in the respective rock and soil. The value of such data is usually affected by unevaluable uncertainties that exceed any effects reportedly

* *Report of the United Nations Scientific Committee on the Effects of Atomic Radiation,* General Assembly Official Records—Seventeenth Session, Supplement No. 16 (A/5216) (1962), p. 416

found; for example, rather interesting observations resulted from efforts made to show a latitude dependence of birth abnormalities since cosmic-ray intensity is latitude dependent. To support the conclusion reached, it was necessary not only to assume that *all* background radiation comes from cosmic-ray debris, which thus ignores completely the contributions of uranium and thorium deposits in the earth, but also to assume that *all* spontaneous mutations result from background radiation. This eliminates the effects of other mutagenic agents, which are thought to account for some 90 percent or more of the mutations produced (mutations are discussed in Chapter 9).

To date no detailed studies have been made of areas in which the level of background radiation is very high, such as that in the Travancore region of India.

SUMMARY OF BACKGROUND INFORMATION

It is beyond the scope of this review to detail all of the assumptions, models, and calculations on which estimates of overall risk may be based. However, it has been pointed out that the genetically significant dose due to natural radioactivity is 125 mrem/yr, that diagnostic and therapeutic medical exposures add respective values of about 30 and 5 mrem/yr to this value, and that the corresponding figure for occupational and other exposures is about 2 mrem/yr. In case of nuclear testing at the 1954–1961 rate an additional dose commitment of about 29 mrem/year of testing may be anticipated.

Accordingly, it appears that the genetically significant dose from all sources of background radiation is on the order of 165 mrem/yr, exclusive of fallout.

It has already been pointed out that the figures given in this chapter are rather general averages of a vast amount of data from which the author has selected what appear to be the more significant items. Thus anyone interested in further information—including the ranges of values for which single averages are given, methods used in estimating the various values given, and original work—is specifically referred to the literature.

REFERENCES and SUGGESTED READINGS

Report of the United Nations Scientific Committee on the Effects of Atomic Radiation, United Nations, New York, 1958, 1962, 1964, and 1966.

Blazt, Hanson, ed., *Radiation Hygiene Handbook,* McGraw-Hill, New York, 1959.

Eisenbud, Merrill, *Environmental Radioactivity*, McGraw-Hill, New York, 1963.
Control of Radon and Daughters in Uranium Mines and Calculations of Biologic Effects, U.S. Public Health Service Publication No. 494, U.S. Department of Health, Education, and Welfare, Washington, D.C., 1957.

QUESTIONS

1. When a material is said to go critical what phenomena have taken place in the material?
2. The Marshall Islanders and *Lucky Dragon* incidents were examples of what type of exposure? What were the results?
3. How dangerous is uranium exposure under normal conditions?
4. What components make up background radiation?
5. From what two principal sources does natural background radiation come?
6. What are the four main sources of radiation exposure to man?
7. Explain the formula $G = \Sigma nDKA/N$ and its implications.
8. What is fallout and what isotopes are of greatest concern therein?
9. What is dose commitment? Will it remain at the same value forever?
10. What is the estimated annual genetically significant dose?
11. Do the facts indicate that therapeutic X-rays may be more harmful than advantageous?
12. List the various sources of radiation exposure to people and comment on the significance of each—the frequency of occurrence and the severity of the exposure.

CHAPTER SEVEN

Effects of Acute Total Body Radiation Exposures

The somatic effects of radiation are those which directly affect the ir-radiated individual—good, bad, or indifferent. Such effects result from short or long term exposures, from high or low level irradiation, and from various types of radiation. Any discussion of the dose dependence of the various effects of radiation exposure as determined from human data is frequently rather qualitatively determined, basically because such data have generally not been obtained under controlled experimental conditions but result from analysis of accidental exposures, medical data, and other conditions in which the radiation exposure is incidental to other activities or purposes. Principal difficulties in data analysis appear to be the imprecision in dose determination for some of the occurrences supplying data and the different radiosensitivities of various individuals. Other factors that complicate the analyses include the relatively unknown effects of exposure rate for given total exposures, the type of tissue or organ which receives the exposure, and the type of radiation concerned. In many cases it has been necessary to extrapolate data obtained from animal experiments, and this introduces special problems. Thus it is necessary to make a careful identification of the conditions of exposure in expressing the somatic hazards of radiation.

Probably the most basic effects—certainly the most widely reported and used—are external radiation exposures to all parts of the human body, identified as total body exposures. The best quantitative figures avail-able refer to such exposures occurring over a short period of time, perhaps a few minutes to something like an hour but certainly less than a day. The effects of such exposures are reviewed in this chapter, and other somatic effects are discussed in Chapter 8.

DOSE-DEPENDENT EFFECTS

Table 7–1 gives the currently accepted general effects in man of gamma ray exposures at various levels; however, by definition these should also be the values in rems for exposures that involve neutrons and other types of radiation. These values are generally based on reasonably well-determined exposure doses from which absorbed doses were evaluated.

It will be noted that an exposure of the order of 400 rems is estimated to cause death in about half the population so exposed, that an exposure of 600 rems will essentially kill all of that population, and that probably no one will succumb to 200 rems. This assumes that no particular medical care is given. Currently such care principally involves rest, use of anti-biotics to limit infection, and treatment to relieve symptoms; it is esti-mated that such efforts may raise these figures by perhaps as much as 25 percent.

A convenient method of indicating dose effects is in terms of the lethal dose (LD) for some fraction of the population concerned. Thus, 400

TABLE 7–1

SHORT-TERM EFFECTS OF ACUTE RADIATION EXPOSURE

Dose (rems)	Effect
25	Perhaps detectable clinically by blood measurements.
100	Nausea in about half of those exposed; fatigue noticed; marked hematologic effects.
200	Nausea in all exposed individuals; fatigue; death possible, especially in the absence of rest and other treatment; increased susceptibility to infection.
400	LD–50 dose in the absence of rest and other treatment; some fatalities even with treatment.
600	LD–100 dose in the absence of rest and other treatment; probably LD–50 to LD–70 with special treatment.
800	Probably LD–100 even with treatment.

Note: Although there is some disagreement on the dose values for the various effects noted, the above figures have been rather consistently quoted. Other effects at the various levels are more completely reviewed in Chapter 8.

rems, the lethal dose for half the population, is frequently expressed as the LD–50 dose; on the same scale 600 rems would be the LD–100 dose. In animal experiments a time factor is also sometimes included. Thus, an LD–40 in 30 days, or LD–40/30, would mean a dose as a result of which 40 percent of the exposed population will succumb within a 30-day period.

From Table 7–1 it may be concluded that for simple purposes of comparison with other forms of injury a total body exposure of 100 rem is the amount which would produce a measurable injury or illness as judged by normal standards. It is, however, customary to consider an exposure of 25 rem as that where clinical treatment should be considered.

This 25 rem dose is the lowest that may be clinically detected. Such a determination requires precise blood analyses to detect minute changes from similar accurate preexposure data; in fact, a common cold changes the blood picture much more than a 25 rem absorbed dose. However, many in the field consider much lower exposures as actually representing injury, even though the body's resistive and recuperative powers are such that exposures to these lower levels have not produced any detectable effect, much less any identifiably injurious one.

The values listed in Table 7–1 would become progressively greater as the time in which a given exposure is received increases. Quantitative factors that relate to this increase have not been given, although the effect itself is very marked, especially for long term exposures and at low continuing levels.

UNCERTAINTY IN DOSE EFFECTS

Since most dose measurements are those of exposure (doses), there are obvious uncertainties in correlating such measurements with any quantitative effects of a given exposure on the cell structure of an individual or the energy released in each of his body cells; for example, a large person with a thick body would be expected to receive less average ionization per gram of tissue than would a slightly built person placed at the same point, although each would be said to have received the same exposure dose. Of much greater importance is the differing overall radiosensitivity of different individuals, as already mentioned. Such uncertainties are obviously even more pronounced for partial body exposures, particularly including those to betas and very low energy gammas; for these the figures of Table 7–1 are not applicable since the principal hazard of external total body exposures to betas is skin burns. In fact, there is no indication that anyone has ever received a lethal dose from betas, either alone or as the principal agent, although it is suspected that one of the criticality fatalities in an experimental facility in the early days of the atomic energy program was significantly hastened by beta burns on the hands.

In view of these factors of uncertainty as well as others that are mentioned subsequently there is little justification for attempting to define dose effects in other than broad ranges; in fact, it is perhaps remarkable that dose effects are so consistent from person to person.

SYMPTOMS OF ACUTE RADIATION EXPOSURE

Injury and eventual possible death from high-level radiation exposure apparently follow a rather consistent pattern of symptoms, the more rapid occurrence and greater severity of which indicate the higher exposure; this takes into account, of course, individual differences. In most cases, if death does not occur within about six weeks, recovery from high but nonlethal radiation exposure eventually occurs and is apparently complete. Death that occurs more than about six weeks after a radiation exposure is normally the result of complications or other factors. Many of the symptoms of high-level irradiation are obviously noted in individuals who do not succumb, and a variety of long term effects to biological systems have been reported to result from such exposures; these are reviewed subsequently.

Nausea

Probably the first observable reaction to high-level exposure is nausea and vomiting, with the time of occurrence depending on the dose. The length of this period, as well as the severity and duration of this symptom, is considered the best and earliest criterion of the extent of the exposure. Early, prolonged, and severe vomiting is a serious clinical sign of a high dose, whereas incidental, late, and infrequent vomiting is probably a sign of low dosage. In all cases, but particularly in this latter case, the possibility that the vomiting is of psychological origin should not be ignored.

Absorbed doses of less than about 100 rad cause nausea in very few individuals. With increased doses, however, the onset of nausea is more rapid and its intensity becomes greater, the change becoming most readily apparent in the 200 to 400 rad range; this reaction is always very rapid and severe at exposures above 800 rads. Nausea due to radiation can last for several days but probably does not continue after about a week.

Diarrhea does not appear to be a radiation-caused effect except possibly at very high doses.

Hematologic Effects

A decrease in lymphocytes in the peripheral blood is one of the more sensitive tests of a radiation exposure, and the amount of this depression as well as its rate of recovery is useful in prognosis, especially in the first 24 to 48 hours after the exposure.

The white blood cell count is also taken as an indication of exposure. There appears to be a slight increase in the total neutrophil count after the exposure and a decrease in the white cell count after the first few days. For about a week, this decrease is rapid, with a continuing decrease for about three to five weeks. After this, if fatality does not occur, the blood cell count gradually returns to normal, but this may take several months. It is thought that the white blood cell count is best correlated with, and considered in conjunction with, other dose indications.

In the fourth or fifth days after exposure bleeding occurs for exposures above a few hundred rad, such bleeding being correlated with a disappearance of platelets from the blood. It may or may not be an important factor at the lower ranges of serious exposures.

Erythema

Erythema, or skin reddening, is another indication of exposure, its early appearance qualitatively indicating a high exposure. However, this can also be produced by low-energy gamma rays or beta particles with no other significant clinical effects. In fact no significant effects other than severe chemical-type burns have resulted either from beta emitters being accidentally in contact with the skin of various individuals for extended periods of time or from work being done where an individual's hands or other parts of his body received high beta exposures. In evaluating erythema indications, however, the possibility of ordinary sunburn or other thermal effects should not be ignored.

Fever

Early fever in an exposed individual is a sign of very high exposure. However, fever occurring much later, several days after the exposure, can be due to infections, since patients receiving high radiation doses seem unusually susceptible to infection.

Fatigue

Fatigue is a function of exposure, its first incidence appearing shortly after the exposure to be followed by a period of well-being, another fatigue period, and so on. The extent of this fatigue and its speed of recurrence may be related to the exposure level.

Epilation

Epilation, which occurs at doses above about 200 rads, is noted 17 to 21 days after exposure. The hair regrows in a few months, but not if the victim is already bald, and the new hair may not be identical with the original. Apparently the time at which this symptom occurs and its completeness are related to the dose received; thus, if it occurs early and is complete, the prognosis is not as good as it is when epilation occurs later and is patchy and incomplete. Epilation may also occur as a result of localized exposures, such as those due to beta emitters remaining on the skin or in the hair for extended periods, as was the case with the Marshall Islanders.

Reproductive System

Also generally noted after high level exposures is a decreased sperm count in males; this may occur rather promptly for high level exposures but usually returns to normal after several months. As described later, this decrease is thought to be loss of the more radiosensitive spermatogonia. Since possible chromosome and gene effects are more serious in the available sperm than in those subsequently developed from the surviving spermatogonia, it is considered advisable that the irradiated individual not procreate for some six to eight weeks after exposure.

Miscellaneous

Not all of the effects described above are observed at all levels of exposure, and the results may not always appear consistent; for example, with exposures in the range of 1200 to 2000 rads, fatality can result from severe intestinal infections and thus occur before epilation or other effects eventually observed at lower doses can occur. For exposures above 2000 rads, fever is a significant factor and there is some indication of irrationality, as would be expected on the basis of serious brain injury or illness resulting in high temperature. It is also possible that this irrationality is caused by interference with the electrical activities of the central nervous system.

CAUSES OF DEATH FROM ACUTE RADIATION EXPOSURE

The direct causes of death at different levels of lethal exposure to gamma radiation are largely of academic or medical interest but perhaps significant in indicating the relative radiosensitivities of various systems. Thus death from exposure in the lethal range from 400–600 up to about 1000–1200 rads results from severe injury to the blood-forming organs in the bone marrow, and death, if any, occurs in about four to six weeks. For exposures in the range of about 1200 to 2000 rads, the principal lethal effect is destruction of intestinal function and the rather radiosensitive intestinal linings so that gangrene of the gut causes death in about one to two weeks. In exposures above 2000 rads, the principal cause of death appears to be disarrangement of the central nervous system, as noted above, and fatalities occur in a period of about 40 hr to 1 week.

No attempt is made to separate neutron exposure effects from those of X-rays or gammas, primarily because there have been no severe neutron exposures in which gammas were not also prominent. However, there is also no evidence that total body neutron exposure would cause effects dissimilar from those indicated above, which are principally based on observations of X-ray and gamma exposures.

It is obviously beyond the scope of this volume to attempt any evaluation of medical treatment following exposure, although there is apparently little that can be done to mitigate the severity of the effects other than rest and treatment of symptoms. There have been some indications that bone-marrow transplants are useful, and this may be considered a possibility for treatment of exposures that will otherwise very probably be lethal.

INTERNAL EXPOSURE

There has been only limited evidence that even very high level internal exposures cause acute effects leading to death in a short period of time, although they may be hazardous as long-term, essentially chronic, exposures.

REFERENCES and SUGGESTED READINGS

The Biological Effects of Atomic Radiation, National Academy of Sciences, Washington, D.C., 1956 and 1960.

Pathologic Effects of Atomic Radiation, National Academy of Sciences, Washington, D.C., 1956.

Effects of Ionizing Radiation on the Human Hematopoietic System, National Academy of Sciences, Washington, D.C., 1961.

The Hazards to Man of Nuclear and Allied Radiations, Medical Research Council, Her Majesty's Stationery Office, London, 1956.

Report of the United Nations Scientific Committee on The Effects of Atomic Radiation, United Nations, New York, 1958 and 1962.

Brucer, Marshall, *The Acute Radiation Syndrome,* ORINS–25, USAEC, Washington, D.C., 1958.

Saenger, Eugene L., Ed., *Medical Aspects of Radiation Accidents,* U.S.A.E.C. 1963.

QUESTIONS

1. What symptoms comprise the radiation syndrome?
2. Erythma can be considered to be an ambiguous symptom. Why?
3. Explain what LD–50 means. What would be the dose that would cause the result without treatment?
4. Indicate how fever and nausea can help to determine what the relative exposure level may have been?
5. What dose seems to be the maximum that a person can survive even with treatment and rest?
6. Is there anything mysterious about the way in which radiation affects the body? Explain why or why not.
7. In what way does high radiation exposure cause a person to be more susceptible to infection?
8. If a person receives a high dose of radiation but because of special treatment manages to recover, what are the effects that may be permanent?
9. As judged by normal standards, what total body exposure would produce a measurable injury or illness?
10. What are the symptoms that accompany exposures of 25, 100, 200, 400, 600, and 800 rems?
11. What are the general causes of death from radiation exposure for doses of 400 to 1200 rads; 1200 to 2000 rads; greater than 2000 rads?
12. What are some of the goals of medical treatment of those who have received acute radiation exposures?

CHAPTER EIGHT

Long-Term Somatic Effects of Radiation

What may be called the long term effects of radiation exposure are the delayed effects of comparatively short term exposures and the eventual effects of chronic, or long term, exposure. Essentially all internal exposures may be considered to be long term ones since, once a radioisotope is taken into the body, it continues its emission until eliminated; such emission, and consequent internal exposure from some materials, may continue for years. Thus, although many of the radium dial painters ceased intake of the radium compound after a comparatively short time, the final effects became apparent only after many years. There is also the possibility that the injuries identified from internal deposits may have been at least partially due to the nonradioactive characteristics of the materials concerned.

Most human exposures to external radiation, except for such events as criticality accidents, have generally had the characteristics of long term exposures as defined above. These, in turn, have a rather consistent set of effects for the various body parts irradiated, although some differences are also introduced by variation in the total dose and the dose rate, as will be noted; similarly, much of the data have been obtained from partial body exposures.

For convenience radiation effects are discussed in terms of the effects on the various body systems, specific organs, and tissues; in addition, the possible role of radiation in inducing various diseases and in causing effects involving the entire body or major parts thereof are briefly reviewed. Obviously much of the information given has been developed from partial body exposures, and the results of experiments with animals have also been rather broadly extrapolated to human analogies. In general such data present the rather obvious difficulties of interspecies differences and have thus been rather cautiously interpreted to give qualitative indications of exposure effects or information on relative effects, such as the relative radiosensitivities of various tissues.

As has been already mentioned, different body parts have different radiosensitivities, with the tissues and systems that consist of rapidly dividing cells being the more radiosensitive. However, radiation applied in sufficiently high doses and for a sufficiently long time will cause injury to any tissue, organ, or system. If the exposure of some specific body part is continued for a long time, carcinomas may eventually develop unless other deleterious effects, perhaps resulting in death, occur first. At the other extreme there are sufficiently low exposures or exposure rates for which no measurable effects may be noted or for which the effects observed may be interpreted as being advantageous.

RADIATION EFFECTS ON BODY SYSTEMS

Muscle and Connective Tissues

Muscle and connective tissues are among the most radioresistant of all tissues. It is thought that comparatively short term doses of the order of 10,000 rems can cause heart disease, perhaps leading to failure, although the basic effect is probably more that of shock than radiation injury. Muscle cancer has not been unequivocally identified, although a correlation between muscle irradiation and creatinuria, an indication of muscle breakdown, has been observed. Fibrosis of connective tissue can be produced by exposures of several thousand rems.

Skin and Epithelial Systems

Skin Burns

Comparatively short-term skin exposures of the order of 700 rad (or reportedly as low as 300 rad) have resulted in erythema, or skin reddening, which appears very much like an ordinary sunburn or chemical burn. For exposures of more than about 1500 rad, blisters appear, and for exposures of 5000 rads, the raw areas produced have difficulty in healing; for somewhat lower exposures the skin may remain rather generally sensitive to mechanical injury, heat, and other radiation. Erythema effects were particularly noted in early medical X-ray experience, the first actual dose-measuring device being the skin and the minimum identifiable dose being called the erythema dose. Today, however, betas with energies above the 70 keV necessary to penetrate the dead outer skin layer are the principal source of such burns, which become apparent after a period of about one to two weeks. There is no good information on a maximum chronic exposure rate at which erythema will not be produced, although it is thought that the skin may become more radiosensitive after continued exposure.

Skin cancer has also resulted from external radiation exposures, although these injuries have occurred only after long exposures to very high radiation levels and have been noted primarily among the earlier X-ray technicians and physicians; they were obviously the result of long term exposures to the very high levels that many unwittingly accepted. It is highly improbable that anyone has developed radiation-induced skin cancer under conditions where he did not receive at least some erythema doses. Thus it is relatively impossible to evalaute the total doses which would be necessary to cause skin cancer, although they are probably in the thousands of rads, with exposures extending over several years. Some late effects of comparatively low level chronic exposure include loss of detail in finger ridge patterns, ridging of fingernails, and dryness of the skin and nails.

Epilation

Another effect of radiation is the loss of hair as a result of radiation exposure doses of 200 rads or more. The returning hair is sometimes of a different color or texture. It is estimated that skin doses of the order of 700 rads might result in permanent epilation. Such effects have not been observed to result from betas or low energy X-rays except as erythema and similar effects have also been observed; for higher energy X-rays or gammas, epilation has been generally observed as an accompaniment to total body exposures.

Probably the most spectacular observation of epilation from partial body exposure occurred in some of the Marshall Island natives after the bomb tests of 1956, when some of the beta-active coarse-particle fallout material landed on their bodies and lodged in their hair where it remained for several days. The natives thus received local exposures literally in the thousands of rad to their skin, with the result that epilation and other results of high levels of radiation were observed. Had the individuals concerned merely washed this dust off their bodies and particularly their hair, the possibility of significant exposure would have been tremendously decreased. Thus, simple personal hygiene would have eliminated much of the actual exposure and injury. Similar action could very probably have alleviated much of the injury received by the Japanese fishermen mentioned previously.

Eye-Cataract Formation

There is evidence that long term eye irradiation at dose levels from 600 to 1000 rads can produce cataracts, although some opacification has been reported at only about 200 rads. Although the effect has been observed for gammas and X-rays, it appears to be most easily induced by fast neutrons with an RBE of 2 to 10, with the RBE increasing with decreasing radiation intensity. The mechanics of such injury is not clear, especially since the opacity resulting from radiation exposure is usually different from that observed in natural cataract formation where the causative factor is probably biochemical in nature.

Digestive Tract

Apparently the epithelium of the small intestine is, next to the blood system, the most sensitive of all tissues to total body irradiation, some 50 rads causing observable effects. It is thus damage to the intestinal mucosa which is the proximate cause of death from acute exposures in the 1000 to 2000 rad range as mentioned.

Bones

In general mature bone tissues are relatively radioresistant. However, the epiphysis, where bone growth occurs, is comparatively radiosensitive, and high radiation exposures will thus inhibit bone growth in the young; the fetal skeleton is particularly radiosensitive. This is one of the main justifications for rather severely limiting exposures of children, one aspect of which has been the restrictions placed on the use of X-ray shoe-fitting machines. It also appears that the healing of a fracture is not significantly altered by irradiation but that the irradiated bone is more susceptible to infection.

Some of the individuals who developed skin cancer as a result of the early high-level chronic X-ray exposures also eventually developed bone cancer, although the causative exposures were generally higher and the bone malignancies occurred much later. There is little experimental evidence on which to base an estimate of the total dose or chronic dose rate that is necessary to induce bone cancer, although it is probably above 3000 rads and thus higher than that necessary to induce skin cancer. Again, there was what would today be called ample warning of the danger.

Many of the principal alpha-emitting radioisotopes—such as radium, plutonium, and uranium—preferentially deposit in the bone, as do other radioisotopes such as strontium-90. The resulting radiation may affect not only the bone but also the bone marrow, which, particularly in the long bones, provides an important blood-forming function; this aspect is discussed in a subsequent section. Although bone necroses have developed at radium-deposit sites, this being particularly noted for the early radium-dial painters whose intake of material is estimated to have resulted in deposits of the order of 5 μg of radium that was being eliminated very slowly, if at all, a similar necrotic effect has not been observed at deposits of uranium, also an alpha-emitter.

Nervous System

The overall effects of radiation on the nervous system are not as clearly established as would be wished. However, it appears that peripheral nerves and receptors are comparatively highly radioresistant in that serious effects are caused only by doses which are several times the 600 rems that, as a total-body exposure, is the minimum lethal dose. In fact doses of the order of 10,000 rems have produced no observable damage to these nerve tissues. Similarly the autonomic nervous system seems to be highly radioresistant.

Data concerning radiation effects on the central nervous system, particularly the brain and especially the cerebral cortex, are rather conflicting. As noted, brain damage is considered to be the proximate cause of death from total body exposures of 2000 to 3000 rems and higher; because of the brain's sensitivity to heat, it is not known what effect, if any, localized heating may have in causing this damage, although it is probable that interference with electrical processes in the brain may be the principal effect. At lower radiation exposures behavioral changes have been observed. For many years western psychophysiologists have generally been of the opinion that LD–50 doses, at least, are necessary for any changes, and these are actually secondary effects. However, Soviet experimenters using Pavlovian conditioning techniques have claimed to observe measurable changes in behavior after total body doses in the range of a few rems or lower. In addition recent experiments in the United States have also indicated that comparatively low level brain exposures have resulted in some test primates having increased concentration abilities. It is also thought that various body reactions to radiation, principally some of those involving the circulatory system, may have been caused by irradiation of the nervous system; although such indirect effects may perhaps be anticipated, they are rather difficult to evaluate.

Blood and Hematopoietic System

Blood Picture

It has already been noted that the blood is the most sensitive indicator of radiation exposure and that lethal effects for minimum exposures result from damage to the hematopoietic system, particularly the bone marrow. The various constituents of the blood have also differing radiosensitivities, the erythrocytes (red blood cells) with their long lifespan probably being the most highly radioresistant and the lymphocytes (a type of white blood cells), which have probably the shortest average lifespan, being the most sensitive. The early reduction in this latter blood component after high-level (some 200 rems or more) acute exposure is one of the major causes of the increased susceptibility to infection of highly irradiated individuals. Not only does radiation produce changes in the normal blood components but it also can produce atypical cells; additional information is readily available in the literature.

Bone Marrow

Injury to the bone marrow is probably the most important overall effect of radiation on the blood system; in fact, animal experiments have indicated the bone marrow to be affected at all radiation levels down to 25 rems. In 400 to 1000 rad acute exposures, it is failure of the bone marrow's blood-forming function that is the proximate cause of death, and it is considered by many that bone marrow from unaffected individuals may be transplated to radiation victims and thus increase their probability of survival. On a long term basis it is thought that the continual insult to the bone marrow by radiation eventually results in anemia, which is aplastic for external radiation but of other types if the exposure is from internal deposits of radium.

Leukemia

A statistical increase in the normal incidence of leukemia has been observed for individuals who have been exposed to short-term radiation doses of 100 rems or greater. The data used include persons who were exposed in the Hiroshima and Nagasaki bomb blasts, certain individuals who received X-ray treatments of the lower spine (and thus essentially of the entire body) to relieve an agonizing disease called ankylosing spondylitis, and X-ray treatment for enlarged thymus in some children. In all cases the exposures concerned have been given in a relatively short period of time, a few weeks at the most for those receiving treatment for a disease and in a single dose at the bomb blasts at Hiroshima and Nagasaki. An evaluation of available data shows that there is little, if any, statistical justification for assuming an increased incidence of leukemia where the dose was less than about 100 rads. Since this is a short term exposure value, it probably contains a considerable safety factor for chronic exposures. Averaged over a 15-year period, the overall leukemogenic effect is apparently about 100 cases of leukemia per million persons per 100 rads for each year at risk.

It should be emphasized that the data indicate only an apparent increase by a statistically significant amount in the normal annual leukemia incidence of about 50 cases per million population. Thus, although there is certainly no evidence of leukemia induction at levels which might not also cause other and more obvious clinical effects, leukemia is one long term effect that, however tenuously, may be related to radiation exposure. As such, radiation exposure is becoming a favorite allegation of leukemia victims who may have been exposed, no matter how briefly or how small the dose.

Medical data also indicate that a minimum period of about 15 months, even after a dangerously high-level exposure, is necessary before clinical symptoms of the disease become apparent, and a drop in this increased incidence of the disease after 4 to 7 years indicates that most of the radiation-induced leukemia cases, if any, will become apparent within 10 to 15 years. In addition, the type of leukemia that is normally induced by radiation is different from the types that most commonly occur spontaneously.

There is no experimental evidence of a chronic exposure level at which leukemia would be induced. Although radiologists, as a group, apparently have a higher incidence of leukemia than other physicians or the general population, this higher incidence is apparently decreasing. This may be the effect of radiation; on the other hand the apparent increased incidence may be merely a statistical effect or it may be because of an unidentified cause. It is observed that other medical specialists also have "special" diseases; for example, the comparative incidence of cancer in psychiatrists and neurologists is even higher than the incidence of leukemia in radiologists; similarly radiologists also suffer from a comparatively high incidence of coronary disease. Thus, although the indicated leukemia effect in radiologists is popularly attributed to radiation exposure—particularly to those few radiologists who perhaps carelessly overexpose themselves—it is entirely possible that acceptance of this rather obvious explanation should await explanations of other similarly observed differences if any of them are to be considered as other than the result of unknown causes or of statistical fluctuations in the rather small samples involved. Although the incidence of spontaneously induced leukemia in the general population has also been rising, the rate of rise has decreased in the past decade; this indicates the rise is probably not related to the increasing rate of radiation exposure in the population.

Reproductive Systems

At best, the effects of radiation on human fertility are sketchily known, and only estimates may be given. Based principally on data from short-term, high-level, and highly localized gonadal exposures of animals it appears that gonadal doses on the order of 200 rads may cause temporary sterility (probably functional sterility) in both males and females, with fertility being recovered over a period of time and the effects on females being less marked than those on males. There are indications from some accident cases, as well as the experience of the Marshall Islanders and the Japanese, that exposures even in the reasonably near-lethal range have no serious permanent effects on fertility, although gonadal doses of more than about 600 rads may produce permanent sterility. There have been few cases, if any, wherein sterility has resulted from long term exposures. Thus a reasonable conclusion would seem to be that reasonably permanent sterility will not be induced by total body irradiation except under conditions where the life of the individual is also endangered.

Animal experiments have indicated that fractionation of a given dose increases its effectiveness in producing sterility; for example, in dogs for which the LD–50 dose is on the order of 325 rads single exposures of 1000 rads to the dog testes will not cause permanent sterility; however, a 475 rad testes dose delivered at a rate of 3 rads/day in 10 min periods for

5 days per week will cause 100 percent permanent sterility, whereas recovery occurs if the total dose is 375 rads. On the other hand, dogs chronically irradiated for life at levels of 0.3 and 0.6 rad/week have maintained their spermatogenic function better than unexposed controls.

Lungs and Respiratory System

As is the case with most tissues, sufficiently high and long-continued exposures to radiation can cause cancerous conditions in the lung. This has become of particular interest since an abnormally high statistical incidence of lung cancer has apparently resulted from inhalation of radon by miners in various locations. As noted, this was first observed in the Schneeberg and Joachimsthal miners for whom an average delay in appearance of the disease was about 17 years and whose lung exposures have been estimated to have been about 1000 rads during that period. Similar results have also been observed in some of the uranium miners of the Colorado Plateau, although similar effects apparently have not been observed in other uranium mines.

Other Organs and Systems

Most tissues and systems not specifically mentioned above are considered to be rather radioresistant, although internal deposits of radioisotopes may result in sufficiently high, and perhaps long-continued, exposures to cause apparent injury. Thus, thorotrast (thorium oxide) used as a contrast medium for X-ray exposures of parts of the reticuloendothelial system has eventually caused cancerous conditions. Similarly, the kidneys have been observed to suffer injury during the process of eliminating sufficiently large quantities of radioisotopes; however, in one set of rat experiments ingested uranium apparently cleared up a congenital kidney infection that was peculiar to the strain concerned.

In general various glands and other body organs and tissues are comparatively radioresistant; for example, the pulmonary lymph nodes of dogs continued their function even when uranium deposits comprised some 15 percent of the weight of the organ itself. Most of this type of information has been developed from animal experimentation, the details of which are in the literature. In addition, most of these localized exposures at the levels described have been produced by internal deposits of radioactive materials, and other types of toxicity may also be involved.

EFFECTS OF CHRONIC TOTAL-BODY EXPOSURE

Longevity

As noted above, a short-term total-body dose of the order of 600 rems or more will shorten the life of any individual to perhaps a couple of months at the most; following survival from somewhat lower exposures recovery is apparently complete. However, there is evidence from animal studies that these somewhat sublethal short-term doses will, on the average, slightly shorten the lives of the exposed individuals, the actual shortening being roughly proportional to the dose; this shortening is small compared with the total lifespan, even for a dose that is only slightly sublethal for the individual concerned. The effects described above are the long-term residual ones following short-term high-level exposures. They are thus obviously not specifically applicable to chronic (essentially continuous long-term, even lifetime) exposures. For these the longevity picture is somewhat more complicated, as will be described.

Data on Human Exposures

The principal data on the effects of chronic, essentially lifetime exposures in humans are comparisons of the lifespans of general or specific populations with those of some individuals or groups who are intuitively believed regularly to have received exposures that exceed background-radiation levels. Such a group is the radiologists, who are estimated to receive radiation exposures of 100 to 1500 mR/year in various nations; these average doses have apparently been steadily decreasing with improved equipment and greater care by the radiologists themselves. The three principal studies of this type—two in the United States and one in Great Britain, where the average annual exposure is about 500 mrem—have indicated that radiologists have a somewhat increased longevity as compared with either physicians as a whole or the general population. The information is best expressed as a mortality ratio, which is the ratio of the number of deaths in a given period from the population concerned to the number of deaths that would have been expected in the same population on the basis of the experience of the control group. The mortality ratio for radiologists is about 90 percent with respect to physicians as a whole and is less in comparison with the general population; however, it is higher than the mortality ratio of most other medical specialists. An age-specific study also indicates an increased average lifespan of about 2.5 percent for the radiologists as compared with physicians generally, even though their average age at death, due principally to their age distribution in the study, was some five years lower than that of the physicians.

Some unexplained results of these analyses include the fact that the pathologists, as a group, had by far the greatest increase in longevity as compared with physicians generally, their mortality ratio being only 62 percent. Also, there is the apparently abnormally high, but decreasing, incidence of leukemia among radiologists; this is frequently attributed to radiation, although no attempt is made to explain similar abnormally high disease incidence in other specialists, such as cancer in psychiatrists and neurologists.

Data from Animal Experiments

Most of the comparatively large amount of data that have been collected concerning the effects of chronic lifetime exposures in animals has been obtained at such high exposure levels that early death or similar deleterious effects have been expected and of course found. Also, the eventual longevity effects of single high-level exposures have been frequently incorrectly interpreted as being identical with those due to low-level chronic exposures.

In chronic exposure experiments with mammals where longevity is reported, an average life shortening is generally observed for subjects receiving chronic exposures of 10 rads/day or more. By far the majority of studies have been carried on at levels above this figure. On the other hand, in studies that also give low-level exposures the data would seem to indicate an average life-lengthening effect at exposures on the order of 5 rads/week (or 1 rad/day) and less. Figure 8–1, although it does not report any particular experiment, does indicate the type of data on which these conclusions are based. There are statistical uncertainties in all of these chronic exposure data, as would be expected. However, there would seem to be little doubt that, overall, they do indicate an effect of increased longevity much more strongly than do they indicate the opposite, which is a decreased longevity.

There is no corresponding information concerning any overall effects on longevity as a result of low-level external exposures to beta emitters. The results of available experimentation with neutrons is unfortunately rather unclear since in most cases such radiation is also accompanied by gamma exposures. However, this is the one type of radiation for which no chronic exposures indicating an average life-lengthening effect have been observed; however, the total exposure levels reported have always been greater than an approximate threshold value of 5 rems/week suggested above.

As in the case of external radiation, experiments have shown levels of injection of plutonium, uranium, and radium which increased the average lifespan of the injected animals. Similarly, there are apparently levels of inhalation and ingestion of uranium compounds for which increased longevity is observed. On the other hand, no injection level for increased longevity was found for the beta emitters of calcium-45, iodine-131, nor strontium-90; however, even the lower levels of injection were rather high for these materials.

The cause of this apparent increased longevity is not well known. However, two possibilities have been suggested; one is that the radiation may selectively destroy injurious bacteria or other parasites and that this reduction of infection, especially in early life, thus increases the probability of survival of the animal concerned. The other possibility is one of stimulation; the "injury" produced by the radiation stimulates the organism to the extent that its overall effectiveness is increased.

The phenomenon of different levels of exposure causing markedly different effects on longevity is not a new or particularly unusual one for biological systems; for example, vitamin A is a necessity for life, but a sufficiently large dose of vitamin A is highly injurious; similarly, large quantities of various elements that in trace quantities are considered to be necessary in the diet may be dangerous. Obviously, such data better fit a stimulation hypothesis than one concerning reduction of infection.

Accelerated Aging

Observations

The effect of sufficiently high radiation exposures in shortening the lifespan of animals has been coupled with an observation that the causes of death, except for an above normal development of cancerous conditions, appear to be similar to those of developing senescence. This has led to the concept that radiation may accelerate aging. Accordingly, the symptoms of old age and eventual death appear at an earlier age than normal. However, there are some rather significant differences in the effects of age and of radiation on various tissues and systems. Some of the rather obviously and grossly apparent overall effects of aging, and related ones of radiation, are the following:

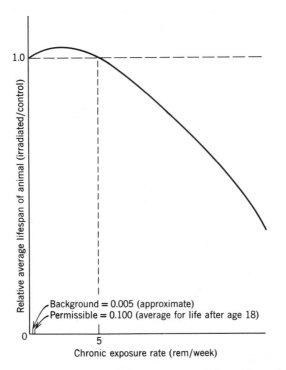

Figure 8–1 Longevity effects of radiation exposure. (*Note:* Figure is not to scale and units are arbitrary. It is not a depiction of a specific set of data but an indication of the type of curve that is obtained from many sets of data. The figure of 5 rems/week as a "no effect" value results from a rather generalized interpretation of the information obtained with a variety of animals, for some of which the value was higher and for others, lower.)

1. Muscles are very radioresistant, although change in muscular function is among the earliest effects of age.

2. The effects of radiation on the skin are similar to those of chemical burns or of long exposure to the sun.

3. Baldness is one symptom of age (in some men), and radiation causes epilation; however, the hair usually eventually regrows, and the epilation effect of radiation is thus probably similar to that of typhoid or other diseases.

4. Radiation can affect the nerve systems, but there is no report of its causing the slow deterioration of the central nervous system or the effects on the peripheral system, such as those of "palsy," which are ob-

served with increased age. It may well be that the early reduction in muscular capability is due to nerve deterioration.

5. Acute radiation exposure, and perhaps chronic exposures as well, apparently reduce resistance to disease and to subsequent radiation exposure itself; however, it has not been shown that the effects are similar to those due to age itself.

6. Acute gonadal exposures can produce sterility (usually temporary), but there is also experimental data showing that animals chronically exposed to low radiation levels better maintain their spermatogenic processes than controls; no relation of radiation and impotence has been suggested.

7. Radiation-induced eye cataracts have been observed, but their structure is usually different from that of spontaneous cataracts; no other eye effects of aging, such as lack of accommodation, have been observed to result from radiation exposure.

8. Leukemia has been statistically linked to certain types of exposure, but the origin of the disease and its connection with age are at best only vaguely understood.

9. There has been no indication that animals mature earlier as a part of an accelerated aging process caused by radiation exposure; if anything, it appears that radiation may delay maturation. In any event it is only in senescence that a connection has been attempted.

Most of the observations of aging have been made on animals that recovered from acute and generally nearly lethal total body exposures, although some have been based on chronic exposures that were high enough to cause an actual life shortening. In many cases it was the experimenter's observation that the animals had appeared to him to have aged. However, similar effects have also been observed in individuals who have recovered from a severe traumatic injury or from a very serious disease, perhaps in their youth. A chronic debilitating disease may also produce an apparent aging effect. In fact, the evidence available would appear to point to radiation as producing effects more nearly similar to those resulting from serious injury or disease than to accelerated aging.

Theories

General. Theories of aging usually fall into the following three categories:

1. Slow accumulation in the cells and tissues of some of the deleterious products of metabolism causes a progressive decrease in efficiency that eventually results in failure of the organ, tissue, or even the or-

ganism. As one example of this theory type, it has been suggested that aging is caused by a gradual and progressive chemical cross-linkage of large vital protein and nucleic acid molecules into pairs and groups which, in some fraction of cases become irreversibly immobilized and form larger groupings. The resulting aggregates eventually interfere with the functioning of the remaining "free" molecules, and the cell dies.

2. "Wear and tear" theories according to which the strain on the body or organ in meeting demands placed on it by both normal living stress and such abnormal stresses as disease eventually so affect the capability reserves of the organ or body that it is finally unable to meet some stress placed on it.

3. Mutation theories wherein it is assumed that somatic mutation of individual cells by various mutagens (unspecified) causes an overall decrease in the cell efficiency, this resulting in degeneration of the organ or tissue concerned and eventually of the organism itself; if nonreproducing cells die as a result of these mutations, their nonreplacement is an eventual cause of serious deterioration. On an information theory model this gradual deterioration is considered to be an increase of noise in the genetic material of the cell. An interesting specific theory of this type assumes that spontaneous mutation of body cells, during an individual's lifetime and throughout his body, results in their actually becoming near strangers to normal cells; the body's allergic reaction to these foreign cells, called transplant immunity, may then produce antibodies against them. However, the mutated cells are assumed to be sufficiently similar to normal body cells that these antibodies actually attack both normal and mutated cells and this results in progressive degeneration of the entire body since the mutated cells are distributed randomly about the body.

Although the above summaries refer specifically to aging, they would also be applicable to the more general effects of radiation exposure or other debilitating effect. Since radiation obviously "insults" the body, such effects would fit with the "wear and tear" theories; since, as is shown later, radiation is also a mutagenic agent, the mutation theories may also be applicable. However, it is of interest that of the mutagenic chemicals only those which are active cross-linking agents appear to shorten the lives of experimental animals on the same order of magnitude as massive, but nonfatal, doses of radiation.

Theories are generally quoted as justification for conclusions concerning the various effects of radiation exposure. However, it may be pointed out that the "models," or basic premises, on which most theories are based assume a linear effect of radiation. In the case of lifespan this would mean a decreased longevity at all levels of exposure, which is apparently the case at the high levels for which definitive data are available. Hence it is not surprising that theoretical results generally indicate a linear relation between longevity and radiation dose. In fact, it appears that little theoretical attempt has been made to take into account a threshold-type effect; the author is sympathetic to the difficulties of such treatment.

Gompertz Function. Various attempts have been made to express mortality and aging relationships mathematically. Of these, the empirical Gompertz function, first stated in 1825 to express the mortality of a normal population, has appeared applicable to certain radiobiological situations. This relationship assumes that the fractional decrease dN/N in a population of N individuals with time is proportional to a factor R. Thus

$$\frac{dN}{N} = -R \; dt.$$

In this relation R is not a constant but is given by the corresponding relation

$$R = R_0 e^{mt},$$

where R_0 and m are both arbitrary constants that are chosen to fit the actually observed data. Thus

$$dN/N = -R_0 e^{mt} \; dt.$$

It has been found that with properly chosen values of the constants this relation will fit observed data over a significant range. Similarly it is obvious that, if this relation does fit different conditions, changes in mortality rates will appear as changes in the values of R_0 and m. Where radiation exposures or exposure rates are sufficiently high to produce significant changes in the mortality rates it has been possible to fit most experimental data with curves of this type with proper adjustment of the constants; however, there has been little theoretical justification for the values of the constants used.

Reversible and Irreversible Effects. In one of the principally quoted approaches to evaluating aging it is assumed that any radiation exposure consists of a reversible component and an irreversible one. By definition the body's reparative function can overcome the reversible effect with complete recovery, but the irreversible one has a long term effect on longevity; it is the accumulation of these long term irreversible effects that produces an overall result. A similar theory has been expanded to show that *all* deleterious occurrences to an individual, whether they be illnesses, cuts and bruises, or what not, have an overall disadvantageous effect on longevity.

SPECIAL ASPECTS OF INTERNAL EXPOSURE

As stated previously, a principal difference between the radiation protection problems of external and internal exposures is that one can move outside of the field of an external source, but there is no getting away from an internal exposure that continues 24 hours per day and 7 days a week. Of course, due to both radioactive decay and biological elimination, the rate of exposure from internal deposits decreases with time, but all forms of radiation, including alphas, are problems.

The most common methods of body entry by radioisotopes are through being breathed (or inhaled), being eaten (or ingested), or being injected (entering the blood stream through the broken skin). There is also a possibility of entry through the unbroken skin as a process of osmosis, and there is some indication that this can occur for some materials, particularly tritium ($_3^1$H). However, by all odds the most common and extensive methods of entry are inhalation and ingestion.

The accepted deleterious effects of internal exposure are based on a few human experiments, unofficially performed to be sure, but capable of providing information all the same, and a huge number of animal experiments mostly based on injection of a radioisotope and frequently showing no actual deleterious effect but only the possibility of eventual deposits.

Whether or not injection data can be well extrapolated to give corresponding information for natural inhalation or ingestion is a good question. Certainly, for other systems the effects are different; for example, the white of an egg, if injected, is as dangerously toxic as rattlesnake venom, but the effects are different if these proteins are swallowed. Similarly, experiments have shown that tumors result from injected strontium-90 but not from inhaled strontium-90, and longevity experiments with animals inhaling strontium-90 show no undesirable effects at much higher levels than had been concluded from injection data.

If it is assumed that it is the radioactive properties of the radioisotopes taken into the body which cause injury, the energy that might be expected to be released in a given organ as a result of the presence of given quantities of these materials can, of course, be calculated, and a level for injury or other effect can thus be stated.

However, such calculations do not give information that is consistent with observations, and available data give some rather puzzling results as related to the actual effects of internal exposures. In particular, conclusions based on the results of inhalation of radon and ingestion of radium, respectively, do not agree with data for similar inhalation or ingestion of certain other radioisotopes, principally alpha-emitters. It appears that better correlation is obtained between the extent of the bone injuries received by the dial painters and the absorbed radiation dose from the radium deposited than does a similar relation for any other factor considered; on the other hand, other materials that also have radioactive properties have not caused the same problem, even though the apparent doses received from them have been much higher than the danger level indicated by the radium data. Chemical or mechanical effects of the materials used have generally been ignored, although it may very well be that the overall effects result from some sort of combined radioactive and chemical toxicity which is much more dangerous in combination than is the sum of the two individually. However, some tests have indicated little, if any, synergistic effect of chemical carcinogens and radiation.

With the exception of radium, radon, and perhaps thorium under the conditions stated, it would appear that injury would not be expected except where the intake levels were rather high, perhaps high enough to provide a significant fraction of the total diet or air breathed. Similarly, it is very probable that injury can be caused by intake of most radioactive materials at sufficiently high levels.

MISCELLANEOUS DATA AND OBSERVATIONS

A variety of effects resulting from external radiation have been recognized in addition to those noted, but in all cases these have resulted only from comparatively high levels of radiation or radiation exposure. Among these are the facts that (a) high level irradiation may have the effect of reducing the body weight of an animal, (b) exposures of 100 to 300 rads can predictably produce malformations in mammalian embryos, and (c) there is a poorly known and direction-indefinite change in the radiosensitivity of mammals as a result of radiation exposure in the LD–50 range.

In addition, the following are some general observations concerning the overall somatic effects of radiation exposures:

1. To produce a given effect, a greater total dose of radiation will generally be needed if the exposure time is long (meaning a low exposure rate) than if it is short (which means a higher exposure rate).

2. A greater exposure to a small part of the body, or an organ, is necessary to produce a given injury level to that organ than is required to produce the same affect where the exposure is applied to the entire body.

PARTS OF A PUZZLE

Although an evaluation of much of the experimental evidence of the somatic effects of various types of radiation exposure is rather straightforward, there are some bits of contradictory data which should be recognized. One of these is the apparent dependence of the internal exposure necessary to cause a given effect on the material deposited and not necessarily on the resultant radiation level. Another is the apparent increased longevity that results from low-level chronic exposures. A third is the apparent absence of a measurable effect due to the background radiation level of different populations. The concept of possible synergistic effects has been suggested as an explanation for the first of these and those of possible stimulation or infection reduction for the second; none has been suggested for the third except that the effect is too small for actual determination.

The intriguing observation that radiation may be necessary for life itself, perhaps as a stimulant, has been suggested both in analysis of some experimental observations of plants and by at least one set of experiments in which a frog heart maintained in vitro continued beating when a radioactive isotope of potassium was a component of its nutrient but stopped when this component was replaced by a nonradioactive potassium isotope; when the radioactive isotope was again used, the heart resumed its beating.

More recently it has been observed that certain plants failed to grow hydroponically in nuclear submarines and that not only did algae fail to grow after a few weeks in the Simplon Tunnel in the Alps but also Artemia eggs produced only some 60 percent of the normal rate of offspring and certain seeds became sterile after 6 months. It was pointed out that the background-radiation level was much below that normally encountered, the Simplon Tunnel specifically being reported free of traces of cosmic radiation.

SOME OVERALL OBSERVATIONS AND SUMMARY

Though by no means a complete analysis, the following indicate some of the rather broad conclusions that may be drawn from radiation exposure data:

1. Radiation can cause injury to any body part or organ if the exposure is high enough and long enough.

2. Radiation can shorten the lives of individuals, but only if applied at sufficiently high levels over a sufficiently long period of time. There are also apparently duration-of-life-exposure levels which are sufficiently low that they may produce a generally increased life expectancy of animals as compared to unirradiated controls; there are also other evidences that low levels of radiation may be necessary for life.

3. With respect to internal exposure only radium, radon, and probably thorium have been observed to be associated with actual injury; in each case malignancies were produced in the bone, lung, and liver-spleen systems, respectively. Radon was inhaled, and the other two were ingested. For other radioactive materials significant effects have generally resulted only from injection of the material, intravenously or otherwise.

4. Most of the effects produced by exposures are neither good nor bad, although unfortunately they have been usually interpreted as being bad. Change does not necessarily mean undesirable change.

5. Many experimental observations are difficult to explain.

6. Too often data taken at high levels of exposure or exposure rate have been extrapolated to lower levels, and the inherent inconclusiveness of such results has apparently not been thoroughly recognized.

REFERENCES and SUGGESTED READINGS

The Biological Effects of Atomic Radiation, National Academy of Sciences, Washington, D.C., 1956 and 1960.

Pathologic Effects of Atomic Radiation, National Academy of Sciences, Washington, D.C., 1956.

Effects of Ionizing Radiation on the Human Hematopoietic System, National Academy of Sciences, Washington, D.C., 1961.

The Hazards to Man of Nuclear and Allied Radiations, Medical Research Council, Her Majestiy's Stationery Office, London, 1956.

Report of the United Nations Scientific Committee on the Effects of Atomic Radiation, United Nations, New York, 1958, 1962, 1964, and 1966.

Long-Term Effects of Ionizing Radiations from External Sources, National Academy of Sciences, Washington, D.C., 1961.

Effects of Inhaled Radioactive Particles, National Academy of Sciences, Washington, D.C., 1961.

Control of Radon and Daughters in Uranium Mines and Calculations on Biologic Effects, U.S. Public Health Service Publication No. 494, U.S. Department of Health, Education, and Welfare, Washington, D.C., 1957.

Henry, Hugh F., *Is All Nuclear Radiation Harmful?* K–1470, USAEC, Washington, D.C., 1961. [Also *Journal of the American Medical Association,* **176**, 671–675 (1961).]

Symposium on Occupational Health Experience and Practices in the Uranium In-dustry, HASL–58, USAEC, Washington, D.C., 1958.

Health Physics, **11**, 76 and 455 (1965).

R. W. Prehoda, *Industrial Research*, **8**, No. 9, 44 (August 1966).

Sacher, G. A., and D. Grahn, "Survival of Mice Under Duration-of-Life Exposure to Gamma Rays. I. The Dosage-Survival Relation and Lethality Function," *Journal of the National Cancer Institute*, **32**, 277–321 (1964).

QUESTIONS

1. It is stated that some exposure to low level radiation may be advantageous. What are some of the advantages that could be attributed to low level radiation?
2. Why is it almost impossible to determine the actual dose that could be necessary to produce erythema?
3. Where and what is the epiphysis and why is it particularly radiosensitive?
4. How would it be possible for mental concentration to be improved by means of radiation exposure?
5. What is the hematopoietic system and how can it be affected to cause death as a result of radiation exposure?
6. What is leukemia and in what way could it be associated with radiation exposure?
7. What is meant by long term effects of radiation?
8. What tissue appears to be most radiosensitive?
9. What type of dose seems to be most effective in producing sterility?
10. What two reasons are given to explain the apparent life-lengthening effect of low-level long term radiation exposures?
11. What are the three general explanations for the process of aging?
12. What three radioactive elements are known to cause damage when taken into the body?
13. Do you feel there exists an unbased prejudice against the effects of radiation?
14. What are the long term somatic effects of radiation to the muscles, skin and epithelial tissues, bones, nervous system, blood and blood-forming organs, lungs, and respiratory system?
15. Rate the body systems as to their relative sensitivity to radiation.
16. Comment on the induction of leukemia by radiation.
17. How do the effects of radiation differ from those of aging?

CHAPTER NINE

Genetic Effects of Radiation

GENETICS AND HEREDITY

Genetics is the study of the transmission of the individual characteristics of living units from one generation to the next. Thus it involves not only inheritance from parent to offspring but also the propagation of the species itself. There are two general types of reproductive processes—asexual and sexual (or bisexual). In asexual reproduction the offspring has only one parent; obviously bacterial reproduction by fission is asexual. On the other hand sexual reproduction normally requires two parents, and in sexual reproduction the highly specialized sex cells from the respective parents—sperm from males and the egg from females—fuse to form the first cell of the new individual.

By a process of rapid cell division, along with differentiation and specialization, this first cell eventually develops into a new individual who has some of the characteristics of each parent. The overall process is extremely complex and requires very accurate and well-timed developments; hence it is not as surprising that abnormalities occur as it is that they are so few. It is calculated that about 6 percent of newborn human infants have inherited abnormalities that are visible and that of these, chromosomal aberrations occur in about 1 percent of all live births; there are also recessive effects, but these do not lend themselves to simple quantitative evaluations.

RANGE OF CHARACTERISTICS INHERITED

The heredity carriers of a cell are its several thousand genes, each of which forms some part of a DNA molecule and each of which is carried

136

on one of the chromosomes in the cell nucleus. The 46 chromosomes of each human diploid cell are paired, one of each pair having been inherited from each parent. The genes of a pair of chromosomes apparently control the same characteristics, and the respective genes for a given characteristic, called alleles, are located adjacent to each other. If the alleles are identical, which means they produce the same form of a given characteristic, the individual's genetic constitution for this characteristic is said to be homozygous, or the individual is a homozygote for that characteristic. If the alleles are different, the individual's corresponding genetic structure is said to be heterozygous, or he is a heterozygote for that characteristic. Any individual may thus obviously be homozygous for many characteristics and heterozygous for others.

In general, each allelic form (or allele) of a given gene produces a specific form of the characteristic controlled by that gene. Many human genes are known to have several allelic forms, and it is also probable that there are genes with only one allelic form. However, since each species has a definite physical form, it is obvious that the various alleles of a specific gene in a species do not produce grossly dissimilar forms of a given characteristic. There appears to be operative what is sometimes identified as a normalizing or stabilizing principle, which in essence implies that there is an overall constancy of species characteristics and that viable variations from the norm usually represent only comparatively slight changes or modifications.

In any event there appear to be definite limits to the number, or range, of physical forms of a given characteristic available even to a species, and in some way a species is protected from incursion of viable variations outside this range from another species. In general there is little successful mating between individuals of different species, especially in animals; in fact success in mating between representatives of different populations would tend to show they belonged to the same species.

VARIABILITY AMONG INDIVIDUALS

The haploid cell produced by meiosis of a germ cell in humans contains 23 chromosomes, one from each pair of the original cell; each of these 23 chromosomes must thus be descended from one or the other of the original parents of the individual producing the germ cell. Since each chromosome pair segregates independently of the other pairs at meiosis, the particular assemblage of chromosomes in a given haploid cell has an original parental arrangement that is selected by chance only. Thus, in one such cell 22 of the chromosomes could be from one parent and one from the other; obviously another germ cell has 1 chromosome from the first parent and 22 from the other. Again, there could be another pair of cells with 10 chromosomes from one parent and 13 from the other, and so on. As has been shown, there are thus 2^{23}, or some 8.4×10^6, different possible chromosome configurations for these haploid cells. However, it is highly improbable that any individual actually produces this number of germ cells; certainly, the number of egg cells supplied by a woman during her reproductive life is much less then these possibilities.

In mating, the two haploid cells from the two parents fuse to form a diploid cell; thus one of each pair of chromosomes comes from each parent. The resulting individual has a specific genetic constitution, or assortment of genes, and is thus a specific genotype. The term, "genotype" is also applied to a specific variation of a given single characteristic or group of characteristics. Since the chromosomes from each parent are a random assortment of those from the two grandparents, the actual genetic constitution of the individual is derived from the parents in equal amounts, from each grandparent in differing amounts depending on chance, and so on into ancestry. A given offspring may exhibit a wide range of final characteristics, and it is not at all surprising that no two humans are identical genotypes, although identical twins may be close to this. There being some 8.4×10^6 different possible chromosomal arrangements for the haploid cells from each human parent, there are $(8.4 \times 10^6)^2$, or some 7.0×10^{13}, different chromosome assemblages possible for the diploid cells produced by one set of parents. Since each of the 46 human chromosomes is estimated to contain some thousands of genes, any individual is thus one specific variation, or genotype, of an almost infinite variety of possibilities available from his parents. As noted, the number of possibilities represented only by chromosome variation is on the order of 7.0×10^{13}. If this value is compared with the total present human population of about 3 billion individuals, it is obvious that only a very small fraction of the genotypes possible to humans can exist in any generation. It has been estimated that the current human population is some 3 percent of the total number that has ever existed. On this basis there have existed only some 10^{11} individuals, and this is less than 1 percent of the possible chromosome variations provided by one couple.

ELEMENTARY MENDELIAN PRINCIPLES

For most species that reproduce sexually, including man, inheritance takes place in accord with principles first stated by Gregor Mendel (1833–1884) and subsequently extended by others. Although it is inheritance in populations that has become of principal concern in evaluating the genetic effects of radiation, it may be of value to review first some of the elements of inheritance for a single gene.

Briefly, Mendel's principle of segregation states that when the two alleles of a single gene, one contributed from each parent, differ, the form of the characteristic that physically appears in the offspring is that caused by one of the alleles; there is no "in between" that represents blending of the two. This allele is said to be dominant, and the form of the characteristic that it produces appears both in its homozygote and in the heterozygote. Hence these two genotypes form a single phenotype, the term used to indicate individuals with the same physical appearance and abilities. The other allele is recessive, and its form of the given characteristic appears only in its homozygote, which is thus not only a genotype but also a phenotype.

Similarly, Mendel's principle of independent assortment states that segregation in accord with the first principle for the alleles of any gene occurs independently of a similar segregation by the alleles of any other gene.

To illustrate the first principle assume two alleles of a gene to be represented by B and b, with the B allele being dominant. Now, if it is further assumed that one parent is the homozygote BB and the other the homozygote bb, their offspring will be the heterozygous genotype Bb, all members of which will be phenotypes with the physical appearance due to the allele B. The offspring of mating among members of this second generation (grandchildren of the original homozygous genotypes BB and bb) will have equal probabilities of being the genotypes B_1B_2, B_1b_2, b_1B_2, or b_1b_2 (the subscripts merely represent the genes inherited from the two parents, genotypes B_1b_1 and B_2b_2). Accordingly the three genotypes of these progeny will be statistically divided in the approximate ratios of 25 percent bb, 25 percent BB, and 50 percent Bb. There will obviously be only two phenotypes; 75 percent of the individuals will have the physical appearance of the dominant allele B, and 25 percent that of the recessive one, b.

Similarly, for a second gene also with two alleles—A (dominant) and a (recessive)—it may be shown that, beginning with homozygous parents of the genotypes AA and aa, the second generation will also have 25 percent of each of the homozygotes *AA* and *aa* and 50 percent of the heterozygote *Aa*. There will also be two phenotypes, 75 percent of the individuals showing the characteristic controlled by *A* and 25 percent that due to *a*.

Now, if it is assumed that the initial parents are genotypes *AABB* and *aabb* for the two characteristics, it may be shown that their offspring will all be the genotypes *AaBb* and thus a common phenotype showing physically the characteristics *A* and *B*. However, each of the next generation will be one of the nine genotypes *AABB* (1), *AAbb* (1), *aaBB* (1), *aabb* (1), *AABb* (2), *AaBB* (2), *Aabb* (2), *aaBb* (2), and *AaBb* (4), with their relative frequencies of occurrence being indicated by the numbers in parentheses.

It may also be noted that there are only four phenotypes for these genotypes, nine individuals with the physical appearance *AB* (genotypes *AABB, AABb, AaBB, AaBb*), three that of *Ab* (genotypes *AAbb* and *Aabb*), three that of *aB* (genotypes *aaBB* and *aaBb*), and only one that of *ab*, this being the homozygous genotype *aabb*. It may be readily seen that the 75 percent of the phenotypes with the characteristic *B* are not identical with the 75 percent that have the characteristic *A*, nor are the same individuals homozygous for the recessive alleles *a* and *b*. It is thus apparent that the respective gene segregations in accord with the first principle occur independently, which is thus an illustration of the second one.

It may be shown that all the progeny that result from random mating of all the members of the second generation as determined above will produce in the third generation the same fraction of each of the second generation genotypes; using the *B* and *b* alleles for example, this would mean 25 percent *BB*, 25 percent *bb*, and 50 percent *Bb*. Similarly, succeeding generations will then show the same distribution of genotypes. This conclusion assumes of course that each of the three genotypes for a particular gene is equally viable and fertile in the given environment, that there is no selectivity in mating for a particular genotype or phenotype, and that there are only the two alleles for a gene.

MENDELIAN POPULATIONS

In general a genetic population consists of a group of the members of a species that has been isolated from other members of the same species for many generations. If inheritance occurs in accord with the principles noted above, such a group is sometimes described as a Mendelian population, and the sum total of all alleles for all the genes possessed by members of this population is called its genetic pool (or gene pool). Population isolation need not reflect physical boundaries but may also result from the mating choices of its members.

EXTENSION OF MENDEL'S PRINCIPLES TO POPULATIONS

Were the descendents of the homozygotes of the elementary illustration given above to prosper and multiply while isolated from others of the same species, they would form a population in which there would be equal numbers of the two alleles B and b in the population genetic pool, and these alleles would thus occur with equal frequencies. In actual populations such equality of alleles does not usually occur, and it may thus be shown that if p and q are the relative fractions of the two alleles of a single gene in the genetic pool (where $p + q = 1$), the relative frequencies of occurrence of the respective homozygotes and the heterozygote in the population will be proportional to p^2, q^2, and $2pq$. If $p = q = 0.5$, the respective percentages of appearance in the population will be 25 percent for each homozygote and 50 percent for the heterozygote.

However, were one allele to make up only 40 percent of the population genetic pool and the other one 60 percent, which means that p and q are 0.4 and 0.6, it may similarly be quickly shown that the population will contain 16 percent of one homozygote, 36 percent of the other one, and 48 percent of the heterozygote. Obviously, if the recessive-allele fraction were 40 percent, only 16 percent of the population would be the phenotype that physically shows that characteristic and 84 percent would exhibit the dominant form. On the other hand, were the recessive allele to comprise 60 percent of those in the genetic pool, this form of the characteristic would occur in 36 percent of the population, whereas 64 percent would show the dominant form.

Eye color is frequently used to illustrate simple genetic inheritance, with brown eyes being dominant and blue eyes being recessive. In accord with the simple notation used the brown-eyed allele would be denoted by B and the blue-eyed one by b. From the above description it could thus be shown that, if 36 percent of a population were blue eyed and 64 percent were brown eyed, there would be 1.5 times as many alleles for blue eyes in the genetic pool as there were alleles for brown eyes.

In a large population there may well be more than two alleles for a given gene in the population pool, and each will comprise some fraction of those available. If these fractions, for example, are p, q, r, and s for four alleles (where $p + q + r + s = 1$), it may be shown that the relative frequency of the respective homozygotes will be proportional to p^2, q^2, r^2, and s^2, whereas the heterozygotes will occur with frequencies proportional to 2 pq, 2 pr, 2 ps, 2 qr, 2 qs, and 2 rs. These genotype fractions are obtained by expanding the relation $(p + q + r + s)^2 = 1$. The number and forms of the phenotypes that appear will obviously also depend on the relative dominances of the respective alleles. In addition to multiple alleles there is also a possibility that certain genes may be represented by only a single allele in the population genetic pool; only a single genotype for this characteristic would thus be available.

CHANGES IN A GENETIC POOL

In the above discussions it has been assumed that the relative fractions of the various alleles in a population genetic pool remained constant. This appears to be generally true for a population which has existed long enough in a relatively unchanging environment for equilibrium or near equilibrium to be established. However, an environmental change for which there is in some way produced a change in the relative viability of the various genotypes may eventually result in a new equilibrium condition, with a corresponding change in the relative frequency of the respective genotypes.

Of perhaps greater import may be the occurrence of a chromosome change (or gene alteration) in a germ cell that takes part in the reproductive process, with the consequent production of another allele for a given gene; such changes are called mutations, and the resultant individual is a mutant. He may thus have certain characteristics (a single characteristic or group of characteristics) that differ from those of the parents, that cannot be identified in either ancestral line, and that are transmitted to succeeding generations. If the bearer of this mutation reproduces and the allele also appears in his descendants, it will thus enter the genetic pool of the population.

If the mutated allele is dominant, it's physical form will obviously appear in any individual bearing it. If it is recessive, its physical form will appear only in the homozygote therefor; in this case the allele may be well distributed in the population, although it may physically appear in only a comparatively few individuals. Similarly, as a result of even limited mating between members of hitherto isolated populations, it is possible for an allele established in one to migrate into the genetic pool of the other where it had not hitherto existed. Here it will have the same effect as a mutation in that population. Any change in the frequency of occurrence of this "new" allele will thus reflect the relative viability of the individuals in which it occurs, whether or not it is dominant with respect to other alleles, and the rate at which it may continue to appear spontaneously in the population by continuing mutation (or migration).

MUTATION FREQUENCY

Since mutation is the generator of possible new genes, it is the ultimate source of the species variability on which the evolutionary process is predicated. It is estimated that approximately one mutation per locus (or type of mutation) per 100,000 individuals per generation occurs spontaneously. For a population of 3 billion individuals this ratio would mean 30,000 such mutants each generation. This figure varies for different mutation loci (or types of mutation).

The relative fraction of alleles of a given gene in the genetic pool may actually reflect a continuing mutation rate that produces one or more of them at a relatively constant rate from other alleles; in fact well-identified alleles are produced at definite rates in various populations by mutation of other alleles, even though all of the alleles of the particular gene continue to exist at some comfortable equilibrium level in the population concerned.

It should be particularly noted that it is highly improbable that any allele, once produced and propagated, will disappear completely from a population even though the individuals having the form of the characteristic it produces are grossly incompetent in comparison to those having the characteristics produced by other alleles of the same gene. In fact many studies of mutation in human populations apparently are concerned with the effects of mutated alleles producing major defects. If a grossly deleterious effect is produced by a dominant allele, perhaps causing early death, it may well continue to exist at a very low fraction in the population. However, if the allele is recessive and its seriously adverse effects thus appear only in the homozygote, it may well comprise a significant fraction of the total alleles of that gene in the population genetic pool even though its presence prevented the homozygote from reproducing, and the homozygous individual's death thus removed those particular genes from the genetic pool.

Certain types of birth abnormalities (also called congenital disabilities), which may or may not be true mutations, are not transmitted to a succeeding generation, perhaps because their carriers do not reproduce. Although they exist at the birth of the individual affected, they are actually of somatic rather than genetic concern as far as the welfare of the species is concerned and are not true mutations as defined.

Few of the mutation-produced deleterious effects are so serious as to prevent reproduction; in fact most of them cause only a minor decrease in viability. Similarly, possible advantageous effects of mutated alleles will probably produce only a slight increase in viability.

It is apparent that an allele must exist in a population in appreciable numbers before there is a significant probability of a mutation affecting it; for example, it would be highly improbable that a mutation which occurs at a rate of 1 per 100,000 per generation would occur in a population in which there were only 30,000 individuals with the allele that would be mutated. However, were the number of bearers of this allele to increase to 10^8, there would be on the average some 1000 of these specific mutants. As a brief example, therefore, if it be assumed that color vision is the result of mutation of a certain type of vision, then it would be concluded that color vision would be very improbable in a population lacking that specific type of vision in a large fraction of its members.

CAUSES OF MUTATION

An important initial effort to account for the appearance of changed characteristics in a population is generally attributed to Chevalier de Lamarck (1744–1829) for his "use-disuse" theory or concept of the inheritance of acquired characteristics. Members of a species "learn," or develop, some adaptation to their environment, and this adaptation then appears in their progeny. In essence, a factor of species variation is supplied by environment, a concept that obviously has a considerable and continuing philosophical and political interest.

Subsequently this concept was experimentally shown to be improbable, and the concept of random change, or mutation, produced by generally unknown factors was then rather broadly accepted. Currently mutation causes, or mutagens, are gradually becoming better identified, and it is of interest that penetrating radiation in the form of X-rays was the first identified mutagen—a discovery made by Herman Müller (1890–1967) in 1925.

The fact that the Lamarckian hypothesis has a continuing philosophical and political appeal is shown by its recent "scientific" acceptance, in principle at least, in the Soviet Union and its wide dissemination in that country. Apparently, however, this acceptance, scientifically at least, has now been foregone, and disgrace has been visited upon Lysenko, its principal protagonist (after whom it has sometimes been identified in certain circles as the Lysenko heresy).

CHROMOSOME ABERRATIONS AND POINT MUTATIONS

Since the physical result of a mutation is a changed genotype, it must result from some change either in the genes or in the chromosomes. In this latter case the most obvious change is a break in the chromosome, and it has been mathematically estimated that about 17 ionizations in the germ cell are necessary for such a break to occur as a result of radiation exposure. However, chromosome aberrations that bring about faulty division in both germinal and somatic cells are thought to be "one-hit" phenomena. For radiation intensities above about 160 rads/min chromosome aberration in some cells which have been studied is approximately proportional to the square of the dose; for lower intensities the power relationship decreases to about 1.5 at 2 rads/min, and it becomes more nearly directly proportional to the dose at still lower exposure rates.

The term "point mutations" refers to mutations that are apparently not associated with recognizable chromosome changes and are thus more probably related to effects on the genes themselves. Many of these have been identified, a particularly important group being used in studies of mice. The fact that some mutations are recognized indicates that they occur with significant frequency, and their results may be easily identified. Early experiments had indicated that mutation frequency is proportional to the dose received, regardless of the exposure rate; however, more recent studies with mice indicate that the mutation frequency produced at lower dose rates, down to 10 rads/day, is significantly less than anticipated on this basis. It appears that the differential radiosensitivities of mature sperm and immature spermatogonia are important to this effect.

Obviously the only mutations studied have been those that occur relatively frequently and can be identified; in humans these are generally highly undesirable. Thus the vast majority of actual mutations are probably almost completely unknown, especially for long-lived organisms such as man.

INDIVIDUALS AS GENOTYPES

To this point attention has been focused on the genetic effects caused by the alleles of a single gene. However, each individual has the overall physical characteristics dictated by his own particular assortment of alleles for the vast number of genes that determine his species. Hence, since the genetic pool of the population may contain several possible alleles for each gene, it is apparent that combination and recombination of these alleles by mating among the members of that population will produce a variety of individuals who are thus essentially also individual overall genotypes. Accordingly, although mutation is the ultimate source of species variability and thus of original genotypes for a single characteristic (and probably also for assemblages of characteristics), these random combinations can produce new genotypes of hitherto unrealized assemblages of genes, even though no alleles that are not already in the genetic pool of the population have appeared. It is apparently frequently accepted today that this is the principal method by which new genotypes are produced in the human species.

On the other hand, mutations do occur, and, if such occurrence is completely at random in the population, they will principally affect the more numerous "successful" genotypes. Hence, selection processes may tend to increase the fraction of a population that has characteristics varying comparatively slightly from the norm—whereas mutation effects tend to increase the fraction of individuals with other genes. Accordingly, if it is assumed that most mutations introduce less desirable characteristics than those of the norm, it would be concluded that most mutations will produce less successful genotypes than the norm. Although such an assumption may underestimate the importance of desirable mutations, it may introduce a perhaps desirable precaution and conservatism of approach to population genetic damage by any mutagens.

The ratio of unfavorable to favorable mutations that are produced spontaneously has been variously estimated, with a probable value of about 50:1 indicated. However, it must be recognized that these refer only to mutations that are readily recognizable. It is probable that such observations would be made even though favorable mutations were actually to occur as frequently as unfavorable ones, because unfavorable effects are usually much more readily recognized than advantageous ones. Thus it is much simpler to identify an idiot than a genius; a circulatory defect would be readily identified, but an improvement to the same extent would probably not be seen.

It is thought that the various undesirable recessive genes that exist as heterozygotes, though individually causing little damage, may, as an overall total, cause much of the genetic damage in a given population. Accordingly, it is concluded that any increase in the mutation rate, such as might be caused by radiation, will correspondingly increase this genetic damage since mutations affect the most frequent alleles in the population, which are those that comprise the norm. There is also evidence that a recessive gene can produce some genetic effects in a heterozygote, although obviously these would probably not be as visible as in the case where it is dominant or in the homozygous condition. However, there are instances, hybrid corn being a specific example, where a heterozygote is more vigorous than either homozygote for its respective genes.

Heredity in living organisms is not as simple as indicated by the above illustrations, since inheritance apparently depends not only on individual genes but also on groups of genes that appear to act as a single factor for no identified reason. In many cases genetic effects differ for males and females, and the characteristic is thus said to be sex-linked. Similarly, crossovers (an exchange of certain alleles between the two adjacent chromosomes) occur, and various genetic aberrations, or "slippages," are also noted. Neither of Mendel's principles is apparently completely followed at all times, and dominance is not always complete, with the result, for example, of blue-eyed parents very infrequently producing a brown-eyed child. However, a decreased viability apparently accompanies the violation of these general principles of inheritance.

EFFECT OF ENVIRONMENT

In addition to the production of new genotypes, whether by genetic recombination or by mutation, a requirement for evolution is a process of selection as a result of which a successful genotype will become a part of the norm for the population, or species, in a given environment, and an unsuccessful one will be eliminated or maintained as a small ratio to the norm.

One of the earliest statements of such a selection principle was the famous Darwinian principle commonly identified (by Herbert Spencer) as that of the survival of the fittest. In its simplest form this principle assumes that a characteristic which is favorable to survival of the species, once it appears, will occur in progressively larger fractions of the population with successive generations and thus eventually become a part of the species norm. If, on the other hand, the new characteristic (or variation of an established one) does not better fit the species for survival, its carriers are less likely to propagate; thus it does not become a species norm, even though it may survive as a small fraction of the population.

It has been noted that a population norm may include several alleles for the various genes, these occurring in the population at different frequencies, which may reflect the relative fitness of their carriers. One theory thus assumes the existence of a variety of "normal" alleles in a population with a corresponding variety of homozygotes and heterozygotes and a similarly differential fitness effect for each in a given environment. In a different environment the relative fitness of the respective variations change, and their frequencies also shift. In fact a specific allele may be advantageous to its bearer in one environment but not in another. This has been observed, a classic example being the sickle-cell anemia prevalent among the natives of certain parts of Africa where malaria is also a problem. The sickle-cell anemia is itself disadvantageous, but as a recessive gene it in some way confers some immunity to malaria on those who have it. Hence, in a malaria-dominated environment it is an overall advantageous factor, but in a nonmalarial environment it has an overall disadvantageous effect.

The environment may thus be considered to be an arena in which the various genotypes, produced basically by mutation, vie for success in setting the species norm. The environment includes obviously the totality of surroundings affecting an organism, and these are not only physical in nature. For humans particularly, various social and sociological factors may not be ignored as selection effects.

RADIATION EFFECTS

It has already been noted that radiation can injure cells by affecting the gene-carrying chromosomes, and there are those who consider chromosomal injury to any cell as being genetic in character. However, it is only by chromosome injury to the germ cells that inheritance is affected, chromosomal effects in other cells causing somatic effects at worst.

It is generally accepted that radiation—including visible light for some single-celled organisms—will cause mutations, but the effect of radiation exposures on human heredity is a very important (and sometimes hotly debated) subject. Although survival of a given viable mutation and its incorporation into a species norm requires many generations and depends strongly on environment, principal attention has been given to mutation production by radiation, and comparatively minor attention has been given to the special aspects of selection; selection studies in a "natural" environment are also rather difficult.

The following would appear to summarize current basic information and conclusions of interest concerning the interaction of radiation and genetics, although it should be noted that not all of the conclusions given are universally accepted.

1. Although some groups of humans have received much greater radiation doses than the average, there is no conclusive evidence of any heredity

effect except for an apparently slightly increased ratio of males born of irradiated parents; similar changes in the birth ratio for unirradiated parents due to a variety of factors, however, makes this conclusion rather suspect. The populations for which data were reviewed included survivors of the Hiroshima and Nagasaki bomb blasts who received doses up to at least 200 rads, some individuals in France receiving radiation therapy with total exposures in the 450 to 1400 rad range, similar therapy in Canada resulting in doses of up to 200 rads, and American radiologists for whom the actual doses received are poorly known. The absence of any identifiable genetic effects attributable to differences in radiation background has already been mentioned.

2. Rather limited observations of unconfined animal populations have also failed to indicate any genetic effects of increased radiation levels. Thus, for example, no change in the overall population of a Texas field mouse colony was observed even with high level irradiation of the gonads of captured males, which were then released, nor were any effects of possible mutations on the colony viability noted. Similarly, small mammals that live in high-radiation level areas apparently show no deleterious effects, and a 1964 radiobiological survey of the Bikini and Eniwetok atolls failed to reveal any definite anomalies resulting from the prior nuclear tests among the thousands of animals and plants examined. In fact the rat population of these atolls, including the roof rat that was accidentally introduced on Eniwetok in the early days of the program, apparently survived above-lethal radiation doses during the tests and appear to produce healthy, normal animals.

3. Radiation can produce mutations. This was first shown in the early 1920s by experiments in which fruit flies, or Drosophila, were exposed to X-rays. Similar experiments with X-rays and neutrons have shown the same effects in various other animals and plants, although there is no direct evidence of mutation resulting from irradiation by alphas or betas.

4. No new mutations have been produced by radiation. Although many biological systems have been irradiated at comparatively high levels, there has been little evidence that a mutation has been produced that had not been previously observed to occur spontaneously.

5. The relative frequency of various recognized mutations is the same for radiation-induced and spontaneously occurring mutations. This has been particularly studied for point mutations, and it has been observed that the relative frequency of a given mutation does not markedly differ from its relative frequency in spontaneous mutation only.

6. Radiation is not the only mutagen, although background radiation

is estimated to cause about 10 percent of spontaneously occurring mutations. Among the recognized mutagens are various chemicals, heat, and so on. Current data indicate that a dose of 30 rads delivered over a short period of time is the spontaneous mutation-doubling dose and that 100 rads is the corresponding doubling dose for chronic exposure. Assuming an average background radiation level of some 165 mrem/yr, a 50-year exposure time (from birth to average end of a reproductive period) would result in an approximate 8.2 rem total. On the basis of a 100 rem mutation-doubling dose this would imply that background radiation produces 12 percent of spontaneously occurring mutations.

7. There is apparently no threshold for the production of mutations; that is, regardless of how low the exposure level or the exposure itself, some mutations will be produced. However, the number of mutations that are produced for a given total exposure is less at lower exposure rates than at the higher ones. Probably the lowest exposure rates for which observations have been made with mammals is about 10 rads/day with mice. Compared with many of those used to study nongenetic effects, these are still rather high.

8. A philosophical viewpoint supports the concept that most radiation-induced mutations will be undesirable because one effect of radiation is chromosome breakage. It is assumed that anything producing as drastic an effect as this must be bad, even though similar mutations occur spontaneously without the agency of radiation.

9. Man's evolution itself occurred during a period when the background radiation level was probably higher than it is today (at least it seems highly improbable that it was ever less). The decay of naturally radioactive material in the earth's crust to nonradioactive daughters has resulted in a lower level of background radiation from these materials than was the case a few eons, or even centuries, ago, and it seems difficult to conceive of a situation that would have resulted in a lower level of radiation from cosmic rays, the other principal source of background radiation, than is now the case.

10. Some geneticists seem to feel that the overwhelming majority of new genotypes are produced by recombination in the population, that the principal mutations affecting man are deleterious, and that these are already existent in his genetic constitution, principally as recessive genes. Accordingly any increase in the mutation rate can only be deleterious and will increase the genetic burden of the population gene pool. However, it appears that overall desirable genotypes will more probably result from specific mutations than from recombination, and it has already been ob-

served that it would be impossible for a mutation-produced change in a given characteristic to occur in the absence of the characteristic itself. Color vision has been used as a specific illustration of this effect.

COMMENT OF HISTORY

Many of the factors noted above involve observable results as measured over a comparatively few generations and without an overall integrating factor, such as would be provided by observation of the actual development (or evolution) of a species. It would appear that man's historical development shows rather strongly that some selection for advantageous characteristics does take place, even though it may frequently not be recognized as such. This conclusion is based on various observations concerning the possible results of evolution itself, as follows:

1. It is a valid philosophical conclusion that today's man is better able to meet his current environment than would be any of his progenitors. It is recognized of course that today's man, or other being, may not be very well equipped to meet immediately the environment of some of his ancestors, and it may thus be argued, as it sometimes is, that any genotype that represents the norm of a species is at essentially optimum competence in its own environment at any time. Nevertheless, if it is concluded that man has evolved in an advantageous way, then the conclusion must also be reached that with time desirable characteristics appeared and that they did become ascendent through some sort of selection process.

2. The improvement of various breeds of animals and plants in agricultural programs has resulted from the careful selection and handling of genetically desirable mutations or genotypes resulting from recombinations.

3. There have been nondesirable survivals (from man's viewpoint), which indicate some sort of a selection process. Of these, it is necessary only to point out that a major problem in the hospitals a few years ago was the appearance of a penicillin-resistant staphyloccocus strain; originally the "staph" bacteria were not resistant to the penicillin or other modern "miracle" drugs. Similarly, the advent of DDT immediately after World War II led some to feel that the days of the common housefly were limited, since DDT was an effective poison that would eliminate them. Literally within a very short time, however, DDT-resistant flies were observed, and studies have even shown how this resistance is actually attained. It is perhaps a fair conclusion that mutations at one time produced flies that had the characteristics necessary for resistance to DDT, and this, plus the fact that the nonresistant flies perished in a DDT-

poisoned environment, permitted this resistance characteristic to become a part of the norm of the species. Again, a selection principle was apparently at work.

4. Medical techniques have permitted survival and reproduction by individuals that have such undesirable mutations as diabetes. Concern has been expressed at the overall eventual effects in the population of the results of these humanitarian actions as well as those of various sociologically-dictated programs which result in the increased propagation and survival of genotypes with various characteristics that may be deleterious to the population in its present environment.

On the basis of the above observations, the author finds it difficult to agree with the widely stated conclusion that all radiation is genetically bad, this conclusion in its broadest aspects, of course, including the genetic effects of even background radiation. Certainly this does appear to be true in view of other activities that, desirable though they may be for various reasons, do not appear to improve the human species survival capabilities.

EVOLUTION AND PHILOSOPHY

It is obviously beyond the scope of this presentation to discuss the philosophy or philosophical implications of evolution other than to note that there is good evidence for the existence of its two requirements—some method for continuously introducing new genotypes, however produced, into a population or even a species, and some process by which favorable genotypes become a normal characteristic of the population or species concerned. Although most changes introduced by new genotypes are probably comparatively minor, a succession of them may eventually produce a significant alteration of a species norm.

Most concepts of evolution apparently imply that new genotypes are produced sufficiently rapidly for the norm of an existing species in general to change more rapidly than the environment requires. Thus, if only a reshuffle of occurrence frequencies of existing variations is necessary to meet a slight environmental change, only a comparatively few generations may be required for adaptation; in fact, it has been reported that a mosquito population required only eight generations to adapt to a chemical poison, and a single mutation is probably all that is necessary to produce a new bacterial genotype. On the other hand, many generations may be required to adapt to a major environmental change; if this change occurs too rapidly or is too great for adequate adaptation, a species may disappear. Similarly, the too frequent occurrence of deleterious genotypes and survival of the unfavorable alleles in the genetic pool may so lower the population survival capability as to cause its demise in a relatively constant environment. It is also possible that an environmental change that is dangerous to survival of one species may so enhance the survival competence of another species that it becomes a dominant form of life in that environment.

Time has largely dispelled the half-century-gone religious controversy over evolution versus divine creation of life, a rather common interpretation of which denies the existence of any evolutionary principle. Were this true, no consideration would need be given to genetic factors since it assumes that all characteristics of a species were fixed at creation and cannot subsequently be changed either for better or for worse. Actually the concept that a population gene pool contains all possible viable genes with new genotypes being produced by recombination only may also suggest a type of divine creation that permits some variation but only within limits established at the creation. This might also include the equilibrium condition where only identified mutations are produced, these occur at some given rate, and they cause generally undesirable effects in their bearers.

OVERALL SUMMARY OF RADIATION EFFECTS

In conclusion it appears that, as in the case for almost all environmental conditions, whether they be "bad" or "good," radiation at almost any level can have an effect on individuals. It has been shown that injury or damage can be produced by external radiation at sufficiently high doses or dose rates; that injury can result from sufficiently large internal deposits of radium, radon, and thoron; and that radiation can produce mutations that are similar to those occurring spontaneously. On the other hand, the only parts of the general field of radiation exposure that have been even reasonably well covered experimentally are those of the short-term comparatively high exposures from which mutations as well as various deleterious somatic effects are usually expected and observed. Also well treated have been the results of the injection of comparatively large amounts of radioisotopes, where, again, rather obvious effects are anticipated and usually noted.

However, it has not been clearly or unequivocally determined whether or not there is actually a threshold level of exposure or exposure rate below which radiation would not be harmful. Similarly, the conclusion that all additional mutations produced by radiation, and radiation alone, must be bad seems hardly a completely scientific approach.

It is thus a not illogical position that the conclusions, frequently so glibly stated, which imply that all radiation is harmful to the individual or to the human species simply have not been validated. Thus, although they may be accepted as convenient, cautious, and conservative bases for radiation protection activities, such acceptance should not mean that they have definite experimental backing, certainly not from the somatic injury viewpoint and probably not from the position of genetic changes either.

It is especially important for those who are working in the field of radiation protection to realize that in evaluating available information, an extrapolation of data, particularly a long one, can give results, but this does not mean that a fact has necessarily been discovered. The conclusion drawn from more or less indirect evidence may, when better information becomes available, be recognized as being founded on a false premise with the result of a different conclusion being indicated. In any event, failure or refusal to consider data that disagree with some predetermined concept is a cardinal sin.

REFERENCES and SUGGESTED READINGS

Any text on genetics.
Report of the United Nations Scientific Committee on The Effects of Atomic Radiation, United Nations, New York, 1958, 1962, 1964, and 1966.
Effect of Radiation on Human Heredity, World Health Organization, 1957.
C. E. Purdom, *Genetic Effects of Radiations,* Newnes, 1963.

QUESTIONS

1. In any population containing dominant and recessive genes, which will *always* be greater—the number of genotypes or phenotypes?
2. Why are animals such as mice and the Drosophila fly used in radiation studies instead of higher animals?
3. Is it possible to say that evolution is a result of mutations and natural selection? Support your answer.
4. Define genetics, genes, alleles, and chromosomes.
5. What are the causes of variability within a species?
6. Explain Mendel's two laws of inheritance.
7. Consider a population of 100,000 people. The frequency of the allele for brown eyes is 0.7 and for blue eyes 0.3. Assuming simple brown over blue dominance, how many people will be blue eyed, how many will be brown eyed, and what fractions of people should there be in each genotype?
8. How do changes in gene pools occur?

9. If a gene is present in a low frequency in a gene pool, will its chances of mutation be greater or smaller than those of a more frequently occurring gene? Why?
10. What is the frequency of spontaneous mutation?
11. What are some of the known mutagens?
12. What is usually considered to be the mutation rate of man?
13. Is there a threshold for genetic effects of radiation? What evidence is there to support your answer?
14. What are the important considerations in radiation-induced mutations? Are there more "good" than "bad" mutations?
15. What effect does the rate of environmental change have on the survival of a species?

CHAPTER TEN

Permissible Exposure Limits

IMPORTANCE OF PERMISSIBLE LIMITS

As noted in Chapter 1, it was only a few years after the discovery of X-rays and radioactive materials that this new scientific marvel was observed to be a good servant but a very poor master as the injurious effects of what would be considered today as tremendous exposure became apparent, first as the short term effects of erythema, and subsequently as the long term effects of skin and bone malignancies.

It was shortly after World War I that the British and Americans made their first attempts to establish permissible limits of exposure, then called *"tolerance values"*. In essence, these values were considered to be the exposures that an individual's body could tolerate without injury.

This general concept of limiting radiation exposure to more or less arbitrarily established values as a method of preventing injury has remained a cornerstone of radiation protection programs at local, national, and international levels. Perhaps obviously, the values have changed as more information has been developed from observation, experiments with extrapolation of results, and theoretical interpretations of data. Currently a permissible dose is defined as the dose of ionizing radiation that, in the light of present knowledge, is not expected to cause appreciable bodily injury to a person at any time during his lifetime.*

It should be noted that actually the justification for these limits and the philosophy on which they are based have changed, as is developed in subsequent sections. In addition, because of the continuing practical

* *NBS Hdbk,* No. 59. P. 27

importance of such limits as well as their intimate relation to the development of the entire field of radiation protection, their bases, history, and the uncertainties involved in their specifications are of corresponding importance to any study of health physics.

INSTRUMENTATION DEVELOPMENT

An important factor in the time at which initial efforts at permissible limit specification could be successful was an ability to measure an exposure or the significant characteristics of its producing agent with a reasonable degree of precision; the development of the Coolidge X-ray tube and later of comparatively sensitive and easily handled detectors was thus important to the specification of exposure limits. In fact, as it became desirable to specify limitation of exposure from various conditions which were difficult to evaluate, a rather simple and general standard of *no* permissible radiation was suggested and adopted. Hence the initial permissible level for many conditions or exposures was actually the limit of detection of the instrument used. It is still possible to relate some limits, particularly such secondary ones as those of surface contamination of alpha emitters, to specific instruments that were the most sensitive of their time.

Although the idea that no exposure is the only permissible exposure remains a significant concept in the field, it is not generally applied to the specification of actual limits, and current instrumentation is generally adequate to measure the quantities used in dose determination at these limiting values.

INITIAL WORK OF RADIATION-PROTECTION COMMITTEES

As far as the United States was concerned the formation of two groups— the National Committee on Radiation Protection (NCRP) in 1929 (as the Advisory Committee on X-Ray and Radium Protection, with the American Roentgen Ray Society and the Radiological Society of North America as its principal proponents) and the International Commission on Radiation Protection (ICRP)—at about the same time were of primary importance in the field of determining acceptable doses or dose rates. This is still true today, and both committees, augmented and with hosts of subcommittees, are active at the present time. Since their inception the members of these respective subcommittees have worked closely together, as have the parent committees; in fact, the ICRP representatives from the United States are also generally members of the NCRP. Subsequently, the NCRP has somewhat expanded its areas of concern as is indicated by its current title of National Committee on Radiation Protection and Measurements.

The initial activities of these groups were concerned with what would today be called good practices, including specifying thicknesses of materials which would provide safe shielding under certain conditions. The first criteria of these types, listed in Handbook 15 of the National Bureau of Standards (NBS), came from the NCRP and were published in May 1931. About three years later, Handbook No. 18 was published, giving recommendations for protection from radium and also specifying a permissible quantity of radium in the entire body of 0.1 μCi. It was in this same year, 1934, that the first use of a permissible limit (tolerance value) of 0.2 R/day (or 200 mR/day) for exposure to external penetrating radiation was adopted by the ICRP (the information was distributed although not formally published). This was considered to be about 0.1 percent of the so-called erythema dose from a short-term exposure. The next year the NCRP dropped this tolerance level to half the ICRP figures, or to 100 mR/day for protection from X-rays. This value was published in 1936 in the revised *X-Ray Protection Handbook,* which was then numbered NBS Handbook 20. Implicit in such specifications is the adequacy of instruments and knowledge of source conditions so that relatively accurate methods of determining doses and dose rates could be established.

It is a tribute to the effective and careful work of the men on these committees as well as the adequacy of the safety factors used that these two basic permissible limits, 100 mR/day for exposure to external sources and a total body deposit of 0.1 μCi, were used throughout the vastly expanded production and use of radiation sources and radioactive materials during World War II, with no evidence of any injury when actual exposure did not exceed these values.

EFFECT OF WORLD WAR II

World War II, with its development of the atomic bomb and the use of a wide range of source materials and radiation fields, made further action almost mandatory, and it was in 1948 that further extensive committee work was initiated to modify and extend the limits established. Also in 1948 the peculiar aspects of certain radiation problems were recognized, and the NCRP and ICRP committees were expanded and subdivided into various groups concerned not only with X-ray protection but also with internal and external exposures, the handling of radioisotopes, waste disposal, neutron exposures, and so on. Various other aspects of protection have resulted from international considerations, including such items as shipment of radioactive materials and the international aspects of waste disposal in streams and the ocean. These have also been taken up in some detail by the subcommittees of these two national and international bodies whose activities are continuing. Accordingly, in subsequent years permissible-limit values of varying applicability have been revised and expanded.

In 1954 the permissible exposure was lowered to 0.3 R/week because it was thought that the development of the newer and higher energy radiation sources provided a greater exposure to deep-seated body organs than had hitherto been the case. At about the same time, the term "permissible dose" was first recommended for general use to underscore the belief of some workers in the field that, because of genetic considerations, there was no longer a tolerable exposure with its implication of a threshold limit and thus the term "tolerance" or "tolerance limit" should be abandoned. The reasons for this proposal were first given in the NBS Handbook 59 in that year.

GENETICS AND THE THRESHOLD

In 1956 a genetics panel of the National Academy of Sciences presented its justification for believing that there was no threshold dose as far as genetics was concerned. Thus, the belief that all radiation is harmful became a basic tenet on which would be based further consideration of acceptable or permissible limits. As a result of this decision and considered risk evaluation the permissible dose was once again reduced, this time to 3 R per quarter (13 weeks), with the further provisio that the average exposure of any individual should not average more than 5 R/yr subsequent to his eighteenth birthday.

DEVELOPMENT OF TECHNICAL DATA

To this point only the NCRP and the ICRP have been mentioned as contributors to radiation protection criteria and information. They are probably still the most influential of the various agencies that have been involved in these activities, both because of their prestige and because their membership is made up of individuals who are or have been leaders in the development and analysis of technical information as well as its application to the problems of radiation protection. However, specific mention should also be made of other groups that have had significant influences on the specification of radiation protection criteria and their usage. A particularly important one is the International Commission on Radiological Units and Measurements (ICRU) which has been active in providing appropriate definitions of importance to the field.

Technical information has come from a vast variety of laboratories both in the United States and abroad. The bulk of work in the United States has been sponsored and financed by the U.S. Atomic Energy Commission (AEC) with the various military groups of the U.S. Department of Defense also providing funds for items of more specific interest. More recently, the U.S. Public Health Service has taken an increasingly active part in obtaining technical data, and significant contributions have come from such governmental agencies as the Bureau of Standards, the Bureau of Mines, and, to a rather limited extent, from some of the states. These various groups have also sponsored technical information and analysis meetings for correlation of data. Comparable organizations in other nations have also contributed similarly to the available information, and such international organizations as the World Health Organization (WHO) of the International Atomic Energy Authority (IAEA), the International Labor Organization (ILO), and, more recently, the United Nations itself have sponsored international groups to analyze information and to make appropriate recommendations.

Other than governmental agencies, various laboratories and hospitals have used private capital to develop radiation-effects information. Important analyses of available data as applicable to the specification of permissible limits have been provided by committees or panels organized by the National Academy of Sciences–National Research Council. In some aspects this work has paralleled that of the NCRP except that no limits as such have been stated, the information provided being instead in such form that it might very well have formed the basis of the NCRP recommendations.

Another organization that has accepted some responsibility for the development of radiation-protection criteria is the United States of America Standards Institute (formerly the American Standards Association). This body, working through several committees and subcommittees in the nuclear standards field—one being specifically Committee N–13 (formerly N–7), Radiation Protection—serves as a clearinghouse for representatives of government, industry, technical organizations, and other groups in the preparation and issue of authoritative standards. Although these groups do not normally become involved in the preparation of permissible limits or similar criteria, they do provide a very important service in obtaining a concensus from the various groups affected of methods and criteria by which permissible limits may be efficiently applied to practical operations. Working with the International Organization on Standardization (ISO), these committees are also bringing some degree of international standardization to the radiation protection field.

Various professional organizations, both collectively and through their individual members, have had an influence on the development of standards. These include the radiological groups, the Health Physics Society, the American Nuclear Society, the industrial hygienists (particularly the American Conference of Governmental Industrial Hygienists), various safety organizations, and industry collaborations such as the Atomic Industrial Forum.

GOVERNMENTAL INTEREST AND ACTION

In additional to the above organizations and operations that have provided technical data and frequently recommendations for safe practices, regulations of governmental agencies have also had a very important effect on the establishment of permissible limits and operating criteria. Probably the highest level of these agencies is the Federal Radiation Council, which was organized in 1959 to advise the President of the United States concerning appropriate guidance in the field of radiation protection for federal agencies and others. Its membership includes the AEC Chairman and the Secretaries of Defense; Commerce; Labor; and Health, Education, and Welfare. As such, obviously, its pronouncements unavoidably have regulatory aspects.

The AEC has been particularly active in preparing such regulations, and its specifications, as well as those of the Federal Radiation Council, appear in the Federal Register. These regulations generally follow the recommendations of the NCRP but also sometimes go somewhat beyond these recommendations in specifying administrative methods and criteria by which it must be determined that NCRP criteria are being met. Since such specifications generally have the effect of law, the conclusion cannot be avoided that they actually also tend to set permissible limits. The AEC also provides additional criteria for its own operating contractors.

Shipment of radioactive materials in the United States, as of all hazardous materials, is under the control of the Department of Transportation which has recently taken over the relevant responsibilities of the Interstate Commerce Commission (ICC), whose specifications for shipments of hazardous materials had been prepared by the Bureau of Explosives of the American Association of Railroads; the Federal Aviation Agency (FAA) which had issued regulations for air transport as did the Coast Guard for shipment by waterways; and the Post Office for its services.

STATE AND LOCAL CONTROL

Certain of the health-and-safety responsibilities for radiation control are also amenable to state action, particularly if these states meet the requirements given in the Atomic Energy Act of 1957 for acceptance of some of the current AEC responsibilities. By early 1969, 19 states (Alabama, Nebraska, New Hampshire, Washington, Kentucky, California, Mississippi, New York, Texas, Arkansas, Oregon, Kansas, Florida, Tennessee, Louisiana, Arizona, Colorado, Idaho, and North Carolina) had made the necessary arrangements for such transfer. In addition, certain municipalities, notably New York City, have also prepared regulations concerning radiation sources, both inside the city limits and those in transit through the city. With the exception of the state-AEC arrangements mentioned, there may be a possibility of confusion from dual jurisdiction in some cases.

It should be pointed out, however, that the responsibility of the AEC, and thus that of the Federal Government, for radiation protection at fixed locations applies only to radioisotopes produced as a result of neutron action (principally in reactors) and to natural radioisotopes that also have fissionable properties, which include uranium and thorium. Thus, except for shipment, such jurisdiction does not apply to other natural radionuclides or to radioisotopes produced by such ion-accelerators as cyclotrons; similarly, it does not apply to operation of X-ray machines or to the high energy particles accelerated by cyclotrons and similar atom smashers.

In this chapter consideration is given only to the criteria developed by the NCRP (or the ICRP) for permissible limits of exposure. The effects of regulatory actions are not considered further nor are the specialized criteria of certain of the NBS handbooks.

SOURCES OF PERMISSIBLE LIMIT SPECIFICATIONS

Todays' rather well-specified permissible limits of radiation exposure and control generally refer to an actual permissible dose from either external or internal exposure. In view of the difficulties in evaluating actual internal doses, values are also given for the permissible *body burden* of a given radioisotope together with *maximum acceptable concentra-*

tions of various radioisotopes in air and water for continuous and for 40 hr/week exposure conditions; these are obviously indirect methods of dose evaluation. In addition, a variety of secondary limits for specific circumstances have also been stated by various organizations to meet what are generally local problems and conditions.

Current values of the primary limits, as well as a variety of practical suggestions and other information, appear in the various NBS handbooks concerned with some aspects of radiation protection or control, particularly Nos. 59, 63, and 69; these are listed in Appendix I. Very similar information is given in publications of the NCRP. Their reports, respectively Nos. 17, 23, and 22, give essentially the same information as the NBS handbooks noted.

Appendix II summarizes the basic external beta-gamma exposure limits as given in the appropriate NBS handbooks, principally No. 59, neutron exposures as given in NBS Handbook 63, and the limits applicable to internal exposure for a few radioisotopes that are given in NBS Handbook 69. It will be noted that the comparatively directly measurable total-body external-dose limits include a quarterly (3-month) limit of 3 rems (or 12 rems/yr) coupled with a 5 rem annual average for years subsequent to attainment of age 18.* These are not only modified by considerations of partial exposure or those of limited duration but also form an important part of the bases for specifying internal exposure limits. In addition, there is an emergency limit of 25 rem. Since an understanding of some of the bases and limitations involved in specifying actual values of radiation protection limits is important to their effectve use, some of these considerations are treated in the subsequent sections of this chapter.

PHILOSOPHY OF PERMISSIBLE LIMITS

In addition to their practical application a specification of permissible limits introduces some rather significant philosophical implications. Probably the most basic of these is the question of the applicability to radiation considerations of a *threshold hypothesis* or a *linear hypothesis*. Under the former concept any exposure below a given noninjury threshold would be permissible; also, a noninjury threshold is not necessarily the same as a noneffect threshold. Thus an exposure that produces an effect which is either noninjurious or perhaps helpful—such as would be the apparent increased longevity at low levels of chronic exposure—should be permissible.

* In 1965 the ICRP suggested 5 rems per year as the total body permissible limit.

However, by the linear hypothesis, which takes the view that all radiation is harmful, there is no such thing as a permissible or acceptable level of exposure. On the other hand, many of its proponents are apparently willing to accept the sociological concept that exposure or possibility of exposure may be balanced against the potential benefits to be derived therefrom, and any usage of radiation must make this value judgment. Hence, radiation control limits cannot be based on a possibility of actual injury but must be based on some more or less well-defined "good" to be accomplished, as well as the ability of this "good" activity to meet some limits selected as permissible for the activity concerned; actual injury is thus assumed to be incurred in all activities, but its significance in any given situation is attributable only to the factor of permissibility. Obviously, in this view someone must decide what is "good" and even the various levels of "good," since these should then be balanced against various levels of radiation exposure reflecting the degree of "good". The nontechnical problems can be terrific.

Fortunately, to date there has apparently been little attempt to reject activities because they are not sufficiently valuable to warrant exposure to the appropriate NCRP limits as are described subsequently, nor have different limits been placed on different activities or parts of an overall activity because of such a value judgment. However, it is possible that some of the legal difficulties encountered in reactor site installation may be attempts by the uninformed to apply such a value judgment; such an application is essentially to the effect that no level of possible exposure to a population, even though not actually anticipated on a routine basis, is low enough to justify the installation. Essentially the same approach is taken by the pacifists and various other fanatics to the problem of fallout from weapons testing; *no* gain, they say, justifies the risk, regardless of how small it may be. On the other hand, practically all medical exposures are considered justifiable, and no concern is expressed over exposures to background radiation, although the NCRP limits do include the statement that "the NCRP re-emphasizes its long-standing philosophy that radiation exposures from whatever sources should be as low as practical." *

ACCEPTANCE OF LINEAR HYPOTHESIS

Regardless of any argument, however, it is a fact that today's limits have been essentially based on the concept that all radiation is bad and that

* *NBS Hdbk,* No. 59, Addendum of April 15, 1958.

any exposure must be balanced against the benefit to be obtained from the resultant "injury" to the individual or population involved. This has probably been the view of a significant majority of the geneticists, although most medical men, other biologists, and many of those working in the field of health physics apparently take a less extreme view, except as an overall concession to ultraconservatism. However, it is an interesting phenomenon that practically all of those who have only emotional interests in the subject and are little acquainted, if at all, with its technical aspects have wholeheartedly endorsed the concept that all radiation is harmful; such groups as labor unions and pacifist organizations have particularly vocalized this opinion.

As a result of the current ascendancy of this view as well as improved means of limiting exposure, there has been a steady decrease in the so-called permissible limits, even to the extent of changing the wording from a "tolerance dose," with its implication of a threshold value as generally defined above, to such nonspecific terms as "maximum permissible dose," "maximum acceptable concentration" (of air or water), and so on. Because of these low levels and their general attainment, more attention has probably been paid to the ability of an operation to meet given arbitrary levels than has been given to an evaluation of the injuries that may result therefrom or even their potentiality. In fact, it is the author's opinion that current levels are so low and the bulk of exposures in the atomic energy industry so much lower than even these levels that it will be virtually impossible to make any determinations of low-level effects, statistically or otherwise, in the future.

It may also be noted that this approach to recognizing what may be described as radiation toxicity is markedly different from that generally used in defining any other type of toxicity, such as that caused by chemical effects from various materials.

COMMENTS ON THE EMERGENCY DOSE

A special comment on the suggested emergency dose limit of 25 rems, which applies only to external exposure, may be in order. Such a limit may very well be appropriate to conditions for which exposure limitation may be planned and for which the gain is commensurate with the very slight risk involved. However, no limit is really applicable to life-saving or similar very stringent emergencies, both because planning is usually impossible for such actions and because failure to take action that may lead to the exposure may have grave consequences. This is no different from the risks that are sometimes taken with respect to incidents in which radiation is not involved. In fact, the statement has sometimes been made that a person should not hesitate to accept an exposure in the 100 to 200 rem range if another's life is at stake. Exposures in military operations are considered to be commensurate with other risks of life taken on the battlefield.

EXPERIMENTAL BASIS OF PERMISSIBLE LIMIT SPECIFICATIONS

The types of information that are available on human radiation exposure have been briefly reviewed, and their weaknesses in providing technical data applicable to a specification of long-term permissible limits are apparent. Thus, the additional information available from animal and plant experiments has been judiciously used, particularly the information from mammal experiments, to fill the rather large gaps in the human data. It has obviously required a monumental effort to distill the great wealth of information thus available, much of it data that appear to be contradictory, into comparatively simple specifications that meet the requirements indicated in the definition of a permissible dose as given above.

FACTORS OF SAFETY

In meeting the specifications for permissible limits, which obviously refer only to somatic effects under present consideration, both safety factors and factors of conservatism have been applied to the available experimental information to reach the published permissible limits.

In general, a *safety factor* represents the factor by which the permissible limit concerned could be multiplied and still remain below a value that according to experimental data would cause no injurious somatic effect to *any* individual so exposed. In addition to these specific factors, the permissible limits are also usually lower than the actual injury values for the condition concerned by additional multiples, called *factors of conservatism*, which are known to exist but for which quantitative values vary widely depending on local conditions and are thus difficult to evaluate; these are generally not considered in a limit specification. Some of these are mentioned in succeeding sections, it being recognized that not all of those identified apply to any given situation.

ESTIMATES OF OVERALL SAFETY FACTORS

Safety factors are usually considered to be in the range of 3 to 10, which means that a permissible radiation exposure dose limit is some 10 to 33 percent of an "injurious" value. However, there are indications that some of the basic safety factors may be much greater for reasons of the type indicated below.

In evaluating the overall factors of safety in the listed permissible limits the accepted position is taken that the permissible exposure for any time period may itself be safely received during any shorter time, provided no additional exposure is received during that same period. One limiting case for the permissible exposure of 3 rems in 13 weeks, and this probably the most injurious one, is for the entire amount to be received in only a few minutes. Comparing this with the 100 rem minimum for actual somatic injury (induction of nausea in essentially all people thus exposed but with recovery certain) indicates an overall safety factor to be in the range of 30 to 40. Obviously this 3 rem dose is also about 12 percent of the "emergency" permissible dose of 25 rems which in turn is about 25 percent of the minimum injury figure.

At the other extreme is the chronic lifetime permissible-exposure figure of 5 rems/yr average. Since longevity data indicate this 5 rem dose may be approximately the weekly exposure above which life shortening may be reasonably expected, it would appear that use of the permissible limit provides an overall factor of about 50 as far as somatic effects are concerned. (If the somatic figure of 12 rems/yr were considered, the overall factor would be of the order of 20.) This figure agrees with estimates made by others that current limits are factors of about 50 below the level at which a deleterious somatic effect might be noted for chronic total body exposures. A few animal experiments have also indicated that injections that produce permissible body burdens are about 2 percent of those causing a decreased average lifespan.

Since the overall safety factor in current permissible limits for total body exposure appears to be in the range of 20 to 50 at the two limits of exposure time, those for acute short term exposures and chronic lifetime exposures, there appears little reason to suspect that the factor will be smaller for intermediate exposure periods. These apparent safety factors all refer to somatic effects only; obviously they have no real significance under the linear hypothesis concept.

FACTORS OF CONSERVATISM

Individual Variations in Radiosensitivity

Since permissible limits are generally based on the effects to the most radiosensitive individual, a significant factor of conservatism for most individuals is thus indicated. The order of magnitude of this factor is indicated by the observation that the minimum short-term lethal dose for the least radiosensitive individual is apparently about three times that for the most radiosensitive one; however, as also noted, similar ratios are not available for other effects.

Animal-to-Man Extrapolations

Obviously no animal is affected by radiation, or anything else, exactly as is man. Again, the best data refer to lethal exposures, and for several animals the LD–50/30 differs only by a factor of 2 to 3. Where similar data have been obtained for more than one animal it has been customary to consider man as no less sensitive than the most radiosensitive animal. Actually, it appears that man is among the more radiosensitive of the mammals tested.

Change versus Injury

Although the basic exposure limitation criterion should obviously be that of actual injury, any recognized change is frequently identified as representing injury, and the limits are set accordingly; for example, it would appear that a short-term dose of 100 rems is about the minimum to cause an actual apparent injury; however, the 25 rem minimum-possible-effect dose is usually considered to be the limiting value. The concept that all radiation is harmful might also be considered under this heading, even though there are levels at which it is difficult to show this to be true; any reduction of exposure limits because of genetic considerations would also come under this general head.

Indirectly Evaluated Exposures

Permissible limits are sometimes applied to measurable conditions that reflect only a possibility (or probability) of actual exposure. This is done particularly for possible internal exposures as determined by environmental measurements. In addition, one method of evaluating the potential exposure due to internal deposits (probably poorly known at best) of a given radioisotope involves a determination of the mass of the radioisotope releasing energy at the same rate as that of a specified mass of radium, ($^{226}_{88}$Ra), ignoring possible synergistic effects. Fortunately, it appears to date that this practice has been conservative, although this may not continue.

Partial-Body Exposures

The consideration that a given exposure to a specific organ is as injurious as would be the same exposure to the entire body is extensively used as an integral part of the so-called *critical organ* theory and approach to the evaluation of permissible limits for internal exposures, as reviewed in more detail later. However, evaluation of any partial-body exposure in terms of total body effects will always be considered to be adequately conservative.

Fractionation of Exposure

Since most basic exposure data involve comparatively short term exposures, the assumption that the same effect will be produced by the same total exposure which was received at a lower rate and thus over a longer time period, normally introduces significant factors of conservatism. To give some idea of the magnitude of this effect, it appears that the LD–50/30 for a variety of animals is increased by a factor of about 3 for a reduction in exposure rate of the order of 10 to 50. A single exception to this concept, that involving the reproduction system, has been previously mentioned.

The probable inadequacy of considering a given exposure to be as dangerous if spread over an extended period as it is when incurred over a shorter one is indicated by some uses, or misuses, of longevity data. In some cases, high-level short-term exposure data have been rather irresponsibly extrapolated to give a figure for life shortening per rad of exposure. The low-level data are rather generally ignored or considered to be statistically inadequate, even though other types of information with similar statistical uncertainties are readily accepted, and the perhaps much larger uncertainties in a long extrapolation of data are conveniently ignored.

An illustration may indicate the type of analysis that may be employed. Consider the case in which a dose of 600 rads produces death in 21 days in a 22 year old man whose life is thus shortened by about 40 years. It could then be concluded that a 15 rad exposure would shorten life by one year, 40 mrad by one day, and so on. Such an extrapolation is obviously nonsense, although it is not too different from the type of analyses that are sometimes given, especially in the popular press.

The rather irresponsible assumption that exposure effects are essentially independent of exposure rates will also obviously ignore the possibility that markedly different overall effects may result from different exposure levels. The apparent increased longevity for chronic low-level exposures as compared to the measurable decreased longevity caused by high-level exposures and exposure rates is one illustration of this possibility.

Another example of conclusions drawn from this type of misuse of data is furnished by rather widely quoted figures specifying an individual's probability of contracting leukemia per rad of exposure. Such evaluations ignore the lack of supportive data at either low exposure rates or short term exposures of less than about 100 rads. Any such probabilities, tenuous at best, should be expressed in units of no less than about 100 rads. Even this does not include a time element and is accordingly very probably conservative for even intermediate term exposures.

This last illustration is actually representative of a special type of data use whereby a given dose is assumed to cause the same *fraction* of an effect as it is a fraction of the dose causing that effect; no consideration is given the possible conservatism in the initial specification of the dose causing the effect itself. This approach obviously is not only rather conservative but may also actually indicate an incorrect result.

Extrapolations of the types indicated above can indicate no exposure threshold; this is especially true because data from even the intermediate exposure ranges into which such extrapolations are frequently extended do not always give statistical support to the conclusions drawn.

Exposure Rates and Proportionality

It has already been observed that there are no officially proposed limits for very short term exposures other than the emergency one of 25 rems and that the shortest exposure period considered for what are essentially chronic exposures is the quarter, with the total body external exposure limit of 3 rems. A corresponding exposure rate extended to longer term exposures would normally thus be 3 rems/quarter. Further fractionation of this exposure rate to provide a permissible value, but for a shorter period, such as the 40 mrem/day, will also obviously provide a factor of conservatism. Such conservatism results from any use of permissible limits for chronic exposures with the proportionate rates applied to shorter term exposures.

Obviously, body deposits of various radioisotopes can be built up in an individual if these materials exist in the environment in such condition that he can inhale or ingest them; these deposits will then decrease with removal of the individual from such exposure, and the time-averaged value of the deposit will be less than the maximum. Accordingly, the use of a long-term limit for a short-term environmental exposure has not only the obvious conservative factor of averaging as described above but, as shown subsequently, it also may be important in the possibility of a permissible body deposit being attained.

Variations in Data Interpretation

With respect to the minimum dose that causes effects, the experimental methods as well as methods of data interpretation have frequently led to considerable inconsistencies; for example, in all of the countries commonly referred to as the West experiments based on clinically observable effects, such as blood changes, are used to test radiation effects. Interpretation of the overall results of a vast variety of such data appears to indicate that the most radiosensitive body tissues are probably the blood-forming organs, with nerve tissues having a very low radiosensitivity. On the other hand, experimenters in the Soviet Union have frequently reported that on the basis of conditioned reflex experiments with animals, nerve tissues are the most radiosensitive, particularly the central nervous system. It is not obvious if acceptance of both approaches would markedly affect current permissible limit specifications, however.

Experimental Methods

Experimental procedures may not accurately reflect conditions of potential exposure. Thus, much of the animal data used in determining internal exposure limits is obtained by injection of the radioisotope, whereas most actual exposure results from inhalation or ingestion. Although this practice is frequently defended as providing "cleaner" data, the fact remains that much of the metabolic process is thus bypassed, and experiments have indicated that this may readily result in significantly different effects not only with respect to sites and amounts of deposit but also in the toxic effects on the various organs themselves; this last probability may actually reflect synergistic effects. However, there has been no indication that criteria developed from injection data are less restrictive than those developed otherwise.

Treatment of Data

Differences in radiosensitivity have already been mentioned as has been the fact that permissible limits are generally related to the most radiosensitive individual. Similarly, internal exposure probabilities are frequently specified in terms of deposit probabilities of the radioisotope(s) concerned. Thus in establishing permissible limits of deposit, confidence limits are determined from data that reflect individual variabilities in deposit probabilities, and the value used is not an average but one that reflects a limiting condition as calculated for some specific confidence interval. Again, human (or animal) variability may be such that rather improbable limits are computed; for example, permissible air concentrations lower than those of background have actually been calculated for some radioisotopes, although not used.

Generalizations on RBE-Values

For external-exposure limits for betas and gamma or X-rays no consideration is normally given to the RBE of the radiations concerned, both being considered as having an RBE of 1 with respect to the effects of therapeutic X-rays (about 200 kV), which are taken as a basis. However, this is not strictly true, the higher energy radiation from cobalt-60 having an RBE of 0.6 to 0.8 as far as lethality effects are concerned; these values could very well be different for nonlethal effects. Similarly, X-rays with energies of up to about 20 kV do not significantly penetrate tissue and are thus essentially skin hazards only, as is the case for betas; on the other hand betas with energy below about 30 keV are considered to have an RBE of 1.7. Generally the possible effects of such RBE simplifications are minor and probably well within measurement uncertainties.

However, the RBE becomes rather important when applied to neutron exposures, which, since neutrons are generally emitted only during a nuclear reaction, are effective only as external sources. Neutrons from such a reaction are not monoenergetic, and thus a wide range of neutron energies is generally involved in an exposure. For chronic exposures as based on eye damage the RBE for fast neutrons is commonly accepted as being about 10. However, for lethal effects from short-term exposures the RBE is probably more nearly in the range of 1 to 2. The RBE of slow neutrons is taken to be about 3.

Multiplication of Factors

Permissible limits of internal exposure include not only safety factors and factors of conservatism of the appropriate types indicated above and in succeeding parts of this chapter but also, and inherently, those that are generally involved in external exposure determination. Thus, since one method of specifying internal limits is based on the limits for external exposure, the overall safety and conservatism factors are the product of factors applicable to the external limit specification and those involved in internal limit analyses.

Although the above analyses do indicate some of the possible causes of uncertainty in our knowledge of radiation effects as well as those inherent in the methods used in their evaluations, they should not be taken to mean that the effects reported from extrapolation of data, even long extrapolations, have no value. On the contrary, such extrapolations may be very valuable, especially in the absence of better information, because they almost invariably lead to caution, which is usually desirable. However, it would appear that such extrapolations should not be stated as being true with the same positive approach as would be the case were the extrapolations not so long, and certainly they should not be used in place of other data that require much shorter extrapolations.

REFERENCES and SUGGESTED READINGS

Taylor, L. S., *Health Physics,* **1,** No. 1, 3 (June 1958).
Taylor, L. S., *Health Physics,* **1,** No. 2, 97 (September 1958).
National Bureau of Standards Handbooks 59, 63, 69 and 78. (See Appendix I.)

QUESTIONS

1. What is the definition of a permissible dose?
2. What was the first tolerance value?
3. What is the age above which the permissible radiation dose is changed?
4. Support either the linear or threshold hypothesis.
5. If an effect is due to radiation, is this effect necessarily bad?
6. What is the safety factor that is built into permissible radiation doses?
7. What were the first limits specified by the NCRP?
8. What general effect did World War II have on the establishment of permissible limits?
9. What effect did the 1956 genetics panel have on permissible limits?
10. What is an emergency dose?
11. What are factors of safety? List several.
12. What are the factors of conservatism? List several.

13. Why was the term "tolerance dose" abandoned? Are today's limits based on the linear or the threshold hypothesis?
14. What are the basic rules for external-source permissible dose?
15. How does the permissible limit concept fit in with the threshold and linear theories of radiation effects?

CHAPTER ELEVEN

Internal Exposure Evaluations

Permissible limits for internal exposure are considerably harder to evaluate than are those for external exposure, just as it is considerably more difficult to determine the existence of an actual internal dose than an external one. Thus, recourse is taken to indirect methods of evaluation, one being based on comparison with the effects of radium-226 deposits and the other on a critical organ concept that involves extrapolation from external exposure limits.

The eventual injurious effects of body deposits of radium-226, principally in bone, have been described. However, in 1934 it was observed that apparently no ill effects would occur to anyone with a body deposit (or *body burden*) of no more than about 1.2 μg of radium (or 1.2 μCi) even though the measurements were made many years after the exposure. Accordingly, a permissible body burden limit of 0.1 μCi was established for radium-226, and this, extended to all radioisotopes, became the second of the basic criteria for radiation exposure limitations during World War II; the other was limitation of external exposure to 0.1 rem/day.

The term "body burden" by which permissible limits for internal exposure are expressed is not precisely defined but it is apparently most frequently used to refer to the amount of a radioisotope deposited in the body at any time. Although the total dose received obviously also depends on how long radioactive material remains deposited and elimination may change the amount of the deposit with time as will be described, the time over which the deposit, and its consequent dose rate, may be averaged in evaluating exposure problems has not generally been included in the concept. As an important illustration of this time effect, the lungs, kidneys, and the gastrointestinal tract are not normally considered to be problems for many radioisotopes that are not deposited in them, even though such materials usually pass through one or more of them because most materials enter the body by ingestion or inhalation and are excreted in the urine or feces.

CALCULATION OF BODY BURDEN

Critical Organ

The basic concept on which internal exposure limits are based is the so-called *critical organ* concept. As a fundamental assumption this concept states that the body burden should not exceed the amount of a given radionuclide that would give any of the following doses to the body organs specified:

1. The dose to the gonads or to the total body during any period of 13 consecutive weeks shall not exceed 3 rems. The dose to the gonads or to the total body at age N years shall not exceed $5(N-18)$ rems in case occupational exposure begins after age 18. If occupational exposure begins before age 18, the yearly dose before age 18 shall not exceed 5 rems and the dose to age 30 shall not exceed 60 rems.

2. The effective RBE dose delivered to the bone from internal or external radiation during any 13 week period averaged over the entire skeleton shall not exceed the average RBE dose to the skeleton due to a body burden of 0.1 μCi of radium-226 (see the next section).

3. The dose to any single organ of the body, excepting the gonads, bone, skin, and thyroid, shall not exceed 4 rems in any 13 week period or 15 rems in 1 year. The dose to skin and thyroid shall not exceed 8 rems in any 13 week period or 30 rems in 1 year.

The body organ that receives the greatest RBE dose from a radionuclide relative to the above values is identified as the critical organ for that material even though not all of the material is deposited in that single organ. Obviously, there will be different critical organs for various radio-isotopes, and the critical organ for a mixture of radionuclides will be that organ which receives the greatest relative RBE dose from the mixture, all exposures being considered additive. This is indicated in subsequent examples.

Based on this concept the permissible body burden q, given in units of microcuries, for a given radionuclide is determined by the following relation:

$$q = \frac{100 \ mR}{3.7 \times 10^4 \times 1.6 \times 10^{-6} \times 6.05 \times 10^5 f_2 \epsilon},$$

or

$$q = \frac{2.8 \times 10^{-3} \ mR}{f_2 \epsilon},$$

where $3.700 \times 10^4 = $ disintegrations/sec-μCi
  $1.6 \times 10^{-6} = $ ergs/MeV;
  $6.05 \times 10^5 = $ sec/week;
  $100 = $ ergs/g-rad;
  $m = $ mass of the organ of reference in grams,
  $R = $ permissible exposure limit for the organ concerned,
  $\epsilon = $ the effective energy (in MeV/disintegration) of a radionuclide (and its daughters) absorbed in the organ of reference,
  $f_2 = $ fraction of body burden in the organ of reference.

If the deposit involves the gonads for which the weekly exposure limit R may be taken as 0.1 rem/week and the RBE* for the radiation concerned is 1, the expression becomes

$$q = \frac{2.8 \times 10^{-4} \ m}{f_2 \epsilon}.$$

If the deposit involves most other organs for which the permitted dose R is about 0.3 rem/week and the RBE is again 1, the expression becomes

$$q = \frac{8.4 \times 10^{-4} \ m}{f_2 \epsilon}.$$

Use of this method for determining the exposure due to 0.1 μCi of radium-226 in the body gives an average dose rate to the bone of 0.56 rem/week (or 0.06 rad/week). Although this value is obviously greater than the 0.3 rem/week considered to be generally acceptable, this deviation is accepted as being based on firmer experimental evidence with humans than are similar data for other radionuclides.

The values of m for various organs are taken from tables of the so-called standard man, which is a compilation of various average attributes of humans. See Appendix III.

For some radioisotopes, of which natural uranium and depleted uranium

* In this usage RBE is defined as the ratio of the equivalent dose and the absorbed dose.

(that with a lower than natural uranium-235 fraction) are probably the outstanding examples, the permissible limits are based on chemical toxicity rather than radiological considerations.

Comparison with Radium–226

Although comparison with radium-226 gave the first values for internal permissible limits for a variety of radionuclides, this approach is currently applied only to those that are deposited in the bone. Thus for radium-226 the bone is considered to be the *critical organ.* For such materials the maximum permissible body burden q for any radioisotope is determined by simple proportionality and is given by the expression

$$q = \frac{q^{Ra} f_2^{Ra}}{f_2} \times \frac{\epsilon^{Ra}}{\epsilon} = \frac{0.1(0.99)}{f_2} \times \frac{110}{\epsilon} = \frac{10}{f_2 \epsilon},$$

where $q^{Ra} = 0.1$ μCi is the maximum permissible body burden of radium-226,

$f_2 =$ fraction of radionuclide in the skeleton as compared to that in the total body;

$f_2^{Ra} = 0.99$ is the value of f_2 for radium;

$\epsilon =$ effective absorbed energy per disintegration of a radionuclide given by $\epsilon = \Sigma EF \text{ (RBE)} n$, with

$E =$ energy (MeV) deposited in skeleton per disintegration;

$RBE =$ relative biological effectiveness,* which is normally 1 for X-rays and gamma rays, β^- (electron), β^+ (positron), and ρ^- (internal conversion electron); 10 for alphas; and 20 for recoil atoms. (However, the RBE is set equal to 1.7 for β^-, β^+, or ρ^- if the maximum energy $E_m \leq 0.03$ MeV);

$F =$ ratio of disintegrations of daughter(s) to disintegrations of parent;

$n =$ the relative damage factor, or distribution factor, which is unity provided the parent element of the chain considered is an isotope of radium or the energy component considered originates as X- or gamma radiation; otherwise, it is 5.

Now $\epsilon^{Ra} = 110$ MeV per disintegration (equivalent) is the value of ϵ for radium.

---

* In this usage RBE is defined as the ratio of the equivalent dose and the absorbed dose.

For a radioactive chain, such as that of radium, the term ϵ actually refers to a sum of the products of the energy; the RBE; and a factor that depends on the distribution, decay, and elimination rates for each of the various components of the chain. Because of these differences appropriate values for ϵ are not usually simply determinable from elementary radiation data; for example, ϵ^{Ra} is given as 110, and this figure differs markedly from the value obtained by simple analysis of the published values of the radiations of the entire radium chain, which is a part of the uranium-238 series (see Table 3–1).

Radionuclide Mixtures

Where more than one radionuclide is concerned their mutual effects on any one body organ are considered to be additive, even though the respective radiations concerned are different. Appendix II gives permissible body burdens together with other data taken from ICRP tables. With reference to these tables it is noted that the body burden for the alpha emitter radium-226 ($^{226}_{88}Ra$) is 0.1 μCi and that for strontium-90 ($^{90}_{38}Sr$), a beta emitter, is 2 μCi. If an individual has a body deposit of 0.08 μCi of radium-226 (or 80 percent of the body burden), a total body deposit of only 0.4 μCi of strontium-90 (which is 20 percent of the strontium body burden) would be permissible since bone is the critical organ for each.

The tables also give figures for a "body burden" based on the total body. This is usually higher than is that for a specifically listed critical organ, although this is not always the case. These values are assumed to apply to all organs not specifically listed. Thus, if instead of strontium-90 the radioisotope concerned in the above illustration were manganese-54 ($^{54}_{25}Mn$), a gamma emitter, a deposit of only 8 μCi (which is 20 percent of the 40-μCi total body body burden of manganese-54) would be permitted in addition to the 0.08-μCi deposit of radium-226. This is obviously less than the 20 μCi that is the smallest manganese-54 body burden listed for which the liver is the critical organ.

DIRECT ESTIMATES OF BODY BURDEN

Although the concept of a body burden and consequent methods of evaluating exposure potentials as indicated above are attractively simple, they are practically rather difficult to apply because of the difficulty of estimating a body burden. For considerable relative accuracy in experimental work the organ of a sacrificed animal is removed and assayed, probably by ashing and counting. However, for humans such measures are not considered to be at all feasible!

Total body counters, or even beta-gamma meters, have been used to indicate the amount of beta-gamma-active material in an individual's body and its approximate location, thus possibly indicating an organ of deposit. In fact such methods, especially those that use sophisticated methods of spectrum analysis and body-scanning techniques, are now generally adequate to indicate the presence of the very small gamma-emitting deposits that are at the limits of permissible body burden values. Even though alpha emitters reportedly have the greatest potential for causing internal injury, the alphas cannot emerge from the body since they cannot even penetrate the skin. Similarly many important body organs are too deeply placed to permit the escape of most betas. However, such emitters may be detected if they also emit gammas (as do U-235 and Pu-239), have gamma emitting daughters (as does U-238), or their betas produce bremsstrahlung (as with Sr-90).

The other possibility for evaluating the amount of a possible deposit is through measurements of excreta, primarily the urine and feces. Nose swabs have also been taken where exposure has been expected, and breath counts have been made, primarily where the exposure was considered to be that from radon. Breath samples were obtained for some of the early radium dial painters, and the literature gives some of the values obtained.

ELIMINATION OF RADIONUCLIDES FROM THE BODY

In general, radioactive material is removed from the body through the two methods of actual biological elimination and of radioactive decay to an eventual stable isotope. It has been pointed out in Chapter 3 that radioactive decay is an exponential process; that is, the number of atoms changing in a given period of time is proportional only to the number present at the start of the period. Thus, if k_r is the radioactive constant of proportionality, N is the number of atoms (which is obviously proportional to the mass) present after a time t, and N_0 is the corresponding figure at $t = 0$,

$$\frac{dN}{dt} = -k_r N \quad \text{and} \quad N = N_0 e^{-k_r t}.$$

Probably the simplest model for evaluating biological removal is to assume it to be also exponential in character, and the calculation is mathematically the same as that for radioactive decay. Accordingly, if k_b is a biological constant of proportionality,

$$\frac{dN}{dt} = -k_b N \quad \text{and} \quad N = N_0 e^{-k_b t}.$$

If both effects are acting simultaneously, as would be the case where the material is being eliminated both by radioactive decay and by biological means, the resulting equations would be

$$\frac{dN}{dt} = -(k_r + k_b)N \quad \text{and} \quad N = N_0 e^{-(k_r + k_b)t}.$$

For an overall or effective removal constant k the above becomes

$$N = N_0 e^{-kt},$$

where $k = k_r + k_b$.

As already shown, an exponential decrease can be best expressed by a half-life T for which $k = 0.693/T$. Thus, for a radioactive half-life of T_r and a biological half-life of T_b the overall or effective half-life T may be determined from the relation

$$k = \frac{0.693}{T} = \frac{0.693}{T_r} + \frac{0.693}{T_b}.$$

Obviously

$$\frac{1}{T} = \frac{1}{T_r} + \frac{1}{T_b}.$$

Thus, the effective half-life T for elimination is

$$T = \frac{T_r T_b}{T_r + T_b}.$$

Effective Elimination Half-Life

The above relation shows the dependence of the effective half-life T on the radioactive and biological half-lives, T_r and T_b, respectively. In some cases, as for iodine-131 in the thyroid, the biological half-life is much longer than T_r; thus the material is removed essentially by radioactive decay. On the other hand, the radioactive half-life of uranium-238 is much longer than its biological half-life, with the result that the principal removal agent is biological elimination. For some materials of interest, such as plutonium-239 or radium-226, both half-lives are extremely long compared with the length of an individual's life; hence the radioisotope, once deposited, appears to be essentially fixed and irremovable.

Although biological removal is probably not exactly exponential in character, especially after a high-level short term inhalation, the rather simple conclusions developed from such a model have proven to be very useful in certain types of analyses, and this concept is rather generally used in calculations of other aspects of permissible limit determination. In many cases the elimination of a radioisotope can apparently be expressed as the result of successive eliminations with differing half-lives, the shorter ones being predominant in the early removal and the longer ones later on. The condition is similar to the apparent decay of a mixture of radioisotopes, as discussed in Chapter 3; for example, analysis of the removal of inhaled hexavalent uranium indicates that an exponential model is applicable only if more than one half-life is assumed.

Power Model for Elimination

In some cases that have apparently been developed from experiments in which the radioisotope was injected, the best fit to an experimental removal curve is provided by a power series of the form

$$R(t) = At^{-n},$$

where $R(t)$ = fractional retention t days after injection ($t \geqq 1$),

$\quad\quad\quad A$ = normalized fraction of injected dose retained at end of unit time,

$\quad\quad\quad n$ = a constant.

The physical concepts of such removal are not as simply justifiable as is the exponential model, and it is less used, although a function of this form apparently better fits elimination of injected hexavalent uranium by humans than does a simple exponential relation. It is not considered further here (the information developed from this model is available in the literature).

Practical Limitations of Excreta Measurements

Although the above analyses seem to provide a rather simple way of determining the amount of a deposit from appropriate measurements of the excreta, the small amounts of radioisotope that are normally excreted from even permissible-body-burden deposits—as well as the metabolism and other characteristics which are variable among individuals—have made such attempts impracticable. Accordingly, although these methods are frequently used to give an indication of the possibility of exposure, no limits based on quantitative measurements of excreta or excreta rates have been specified for any radioisotope. However, most operating facilities have taken the desirable step of using excreta measurements as a part of their control programs and have thus established their own local limits for taking action.

USE OF ENVIRONMENTAL LIMITS

As indicated in the preceding sections, it is generally impracticable to attempt a measurement of internal deposits and thus to determine directly the actual body burden of an individual and the rate at which the material in such deposits is being eliminated. It has become customary to attempt to limit the possibility of accumulation of deposits above the maximum permissible body burden by controlling the amount of such material in the environment to such limits that excessive amounts will not be introduced into the body. Since intake is most commonly accomplished by breathing or ingestion, permissible limits concern the concentration of some radioisotopes in air and in water; some of these values are listed in Appendix II. The relative hazards of the respective radioisotopes may generally be considered to be proportional to their respective maximum permissible concentrations (MPC). These factors are defined and considered subsequently.

Basically, the MPC values in these tables assume continuous breathing or ingestion by water drinking. However, for occupational exposure it is also assumed that the 40-hr/week exposure period may be averaged over the entire 168 hr of a week and that the 50-week normal occupational work period (with 2-week vacation) may be averaged over an entire 52-week year.

RADIONUCLIDE BUILDUP IN CRITICAL ORGAN

The simplest method of determining buildup of a radioactive nuclide from environmental conditions assumes continuous intake of material and an exponential model for excretion. For simplicity it is also usually assumed that transit time in the body is short and that the amount of material that reaches the critical organ is *always* a constant fraction of the intake. This fraction is not necessarily the same as f_2, which is the fraction of the total deposited body burden that is found in the critical organ. The metabolism involved in the deposit mechanism accounts for any difference. In any event, if q is the total body burden, then $f_2 q$ is the burden in the organ of concern, usually the critical organ.

Under these conditions the time rate of change of a given radioisotope in a critical organ is somewhat analogous to the buildup of a radioactive daughter as described in Chapter 3 and is thus determined by the relation

$$\frac{d(qf_2)}{dt} = -k(qf_2) + P,$$

where P represents a continuous rate of uptake of the radionuclide by the critical organ as would result from continuously breathing air containing a given radioisotope concentration. If $qf_2 = 0$ when $t = 0$, the solution of this equation is

$$qf_2 = P(1 - e^{-kt})/k = PT(1 - e^{-0.693t/T})/0.693,$$

where $f_2 =$ ratio of radionuclide in critical organ to that in total body,

$k =$ effective decay constant $= 0.693/T$,

$T =$ effective half-life $(T_r T_b)/(T_r + T_b)$,

$T_r =$ radioactive half-life,

$T_b =$ biological half-life,

$t =$ period of exposure (in determining maximum permissible values for occupational exposure t is assumed to be 50 years, or about 18,000 days, and for most materials is thus effectively infinite).

If qf_2 is expressed in microcuries, P is similarly expressed in microcuries per day if the times listed are given in days; this is usually the case. The rate of intake, P, obviously depends on the concentration of material in the environment and its transfer to the critical organ.

Thus for breathing

$$P_a = M_a Q_a f_a,$$

where $M_a =$ the radioisotope concentration in the air,

$Q_a =$ the average rate of air intake,

$f_a =$ the fraction of the material taken in that eventually goes to the critical organ.

Similarly for water drinking

$$P_w = M_w Q_w f_w,$$

where $M_w =$ the concentration of the radioisotope in water,

$Q_w =$ the average water intake rate,

$f_w =$ the fraction of material taken in that gets to the critical organ.

In general then

$$P = MQf.$$

(*Note:* The quantity P above is equivalent to the quantity $(M)S$ used by the ICRP, and f may frequently be considered to be nearly equal to f_2.)

Since it is obviously impossible to make such determinations for every individual, usable values of Q and f are usually determined from appropriate figures of the "standard man," as mentioned previously. For populations with different habits or average dimensions, including those of specific organs, the values will be somewhat different. A compilation of such figures for natives of India does indicate different values and correspondingly different permissible limits.

Where q is the maximum permissible body burden the values of M are the appropriate MPC values, indicated by $(MPC)_a$ for air and $(MPC)_w$ for water.

DETERMINATION OF MAXIMUM PERMISSIBLE CONCENTRATION FOR CONTINUOUS EXPOSURE

For continuous exposure, the methods used to relate the MPC values to the other identified constants are obvious. However, those specified for occupational exposure are generally of more interest. For this purpose it is assumed that the exposure occurs in a 40-hr week of five 8-hr days and that the individual works 50 weeks per year. It is also assumed that because of greater activity during the work period he takes in half his daily allotment of air (or water) in his 8-hr workday. From the standard man figures, which give the water intake at 2200 cm³/day and the air intake at 2×10^7 cm³/day, the respective intakes during the workday are thus assumed to be 1100 cm³ of water and 10^7 cm³ of air. However because an individual does not work steadily, the average intake of water or air per day over an entire work year, Q, is given by

$$Q = \tfrac{5}{7} \times \tfrac{50}{52} \times Q' = 0.687 \, Q',$$

where Q' represents the actual intake rate during the individual's work time, which is also the exposure time; Q' is thus 10^7 cm³ of air or 1100 cm³ of water. Accordingly the average intake rate of a radionuclide at the MPC level in air is

$$P_a = (MPC)_a \, Q_a f_a = (MPC)_a f_a \times 0.687 \times 10^7 =$$
$$6.87 \times 10^6 \, (MPC)_a f_a.$$

Similarly, the rate of water intake at the MPC level is

$$P_w = (MPC)_w Q_w f_w = (MPC)_w f_w \times 0.687 \times 1100 = 754 \, (MPC)_w f_w.$$

It has previously been shown that

$$qf_2 = P(1 - e^{-kt})/k.$$

Thus

$$P = qf_2 k/(1 - e^{-kt}).$$

Since $k = 0.693/T$ and $P = (\text{MPC})Qf$, the above expression becomes generally

$$\text{MPC} = \frac{0.693\ qf_2}{QfT[1 - e^{-(0.693/T)t}]}.$$

For the maximum permissible air concentration $Q_a = 0.687 \times 10^7$, and thus

$$(\text{MPC})_a = \frac{10^{-7}\ qf_2}{Tf_a\ (1 - e^{-(0.693t/T)})}\quad \mu\text{Ci/cm}^3$$

since $0.693/0.687 \simeq 1$.

Similarly the approximate value of the maximum permissible water concentration is given by

$$(\text{MPC})_w = \frac{9.2 \times 10^{-4}\ qf_2}{Tf_w(1 - e^{-0.693t/T})}\quad \mu\text{Ci/cm}^3.$$

As shown by Appendix II the MPC values for continuous exposure are always approximately one-third of those for occupational exposure. This comes rather obviously from the fact that the daily intake rate during the work period is assumed to be 50 percent of the total daily intake, and, because of the 40-hr week and the 50-week work year, the average concentration inhaled (or drunk) is 0.687 times the actual concentration. Thus $0.5 \times 0.687 = 0.343$, which is approximately one-third. It should particularly be noted that f_2 is in general different from f_w or f_a.

TIME DEPENDENCE OF BODY DEPOSITS

The $(\text{MPC})_a$ and $(\text{MPC})_w$ values are selected so as to meet the appropriate exposure criteria for the organs of deposit. They are chosen so that, if they are an individual's maximum rate of radioisotope intake, he will never have a greater deposit than the maximum permissible body burden even though his exposure continues for 50 years. In general, his actual body burden at any time will be less than the maximum value that it is approaching at a time of 50 years. The extent of this difference depends on the effective half-life of the material in the body; this is shown mathematically from the previously developed equation

$$qf_2 = \frac{PT(1 - e^{-0.693t/T})}{0.693}.$$

Thus q, the body burden after time t, will be

$$q = \frac{PT(1 - e^{-0.693t/T})}{0.693f_2}.$$

If the time t and the effective half-life T are given in years, the maximum permissible body burden q_0 is assumed to be attained when $t = 50$ years, and thus

$$q_0 = \frac{PT(1 - e^{-0.693 \times 50/T})}{0.693f_2}.$$

Accordingly, after a time t in years, where $t < 50$, the ratio of the individual's actual body burden to the maximum permissible one is given by

$$\frac{q}{q_0} = \frac{PT(1 - e^{-0.693t/T})/0.693f_2}{PT(1 - e^{-0.693 \times 50/T})/0.693f_2} = \frac{1 - e^{-0.693t/T}}{1 - e^{-0.693 \times 50/T}},$$

where $T \ll 50$ years, $e^{-0.693t \times 50/T} \to 0$, and $q/q_0 \to 1 - e^{-0.693t/T}$.

For the condition in which t becomes much greater than T the term $e^{-0.693t/T}$ also approaches 0, qf_2 approaches $PT/0.693$, and q approaches q_0. Where the effective half-life is very short compared with 50 years an exposed individual will have approximately the maximum permissible body burden for a major part of his life if he is continually exposed at the MPC, since a deposit reaches about 90 percent of its maximum after a period of three half-lives. However, once intake ceases an individual's body burden will also decrease rapidly, and at a rate comparable to that of the buildup, as the radioisotope is eliminated. On the other hand, if the effective half-life of the radioisotope is long compared with 50 years (or the individual's life), the buildup will be very slow, with the result that his body burden for the most part will remain far below the maximum permissible value; similarly, after cessation of exposure the actual body deposit will decrease slowly.

It is obvious that the above consideration introduces a significant safety factor when the exposure time is limited; for example, the actual time that material was ingested by the radium dial painters was comparatively short and the amount of material in the bones of those showing no signs of injurious effects was not determined until many years later. The deposits thus measured were then used to specify a maximum permissible limit. The factor of conservatism thus introduced is indicated by Figure

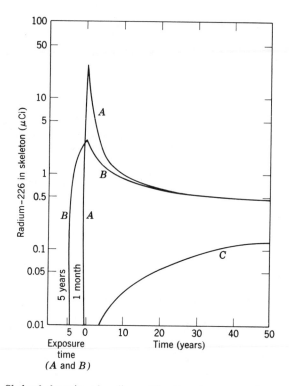

Figure 11–1 Skeletal deposits of radium. The time in years refers to time elapsed since exposure for curves *A* and *B*, and to duration of constant exposure for curve *C*. Curve *A:* deposit period of 1 month; curve *B:* deposit period of 5 years; curves *A* and *B:* 0.5 μCi remains after 35 years; curve *C:* continuous deposit necessary to reach 0.1 μCi after 35 years.

11–1, a modification of one prepared by Evans,* to show actual body deposits of an individual if he were exposed for two specified periods of time and were then to have a terminal maximum permissible body burden (after 35 years). These are compared with the corresponding body burdens of an individual ingesting radium at the current permissible limits for 35 years. Since the total dose received under any exposure condition specified is proportional to the area under the appropriate curve, a significant factor of conservatism exists in use of the MPC intake values as compared with either of the short term deposit conditions indicated. Thus it is obvious that the current MPC value is much safer than indicated by the specified safety factor of 12 obtained from the ratio 1.2 μCi/0.1 μCi.

* R. D. Evans, *Health Physics,* **8,** No. 6, P. 751, 1962

The above sections consider intake to be continuous. However, a more realistic exposure condition for inhalation, especially for materials with a short effective half-life in the body, may be developed for normal work conditions whereby an individual is assumed to be exposed for some fraction of the time each day. Under these conditions it may be shown that the greatest body burden for each day (or cyclical unit concerned) occurs at the end of the day and is given by the relation

$$q = q_1 \sum_{s=1}^{s=n} e^{-0.693(s - 1)rt/T},$$

where t = the time of exposure (in the same units as the half-life T),

r = the fraction of the day (or cyclical unit) during which exposure occurs, and rt is exposure hours/day.

n = the number of days (or cyclical units) since exposure began,

q_1 = the body burden at the end of the first day, which is given by $q_1 = q_0(1 - e^{0.693t/T})$, where q_0 is the greatest body burden attained after an essentially infinite exposure time at the given rate.

Thus for a five-day week during which exposure at some rate is continuous for t hours per day the greatest body burden attained, which will be that at the end of the fifth day, is given by

$$q = q_0 (1 - e^{-0.693t/T}) \sum_{s=1}^{s=5} e^{-0.693 (s-1)t/T} =$$

$$q_0(1 - e^{-0.693t/T}) \times$$

$$(1 + e^{-0.693t/T} + e^{-0.693(2t/T)} + e^{-0.693(3t/T)} + e^{-0.693(4t/T)}.)$$

Figure 11–2 shows the ratio q/q_0 for an exposure time of 8 hr/day, with the effective isotope half-lives being 1, 8, and 64 hr, respectively.

The $(MPC)_a$ for the 40-hr work week is about three times that for continuous exposure. Thus, if it is assumed that the value of q_0 indicated by Figure 11–2 is based on inhalation at the $(MPC)_a$ for the 40-hr week, the resulting dependence of q on time is shown in comparison with both the q_0 for work-week inhalation and that for continuous inhalation, which is one-third as great. It will be noted that, although the body burden fluctuates with time, its average remains below that for continuous exposure; in fact for long-lived radioisotopes during this first week it may never even reach q_0 for continuous exposure.

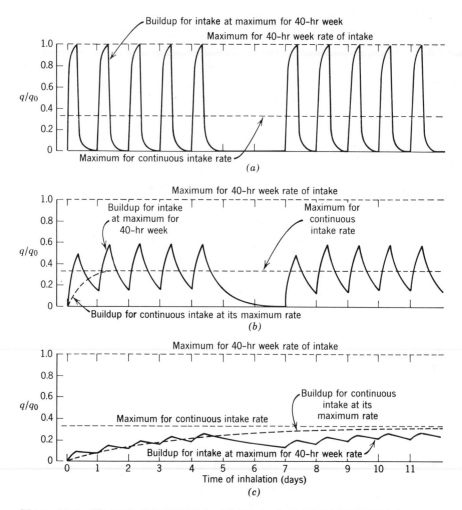

Figure 11–2 Theoretical buildup of radioisotope in body after inhalation—work week and continuous rates. Effective half-life of radioisotope: (*a*) 1 hr; (*b*) 8 hr; (*c*) 64 hr.

For longer term exposures the exposureless weekend must also be considered, and it may be shown that the Monday evening peak of the second week is given by the relation

$$q = q_1(1 + e^{-0.693(9t/T)} + e^{-0.693(12t/T)} + e^{-0.693(15t/T)} + e^{-0.693(18t/T)} + e^{-0.693(21t/T)},$$

Analyses may similarly give the peaks for succeeding days and weeks. In each case it is obvious that the exponential terms represent the fractional difference between the body burden on some successive day and that following the first day of exposure given by $q_1 = q_0(1 - e^{-0.693t/T})$. For very long half-lives the fluctuations shown become negligible, and the curve would approach that for continuous exposure.

SYNERGISTIC EFFECTS

The possibility that chemical toxicity effects, or perhaps combined chemical and radiotoxicities, may be of importance in internal exposure has been mentioned as a factor that should not be completely ignored. It has been noted that apparently only radium-226, radon-222, and thorium-232 have caused identifiable human injury, and it was assumed in each case that the radiation caused the injury. This is supported, particularly in the case of radium, by the observations that the radiation levels and consequent energy release rates in the damaged bone correlated better with the extent of injury observed than any other factor, even though more than one radioisotope of radium was determined to have been involved. On the other hand there is no evidence that what must have been tremendous exposures to uranium materials of a comparatively large number of people in the early days of the Manhattan Project have caused any detectable injuries or effects even after 20 to 25 years; the injurious effects of radium to the dial painters were identified within about 7 years, and those caused by thorotrast some 20 to 30 years after intake and deposit.

The relative nontoxicity of uranium is further pointed out by animal experiments wherein the lungs of dogs and monkeys have accumulated as much as 3 to 15 percent by weight of uranium as a result of inhalation. No obvious ill effects have been reported from the consequent huge dose rates and large total doses. In fact, far from being the relatively toxic material uranium's limit figures (based on chemical effects for natural uranium but radiation for other isotopes) indicate it to be, this radioactive element would be considered essentially nontoxic were its effects to be based on the toxicity criteria that are normally applied to food or drugs. Similarly a few individuals have been exposed to plutonium and maintained plutonium body deposits that were much higher than permissible limits for comparatively long periods without adverse effects being observed. Thorium exposures apparently give rather paradoxical information, with different effects observed for various intake modes.

OTHER INTERNAL EXPOSURE CONSIDERATIONS
Particle Size

Several comparatively minor items may also be noted as being perhaps related to determination of permissible limits for internal exposure. One of these, for inhalation, involves the so-called large-particle-exposure theory. According to this concept the actual injury would be greater for a given total activity if a single large particle were inhaled than would be the case if a large, diffuse group of small particles were inhaled. Similarly the point at which the particles are deposited in the lungs and their mode of removal are also sometimes considered important, although the effects of such differences are not well evaluated experimentally.

Mode of Body Entry

It has been noted that different internal deposits, and thus different potential hazards, occur for a given radioisotope for various methods of body entry. Thus if uranium is inhaled it is considered to be from 100 to 1000 times as hazardous as it is when ingested; similarly ingestion of radium-226 is probably its most hazardous method of intake.

For ingestion the possibility has been considered that the mode of intake, whether by drinking water or by eating differing kinds of food, might produce different effects as far as the eventual site of the deposit is concerned. However, it is considered that any effects resulting from these factors would probably be negligible compared with the overall effects.

In addition to inhalation and ingestion there are normally two other methods by which internal exposure can result. First is the entry of material through a wound or other break in the skin. Since such an event is usually a rather unpredictable occurrence, no attempt can be made to specify a limit for this sort of body entry, and the only actions indicated are efforts to prevent the accident, limitation of environmental surface contamination, and after-the-fact treatment.

The other method of entry is the diffusion or penetration of material through the skin; for example, uranyl nitrate solution will rather readily penetrate the skin and get into the bloodstream. However, it is usually difficult for sufficient material of this type to get into the bloodstream to be a problem unless it is in such concentration as to cause other effects; for example, sufficient exposure to uranyl nitrate solution would also result in chemical burns of the skin. On the other hand, the rather rare tritium ($^3_1$H) may comparatively readily diffuse through the skin in a variety of forms, and carbon-14 may also enter as a component of a readily diffusible compound.

Second-Order Effects

Surface and Clothing

In addition to environmental contamination, such as that of air or water, which may readily be taken into an individual's body there are also environmental conditions, such as the presence of radioactive materials on work surfaces or on clothing, in which a good possibility of eventual body intake exists but is much less susceptible to quantitative analysis. Thus, if such materials are stirred up into the air by intermittent operations or conditions, they may then be inhaled even though the air is not continuously contaminated. The initiating factors may be so irregular in time or space that a good average for the air contamination over the period concerned is not predictable. This may be of particular importance with alpha emitters, which constitute essentially no problem as long as they remain outside an individual's body but are considered to be the most hazardous of internal emitters. Criteria for permissible limits of such environmental conditions have not been widely suggested, certainly not in the same way that the criteria discussed heretofore have been disseminated.

With respect to clothing contamination, it is possible for the material to be shaken off and then inhaled. However, there is the additional interesting consideration that if the alpha-emitting contaminant is so tightly bound to the clothing that it cannot escape into the air, it cannot be a problem—and if it is so loosely held that it does shake off and become an air contaminant, then the actual removal of the material will cause a decrease in the amount of material on the clothing and consequently of that in the air; this of course assumes that there is no continual recontamination of the clothing. Such considerations also apply to the possibility of contamination being taken home on employees' personal clothing. Laundering of such clothing, however, may present its own problems.

Primarily on the basis of certain tests in England, the IAEA and various European nations have suggested permissible surface contamination limits for beta emitters and for the alpha emitters of natural uranium, thorium, and radon daughters of 10^{-3} μCi/cm^2 (2×10^3 disintegrations/min/cm^2) for controlled areas and 10^{-4} μCi/cm^2 (2×10^2 disintegrations/min/cm^2) for inactive areas. For materials of very high radiotoxicity, such as plutonium and radium, the respective limits are 10^{-4} μCi/cm^2. The similar limits for clothing contamination are 10^{-4} μCi/cm^2 for beta emitters and most alpha emitters but 10^{-5} μCi/cm^2 for alpha radiation from radium and plutonium.

Otherwise, various installations have provided their own criteria for evaluating the possibility of an air-contamination problem from surface contaminants. As one example, it was the author's experience with uranium materials that, if the average level of transferable contamination (material removable by rubbing) over an individual's work surface averaged no more than about 1.4×10^{-5} μCi/cm^2 [or 30 disintegrations/min/cm^2, where the overall counting geometry of the instrument (ratio of activity measured to that present, as is described in Chapter 12) was about 0.3 and 100 cm^2 areas were measured], it would be possible, under rather severe conditions that include the use of fans, to produce an air concentration at approximately the maximum permissible value of 10^{-4} μCi/m^3. Permissible levels (those for which no special action, such as some required use of respiratory protection, is taken) was then set at 1 percent of this value. This value is obviously much lower (and thus more conservative) than the values listed by the IAEA. Similarly a uranium-alpha clothing-contamination limit of 3×10^{-4} μCi/cm^2 as based on experimental data was established.

Hand and Skin Contamination

As has been mentioned, alpha emitters in contact with an individual's skin do not cause injury directly, whereas beta emitters do. In addition, there is a possibility that any radioisotope so placed can be inhaled, ingested, or, in rare cases, absorbed through the skin. Accordingly, permissible values listed by the IAEA for skin contamination are on the order to 10^{-4} μCi/cm^2.

Possibly the greatest probability of intake as a result of body contamination results from hand contamination. On the basis of experimental data it is the author's opinion that a possible permissible limit of hand contamination of about 3×10^{-3} μCi/hand (or 7000 disintegrations/min-hand surface using an instrument with an overall geometry of about 15 percent) is adequate for protection of even the most avid cigarette smokers. These figures are probably only guides, because they depend on the test conditions for which they are derived and, to some extent at least, their administrative interpretation; for example, the author's observations specifically apply to the conditions encountered in a group of operations with uranium solutions and dusts at various levels of uranium-235 enrichment. Similar data may also be obtained for other types of operations, and different criteria may be developed.

COMBINED EXPOSURES

The above analyses have been concerned with radiation exposures of specific types, such as that from an external source or from inhalation of radionuclides. However, the considerations developed for their evaluation may be rather readily used for radiation from a variety of sources and involving more than one type of exposure. Basically the criterion for any radiation limitation is that no organ should be irradiated above an overall permissible level, regardless of the radiation source. External radiation is assumed to expose every organ equally. Thus, if an internal deposit is half the maximum permissible body burden, the concurrent radiation dose from an external source can be only half the normal permissible value. This may be more precisely expressed by the relation

$$\Sigma F_a + \Sigma F_w + \Sigma F_i + \Sigma F_e < 1,$$

where ΣF_a represents the fraction of the permissible exposure which is received from inhalation of any variety of radioisotopes in the air, m in number, with each existing at a fraction F_{am} of its individual MPC, and the total fraction given by

$$\Sigma F_a = F_{a1} + F_{a2} + F_{a3} + \cdots + F_{am};$$

ΣF_w is the similar fraction of the permissible exposure which is received by drinking water that contains a variety of radioisotopes, m in number, and given by

$$\Sigma F_w = F_{w1} + F_{w2} + F_{w3} + \cdots + F_{wm};$$

ΣF_i is the fraction of the permissible exposure which is received internally otherwise; this would include ingestion of food that contains radioisotopes, being wounded by contaminated items, etc. ΣF_e is the fraction of the permissible exposure received from all external sources, such as gamma- or X-rays, neutrons, etc., with a total fraction given by

$$F_e = F_\gamma + F_n + F_X + \cdots + F_{em}.$$

The above basic relation may be applied individually to each organ when internal exposure is involved.

The respective fractions may be easily determined. Thus $F_a = M_a/(\text{MPC})_a$, where M_a is the actual air concentration of a given radioisotope and $(MPC)_a$ is its corresponding maximum permissible air concentration. Similarly, $F_w = M_w/(MPC)_w$, where M_w is the actual concentration of the radioisotope in the water and $(MPC)_w$ is the maximum permissible concentration. Similar approaches may be used for other internal exposures. The effect of exposure time on body burden buildup has been mentioned.

For external exposure the fractions depend on the time interval involved and may involve either a dose rate or a total dose for the time of concern. Thus, for the period of a quarter $F_e = E_1/(3 \text{ rems/quarter})$, where E_1 is the total exposure during the quarterly period. Similarly, if the period accepted is a weekly value and E_2 is the exposure for the week, $F_e = E_2/(0.24 \text{ rem/week})$. If such exposure is further divided to give an hourly rate, the corresponding value is $F_e = E_3/0.006 \text{ rem/hr})$, where E_3 is expressed in terms of rems per hour. Ordinarily, however, the smallest time period considered is that of a quarter.

Because of general uncertainties in dose determinations and individual variabilities, it is usually accepted as a general criterion for mixed exposures that any single exposure or exposure rate that contributes no more than 10 percent of the exposure produced by a primary exposure source, internal or external, may be ignored either in specifying exposure limitations, such as time of work, or in determining maximum permissible concentrations. Two or more such components may not be so ignored, however.

The ICRP has recently suggested no limitation of exposure to external radiation when the estimated body burden is no greater than 50 percent of the maximum permissible value.

GENERAL POPULATION LIMITS

The above values and discussions have all been concerned with so-called atomic workers. These are those individuals whose normal work involves the possibility of exposure to penetrating radiation or radioactive materials, and the limits provided are presumably those which provide a maximum amount of safety consistent with the possibility of doing the work assigned. However, due primarily to assumed population genetic effects, it has been customary to consider permissible exposures for people outside the above population to be held at a factor not more than 10 percent of the values applicable thereto. The same 10 percent safety factor is also to be used for all exposures of individuals under 18.

OVERALL PRECAUTION

In view of the uncertainty of the available data, it is probably wise to use caution in limiting exposures. However, it is equally important to realize that the data are not particularly conclusive on the subject, and it is probably comforting to know that the safety factors inherent in the limits currently suggested may be rather large. Certainly this is true for exposures that have no genetic applications.

REFERENCES and SUGGESTED READINGS

National Bureau of Standards Handbooks 59, 63, 69, and 78.

Report of ICRP Committee on Permissible Dose for Internal Radiation, 1959. (Also given as vol. 3 of *Health Physics,* June 1960.)

Various recommendations of the ICRP as amended in 1959 and revised in 1962 and 1965. Published for ICRP by Pergamon Press, London, 1964 and 1966.

Symposium on Occupational Health Experience and Practices in the Uranium Industry, HASL–58, USAEC, Washington, D.C., 1958.

Bernard, S. R., *Health Physics,* **1,** No. 4 (1958).

Evans, R. D., *Health Physics,* **8,** No. 6, 751 (1962).

Safe Handling of Radioisotopes, IAEA Safety Series No. 1, 1962.

"Health Physics Addendum," Safe Handling of Radioisotopes, IAEA Safety Series No. 2, 1960.

"Medical Addendum," Safe Handling of Radioisotopes, IAEA Safety Series No. 3, 1960.

Basic Safety Standards for Radiation Protection, IAEA Safety Series No. 9, 1967.

Radiation Hazards in Perspective, Technical Report Series No. 248, World Health Organization, Washington, D.C., 1962.

QUESTIONS

1. What are the two basic methods of internal exposure determination?
2. What type of radiation has the potential to cause the greatest internal injury?
3. What is the main method of determining internal exposure?
4. Of the two removal methods (biological plus radioactive decay) which is more important for radioactive material with a long half-life (compared to human lifespan)?
5. What is the time length (period) that is considered when calculating maximum occupational exposures?
6. What effect does the deposit of several radionuclides have on the dose limit?
7. How are radioisotopes eliminated from the body?
8. What are synergistic effects?
9. What is the biological half-life of a radioisotope?
10. The limits set up in Chapter 11 apply to atomic workers. How can this be related to the general population?
11. Define "body burden" and "critical organ."

12. Give the maximum permissible body burden for a certain radioisotope as based on some critical organ. Does this mean that such a deposit would give a maximum permissible dose to the entire body or only to the critical organ? Would this be the amount deposited in the critical organ?

13. If a radioactive isotope has a half-life of greater than 50 years, what is the chief means of elimination?

14. Would you expect there to be greater danger from material with a long (50 years) or short (hours and days) half-life?

15. What is the most important method of intake of uranium? of radium?

16. Given 0.05 μCi of radium-226 in the body, what is the permissible concentration of lead-210 that can be present? What is the overall critical organ in this case?

17. *a.* If 5 μCi of platinum-197 and 20 μCi gold-196 are deposited, what fraction of the maximum permissible body burden of any other radioisotope would be permissible with respect to each of the identified critical organs?

    b. In addition to the 5 μCi of platinum-197 and 20 μCi of gold-196, what body deposit of tungsten-181 would be permissible?

18. What would be the external exposure considered permissible for an individual known to have body deposits of 5 μCi of platinum-197 and 20 μCi of gold-196? It is assumed that it is not possible for him to inhale or ingest any other radioisotope.

19. If 0.01 μCi of radium-226 were deposited, how much thorium-228 could be deposited?

20. If a person has 0.01 μCi of lead-212 deposited in his body, how much iodine-132 can be deposited?

21. List the fractions of permissible burdens of the respective organs in the body as produced by an overall body deposit of 1.0 μCi of tellurium-132. If the remainder of the maximum permissible body burden is made up of zinc-69, what is the overall critical organ, and what is the fraction of a permissible deposit in each of the other organs of concern?

22. An individual has a beryllium-7 body burden of 400 μCi. What would be the maximum permissible body burden for potassium-42? What would be the MPC in water of potassium-42 in this case? If the body burden of potassium-42 were half the permissible value for this individual, what would be the maximum permissible body burden of selenium-75?

23. What are the synergistic effects of internal deposits? Give an example.

24. Explain the mathematical model for the removal of radioisotopes from the body.

CHAPTER TWELVE

Radiation Detection and Measurement

Radiation cannot be detected by any of the normal five senses, with the result that no one has a built-in method for immediately determining that radiation levels are high enough to cause injury, except possibly for levels that are lethal in literally a few seconds. In this, penetrating radiation is somewhat similar to sun exposure in that one can receive a severe injury without being aware of it. Thus, a method of detecting and accurately measuring radiation levels is a prime need in the study of radiation and its effects as well as in implementing radiation protection efforts.

Most radiation detection and measurement methods depend on ionization which, as already described, is also a principal cause of radiation's biological effects. It has long been known that a gas will be ionized by X-rays or charged particles; if this ionized gas is in the electrical field between two electrodes charged to some potential difference, the charged particles will be removed by recombining with each other or by traveling to one or the other of these electrodes. Similarly, if this ionized region is included in an electrical circuit, an electrical current that can be observed by a sufficiently sensitive detector will be produced.

POCKET CHAMBERS AND DIRECT-READING DOSIMETERS

One of the first devices used to detect ionization is the electroscope, and instruments of this type are still used for that purpose. Briefly, when an electroscope is electrically charged, as shown in Figure 12–1, its leaves will diverge because of electrostatic repulsion. If the gas around these leaves is then ionized, the instrument will be discharged as the charged

199

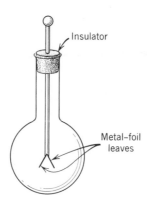

Figure 12–1 Simple electroscope.

leaves attract charges of the opposite sign, and the leaves will fall together. A similar instrument is the electrometer, and the widely used direct reading dosimeter of Figure 12–2 is merely a fountain-pen-sized electrometer. For use, the dosimeter is charged by an auxiliary charger to a specific potential as shown by the position of a pointer on a scale calibrated in dose units and viewed through the instrument. If placed in a radiation field, the device will be discharged an amount proportional to the ionization produced therein by the radiation, and the pointer will move across the scale, which is calibrated in dose units. In theory, a given exposure dose will produce a given amount of ionization in the chamber, and this will in turn cause a specific discharge which is shown by displacement of the movable vane.

Another simple type of detector employs an ion chamber (or ionization chamber), which is merely a modification of "Faraday's Ice Pail." In essence this is merely a piece of metal or a metal container that is insulated from its surroundings and may be electrically charged with respect thereto.

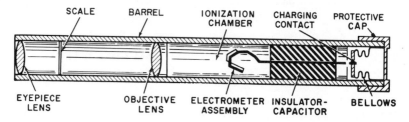

Figure 12–2 Direct-reading dosimeter. (Diagram courtesy Dosimeter Corporation, Cincinnati, Ohio.)

In a practical adaptation of this device a metal rod is inserted into a metal container and insulated from it. The system is then charged so that an electrical potential difference, or voltage, exists between the rod and the surrounding chamber, the magnitude of this voltage depending on the capacity (electrical capacitance) of the system and the charge placed on the rod; the unit itself is thus a simple capacitor, or condenser, with the gas in the ionization chamber forming the dielectric. The chamber may itself be grounded, which means that it is maintained at the same electrical potential as its immediate surroundings. The system may be discharged by making a contact between the rod and the surrounding container; similarly, if the medium separating the rod and container is a gas, the system may be discharged by ionizing the gas in this ionization chamber.

The pocket chamber of Figure 12–3, which is also a fountain-pen-sized instrument, is a commonly used ion chamber that is calibrated to determine dose. The device is charged by an auxiliary source so that a definite potential difference exists between the rod and its container. When placed in a radiation field, the chamber will be discharged an amount dependent on the total ionization produced therein, which is proportional to the exposure "dose" received; this discharge—or rather the charge remaining on the chamber—is subsequently measured with an auxiliary unit (usually a special connection on the charger itself), with the reading being calibrated in dose units.

Pocket chambers and dosimeters are widely used radiation monitoring devices, especially for short term exposures, and their usage in dose determination (second only to that of film, which is mentioned later) has been recommended by some for even more general application. Conditions of usage generally dictate that they be of rigid metal construction, and this makes them peculiarly appropriate for gamma measurement, although special walls of low area density have also made possible some beta detection. However, a variety of attachments and modifications have also

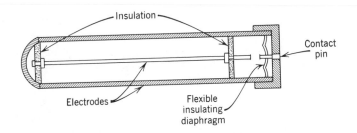

Figure 12–3 Pocket chamber.

been developed in an attempt to make it possible for them to measure neutron flux as well as to differentiate between gammas, X-rays, and betas; such efforts have not usually been too successful.

ION CHAMBERS

Ion chambers are also frequently used for precision radiation field and dose measurements. In essence, the ions produced by ionization in the chamber itself are collected on its walls, and the resulting charge thereon is measured by an auxiliary circuit. For precision measurements of radiation fields ion chambers of various sizes and construction are used for different dose ranges and gamma energies.

Two varieties of ion chambers are of particular importance in dose determinations; these are the so-called air-equivalent and tissue-equivalent chambers devised for precision measurement of the exposure and absorbed doses, respectively. As previously mentioned, the size of the former, as well as its materials of construction, are chosen to duplicate as nearly as possible a volume of air sufficient to capture all of the secondary products of initial ionization in an interior gas region and thus to detect essentially all of the energy released by the ionization produced in this interior volume.

In a similar manner, the tissue-equivalent chamber is designed to simulate the radiation stopping effect of human tissue and the consequent energy release therein. Although this, theoretically at least, should give data that are more nearly related to actual effects on an individual than the exposure dose, the former is probably the more widely used because of its comparatively simple construction and interpretation.

PRINCIPLE OF THE RATEMETER

Instruments that use the condenser-type ion chamber, such as the pocket chamber, are also peculiarly adaptable to measuring a continuing ionization or ionization rate. In this usage a source of potential remains continuously connected across the electrodes of the capacitor-type chamber. This "capacitor" will not be discharged by ionization in the gas of the chamber since the potential source will continuously recharge it, this necessitating a flow of current in the external circuit supplied by the potential source, usually a battery. Since this current is only that necessary to keep the device charged, it is equal to the time rate of neutralization of the charge of the ionized particles that move to the electrodes, thus producing an ionization current in the chamber. Figure 12–4 gives an elementary indi-

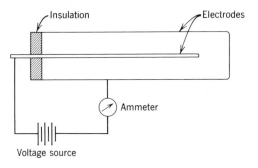

Figure 12–4   Simple ratemeter circuit.

cation of the essential components of such a circuit. The current in the external circuit, which is equal to the rate of ionization in the chamber, is measured by the meter, and the greater the rate of ionization in the chamber, the greater this current. In this usage the instrument indicates an exposure rate instead of a total dose.

Since ionizing events occur at irregular intervals, the resulting meter response is not constant but appears as pulses. However, if the ionizing events are frequent enough for their statistical appearance to be essentially constant, then a correspondingly constant current is observed. Similarly, if the circuit meter response has a sufficiently long time constant with respect to the irregular appearance of the ionizing particles so that the pulses are smoothed out, an apparently constant current is observed. The details of such analyses are obviously beyond the scope of this presentation, although they do indicate why the ratemeters of many radiation detection instruments give irregular jumps at low levels of radiation but provide constant readings at higher levels.

ELECTRICAL CONDUCTIVITY OF GAS

The various factors that affect the flow of current in gases have long been known. Thus, if external X-rays are used to produce a given ionization rate in the gas between two electrodes, the dependence of the resultant current on a change in the voltage applied across the electrodes, as determined by a ratemeter, is shown by Figure 12–5. Curve $ABCDE$ represents qualitatively the response for one level of X-ray intensity (and corresponding constant ionization rate in a chamber), whereas curve $AB'C'DE$ represents another level of X-ray intensity and constant ionization rate at a different level.

In the region AB (or AB') the voltage is so low that many of the ionized particles that are produced do not reach the electrodes but recom-

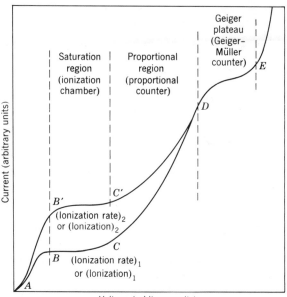

Figure 12–5 Electrical conductivity of gas.

bine in the gas. However, in the region BC (or $B'C'$), all of the particles produced are collected so that despite the increased potential there is no increased current. This plateau is called the *saturated region*, and the current is the saturation current.

As the potential is still further increased, the ions produced, primarily the electrons, gain enough energy as they move toward the positive electrode to ionize additional gas atoms in the interelectrode space, and the current correspondingly increases with increased voltage to point D, which is the start of another plateau that extends to point E. In this region CD (or $C'D$) any ionization that is initially produced is thus multiplied, or amplified.

In the region DE, called the *Geiger plateau* or *Geiger region*, the current in the tube is limited not by the gas ionization rate but by the tube geometry, constants of the external circuit, etc. At potentials above E electrical breakdown may occur in the tube, this being a self-sustained electrical discharge which no longer depends on an external ionizing event for its initial supply of electrons and other ions. One result of such breakdown is usually damage to the electrodes.

The characteristics of the gas in the tube—including its pressure and type—the geometry and material of the tube itself, and the electronic response of the associated circuit are the principal factors that have an effect on the shape and quantitative values of the response curves indicated in Figure 12–5. Obviously, at a given voltage below the Geiger region the actual current shown by a specific meter depends on the ionization events—either their number, or their energy, or both.

RATEMETERS WITH ION CHAMBERS

Ion chambers as used with ratemeters normally operate in the saturation region (BC or $B'C'$ in Figure 12–5), with all of the primary source ionization being collected and measured. The associated meter circuitry usually provides current or pulse amplification. Such meters generally indicate the total energy of the incident radiation or the part that produces ionization; for example, fewer high energy betas may be necessary to produce a given initial ionization than is the case for lower energy betas. In addition to measuring the total ionization produced in the tube, pulse analysis and possible discrimination between types of ionizing sources may also be provided in this detection region. In this case the *pulse height*, or instrument response to an ionizing event in the tube, depends on the amount of ionization that is initially produced by the event itself. Thus, since a beta particle will generally make more impacts in traveling a distance equal to the chamber dimensions than will a gamma photon of the same energy, discrimination between types of radiation, or even the energies of a specific type, may be accomplished by adjusting the external circuit constants so that the meter will respond only to pulses above or below given levels—or even between two levels. Such usage is particularly useful for alpha particles (if the chamber is designed so that they can enter), because they will lose essentially all of their energy in a comparatively short distance, which can be less than the actual dimensions of the chamber itself. Laboratory counting of alpha activity is frequently accomplished by placing the source itself inside the chamber.

PROPORTIONAL COUNTERS

Proportional counters operate in the region CD of Figure 12–5, in which pulse magnification occurs as a result of ionic multiplication in the gas. Gas-multiplication factors as high as 10^5 to 10^6 have been obtained by such operation, although these effects are generally rather sensitive to the gas pressure itself. Proportional counting has the advantage over counting in the saturation region in that lower energy pulses may be detected and less complicated circuitry is sometimes feasible because the circuit amplification can be less.

Such counters filled with ethylene and lined with polyethylene may be operated in such a way that only protons recoiling from fast neutron impact are counted. Under these conditions, the instrument should count fast neutrons in the presence of gammas.

GEIGER-MÜLLER METERS

In the Geiger region (*DE* in Figure 12–5) all detected pulses, regardless of their initial energy, have the same energy output since the overall amplified energy caused by a given pulse, regardless of its initial magnitude, is limited only by circuit and tube constants. Such a tube cannot be used as a radiation or pulse discriminant, although counting in this range permits detection of the least amount of energy released per pulse and is thus the most sensitive method available for gas-counting devices.

Since each ionizing event eventually results in ionization of a significant fraction of the gaseous molecules in the detector, the ions thus produced must be swept from the detector before a succeeding ionizing event can produce a pulse. This time delay, sometimes called dead time, determines the frequency of the ionizing events that can be detected and, as an average, the resolving time of the unit.

Most low-level beta-gamma detectors use the Geiger-Müller (G-M) tube that operates in this range as their detector unit. Measurements with such detectors reflect only the number of incident particles without regard to the total ionization energy released thereby in the chamber.

MODIFICATIONS FOR SPECIAL PURPOSES

Detectors are generally designed for their particular planned use since, in most cases, changes that provide an increase in one desirable characteristic will usually cause a corresponding decrease in another. Optimum electrical response design may thus need to be sacrificed for an arbitrary desirable property, such as portability or compactness. Various shields of the detecting chamber may be designed to permit discrimination between the types of radiation.

By suitable circuit modifications, alarms may readily be incorporated as parts of these instruments, and the alarm points may be readily adjusted to specific requirements. With miniature circuit elements, Geiger-Müller tube devices, including alarms, have been produced in sizes little larger than fountain pens and may be readily worn.

Special gases and chamber coatings are used in attempts to detect neutrons but generally with success only for thermal neutrons. Of these, boron-10 ($^{10}_{5}B$), which has a comparatively high thermal neutron cross section (see Figure 12–9) in the reaction

$$^{10}_{5}B + ^{1}_{0}n \rightarrow ^{4}_{2}He + ^{7}_{3}Li$$

is used for thermal-neutron detection, the boron being introduced either as the gas boron trifluoride (BF_3) in the chamber or as a solid coating, usually boron carbide (BC), on the walls or elements in the chamber. The latter has the advantage of permitting a desirable counter gas to be used.

For higher energy neutrons the thermal-sensitive units are usually encased in thick paraffin to thermalize the neutrons before detection. Special electronic neutron-detection devices, generally for laboratory-type use (except for those that are subsequently mentioned) have been developed; these are described in the literature.

It has already been noted that actual ionization is not generally produced by neutrons directly. Their detection and measurement depend on secondary ionization-type effects, such as proton recoil in paraffin or other suitable gaseous or solid hydrogen-rich materials serving as the proton source in the chamber, nuclear transmutation with consequent emission of charged particles or electromagnetic waves, and so on. Of particular importance in this method is the fission chamber in which uranium-233, uranium-235, or plutonium-239 are incorporated into the ion chamber, usually in the form of thin foils; fission is caused by incident neutrons, and ionization by the fission products is detected. A particular advantage of this method is the ease of discrimination against gamma photons. These phenomena have been and will be mentioned elsewhere and are discussed in the literature.

Figure 12–6 shows several instruments that are typical of the types of detectors described.

EFFICIENCY OF GAS DETECTORS

A very important characteristic of any radiation detection or measuring device is its efficiency or relative response to incident radiation. This is usually expressed as its overall detection (or counting) geometry, which may be defined as the ratio of the counts (or response) shown by the instrument under a given condition to the number of events occurring (or activity involved). As is indicated later, this ratio depends upon a variety of effects other than actual geometrical dimensions. Obviously, knowledge of the overall factor is necessary in measuring the activity, or

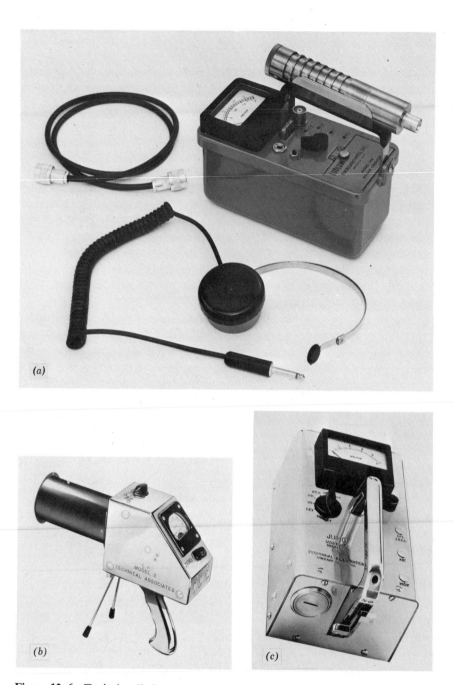

Figure 12–6 Typical radiation detection instruments. (*a*) Geiger-Mueller meter (low-level beta-gamma detector). (Courtesy of Ludlow Measurements, Inc., Sweetwater, Texas.) (*b*) Cutie-pie meter (high-level alpha-beta-gamma detector). (Courtesy of Electro-Neutronics/Technical Associates, Oakland, California). (*c*) Juno meter (high-level alpha-beta-gamma detector). (Courtesy of Electro-Neutronics/Technical Associates, Oakland, California).

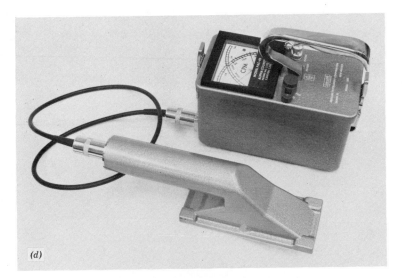

(d)

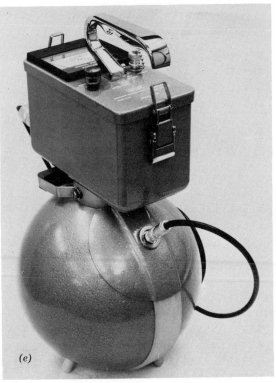

(e)

Figure 12-6 *(Continued)* (d) Scintillation alpha counter. (Courtesy of Eberline Instrument Corp., Santa Fe, New Mexico.) (e) Neutron rem counter. (Courtesy of Eberline Instrument Corp., Santa Fe, New Mexico.)

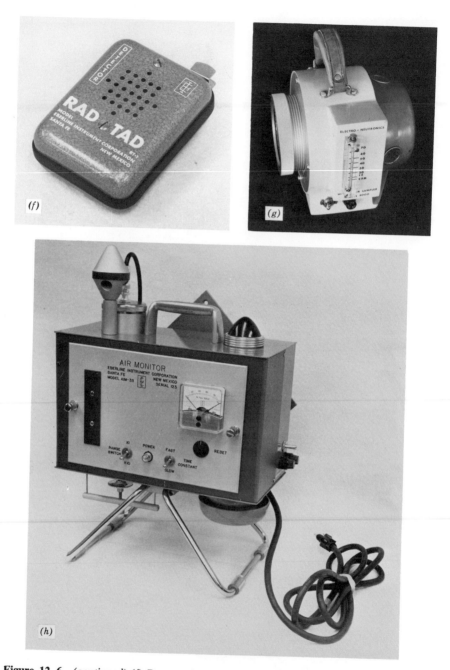

Figure 12–6 *(continued)* *(f)* Personnel monitor and alarm (gamma detector). (Courtesy of Eberline Instrument Corp., Santa Fe, New Mexico.) *(g)* Portable air sampler. (Courtesy of Electro-Neutronics/Technical Associates, Oakland, California.) *(h)* Alpha air monitor. (Courtesy of Eberline Instrument Corp., Santa Fe, New Mexico.)

strength, of a radioactive source, although it is also used for evaluating radiation fields.

For an ion chamber placed nearly in contact with an extended surface covered with an alpha emitter the overall counting geometry may be near 50 percent. Because of their short path length, essentially all alphas that enter the chamber will cause ionization and thus produce a pulse. The alpha source emits particles isotropically; that is, there is no preferred direction of emission. Hence, as shown by Figure 12–7, no more than half the emitted particles even start in the direction of the chamber, the remainder being ejected away from it and thus not being captured. In practical cases such a chamber is not in direct contact with the surface, and thus less than 50 percent of the particles that are emitted will enter it. There will be additional minor losses due to those which are not captured in the chamber because of their statistically long path length, their entry near a corner and consequent exit, and possibly other factors. However, this is a condition where the counting geometry may be nearly 50 percent.

Another special condition is posed by measuring betas or gammas that enter a cylindrical Geiger-Müller tube, which generally consists of a positively charged wire down the axis of the negatively charged coaxial cylindrical electrode. Even if placed in contact with a surface covered with a beta emitter, as shown in Figure 12–8, much less than half the beta particles emitted from the region even directly under the tube will enter it. Of these, some may not penetrate the chamber wall, and some may release too small an amount of energy to trigger the instrument; in addition some particles may actually pass entirely through the chamber with no ionization since the range in a gas of many betas is so great that few of them will be completely stopped in the chamber. Additional losses result from such factors as two particles entering the chamber at so nearly the same time (the interval being less than the dead time) that only one

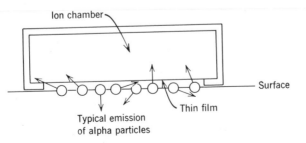

Figure 12–7 Surface alpha counting.

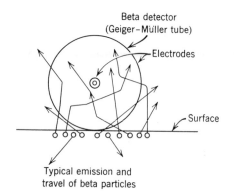

Beta detector
(Geiger–Müller tube)

Electrodes

Surface

Typical emission and
travel of beta particles

Figure 12–8   Surface beta counting.

pulse is produced; thus the overall geometry of most Geiger-Müller monitoring devices for betas is on the order of 10 percent. For gammas, this
overall geometry is even lower, even though the overall total radiation
energy loss in penetrating the chamber walls is much less. However, a
much smaller fraction of the incident gamma energy is deposited in the
chamber gas, this fraction being strongly dependent on the energies of the
incident photons.

From the above examples, it is apparent that the overall detection
geometry of various radiation-detecting meters is a function of the unit's
dimensions, the construction materials of both case and detector itself,
circuit constants, and the type and energies of the radiation involved.
Overall counting geometries are important in all precision work, even
though the comparatively rough averages that are obtained with monitoring devices are generally adequate for field work. In addition counting
geometries of this type are probably more important for instruments using
gas-detection methods than for others.

SCINTILLATION DETECTION

Probably the most widely used counting and dose-rate measuring instruments today, however, are scintillation counters, which are actually modifications of the earliest method of detecting the presence of radioactive
materials and measuring the intensity of the resultant radiation. Such
detection depends on the fact that when "invisible" radiation strikes certain
materials luminescence results so that a visible spark of light is observed.
Originally, the dark-adapted human eye looking through a low-power
microscope was the detector of the sparks produced by alpha particles

striking an appropriate screen, usually zinc sulfide (ZnS). Today, however, these sparks are detected by a phototube, and the resulting energy is highly amplified in a photomultiplier so that the final pulse is of sufficient energy to operate an electronic detector, such as a scaler or other recorder.

Although their initial pulses are small compared with those from gas-tube detectors, counters using scintillation detectors have several particular advantages over other types of detection and measurement instruments. Among these are the facts that (a) since these detectors are usually solids or liquids, their densities are such that charged particles with very high energies are usually stopped in small distances and even high energy gammas may be counted with comparatively small detecting units, (b) the pulse produced is proportional to the total energy released by a particle as it is stopped, and (c) their resolving times are so fast that pulses in the microsecond (10^{-6} sec) to nanosecond (10^{-9} sec) range may be easily distinguished.

Scintillant Materials

A wide variety of scintillant materials, or phosphors, are now available, and they are useful for a correspondingly wide variety of applications. Probably the most generally used units are those that are prepared from inorganic crystals which usually contain a slight amount of an impurity as an activator. Among the more popular of these inorganic crystals are zinc sulfide with a thallium activator [ZnS(Tl)] for alpha detection, sodium iodide with a thallium activator [NaI(Tl)] for gamma detection, and lithium iodide with an europium activator [LiI(Eu)] for slow neutron detection.

In addition to these crystals other crystals prepared from organic compounds, mostly in the class of the aromatic hydrocarbons, are also frequently used, although their light output is generally less than that of the inorganic phosphors. These organics are usually very pure materials and do not require an activator impurity, as is the case with inorganic crystals. Anthracene ($C_{14}H_{10}$) with a high sensitivity to betas is one of the most widely used of these organic crystals.

Plastic crystals are actually organic scintillators which are dissolved in either polystyrene or polyvinyltoluene, and the liquid monomer-fluor solution is then polymerized. By this method a scintillator of almost any dimensions may be obtained. For really huge scintillators, however, liquid organics are used, these having very much the same characteristics as the plastic scintillators. In fact, such detectors have been used in attempts to detect the neutrino.

In addition to solids and liquids essentially all of the noble gases may also be used as scintillation detectors when very short dead times (of the order of nanoseconds) are required and it is desirable that the light output per MeV be insensitive to the charge and mass of the particle being stopped and detected.

Photo-Luminescence

A special type of scintillation dosimetry, called photo-luminescence, involves the use of silver metaphosphate "doped" glass rods or disks for gamma measurements. In this case exposure to ultraviolet radiation following exposure to penetrating radiation causes the glass to fluoresce (or, more properly, to phosphoresce) in an amount proportional to the dose when placed in an appropriate electric field. The glass rods also have a "memory" for radiation; that is, the fluorescence produced by any radiation exposure becomes a permanent part of the fluorescence of the rod itself so that in determining a given exposure it is necessary to measure the fluorescence both before the exposure and after it. Silver phosphate glass rods, especially if encased in lithium cans, show less effect of neutron exposure in neutron-gamma fields than most of the devices that have been tested. Typical dosimeters of this type used for personnel monitoring are in the form of glass rods about 6 mm long and 1 mm in diameter, and can indicate doses down to about 100 mrad. Claims are also made for greater dose sensitivities but this usually involves very careful laboratory-type treatment and handling.

Thermo-Luminescence

The thermoluminescent property of certain materials, particularly calcium fluoride or lithium fluoride, in solid, powdered, and extruded forms has also been widely used for dosimetry in a wide range of dose levels. Lithium fluoride is nearly tissue-equivalent, and commercial lithium fluoride units can easily indicate doses of the order of 5 mR, and has a remarkably flat response to gamma-rays in the 20 to 1200 keV range. After exposure to ionizing radiation the lithium fluoride is heated, and the light emitted as a result is calibrated in terms of the exposure. Adequate heating, called annealing, removes the effects of prior radiation exposure and thus permits the TLD-dosimeter to be reused. Gamma dosimetry in a mixed neutron-gamma field is possible with use of rather pure lithium-7 ($^{7}_{3}Li$) since its cross section, even for thermal neutrons, is rather small. On the other hand the comparatively high thermal-neutron cross section of lithium-6 results in its ready activation by thermal neutrons, though very slightly by fast neutrons. Thus, use of lithium fluoride with and without lithium-6 permits, certainly in principle, a determination of gamma-ray and neutron exposures. Currently thermoluminescent dosimeters have been adopted by at least one major installation in the United States as the basic gamma-measuring device; accordingly this is apparently the only dose-measuring method, other than film, for which such primary usage has been accepted.

CHEMICAL DOSIMETER

Another device for measuring gamma exposure is the so-called chemical dosimeter, one form of which consists of a halogenated hydrocarbon, such as tetrachloroethylene, in which acid is produced by radiation. This is detected by a color change in an overlayering dye-water solution; the extent of the color change as shown by the resultant color should then indicate the extent of the exposure. One such device uses reagent-grade tetrachloroethylene, chlorphenol red, Ionot[R], McIllvaine's buffer, and quadruply distilled water; another employs ferrous sulfate. Such devices have appeared to be rather successful for quantitative measures of very high radiation exposures of the order of thousands of rad, which are obviously much higher than those that are of interest from personnel considerations, and it appears that exposures above about 50 rad may be measured with reasonable precision; at lower doses there is more uncertainty in the doses indicated.

NEUTRON DETECTION BY ACTIVATION

The fact that neutron collision may readily activate a wide variety of materials is the basis for a neutron detection and measurement method whereby appropriate substances are exposed to the neutron flux. The beta-gamma radiation from the resulting activation is then read and interpreted in terms of a neutron dose. As discussed previously and shown by Figure 5–1, the neutron RBE is energy-dependent; hence dose determination by the activation method depends on some method of determining the incident neutron spectrum as well as the total neutron particle fluence. Because of their higher RBEs, it is obviously important to determine the flux of the higher energy neutrons; this very greatly limits the materials available since, although slow and thermal neutrons are readily captured by many substances, the same is not true for the faster particles (particularly of concern are those which originate from fission).

The cross sections of certain elements which have proven useful for dose determination for neutrons in the fission range (up to about 7 MeV) are shown in Figures 12–9 (*a–d*) and 12–10; materials that are useful for high-energy neutrons are listed in the literature.

From Figure 12–9 it may be seen that neutrons of thermal energies (about 0.04 eV) will readily activate gold, indium, manganese, and copper, and will easily cause fission in uranium-235 and plutonium-239. Similarly, gold and indium respond to neutrons with energies of up to about 1000 eV, with particularly high capture cross sections for energies below about 10 eV; fission may be produced in uranium-235 and plutonium-239 by neutrons of a wide range of energies but particularly those with energies below about 1000 eV; copper and manganese have comparatively high cross sections for neutrons of certain energies in the 100- to 100,000-eV range.

Figure 12–10 shows the cross sections of materials in which activation is produced by neutrons with energies above some threshold value; unfortunately, only a few substances exhibit this characteristic, especially at energies that are useful for personnel monitoring. Neutron capture by neptunium-237 and uranium-238 results in fission, and the threshold reaction shown for indium is an isomeric one with a half-life of about 4.5 hr; the thermal and low-energy reaction of indium has a 54-min half-life. As may also be seen from Figure 12–9c and 12–9d a pseudothreshold fission reaction may be obtained by shielding plutonium-239 or uranium-235 with a suitable thickness of boron-10 (or a greater quantity of natural boron).

In one version of an activation monitor, sulfur, uranium-238, and neptunium-237 respond to high- and high-intermediate-energy neutrons; boron-shielded plutonium-239 and cadmium-shielded gold are activated by low-intermediate and low energy ones; and the difference in reading between cadmium-shielded gold and bare gold foils reflects the thermal neutron flux since the cadmium shielding removes these thermals. Because uranium-238 is almost invariably contaminated by uranium-235, with its large cross section for thermal and low energy neutrons, it must also be enclosed in the boron shield.

As alternate materials indium may readily replace the gold, uranium-235 the plutonium-239, and indium (using its threshold reaction) the uranium-238; cadmium-shielded manganese (or copper) may also be used for intermediate energies since the cadmium shielding absorbs the thermal neutrons.

Since the thermal and low-energy neutron capture cross sections of indium are much greater than that for the high energy threshold reaction, most of the initial activation resulting from exposure to a fission source will be due to these low energy neutrons. However, its 54 min half-life activity also decays much more rapidly than that of the 4.5 hr half-life threshold reaction, with the result that after some period of time the latter will predominate; for example, figure 3–4 shows that after 16 hr the threshold activation will predominate even though the thermal activation was initially 1000 times as great. This time may obviously be reduced by encasing the indium in a shield of indium and cadmium, or cadmium alone, since such shields capture a greater fraction of the incident low-energy neutrons than of high-energy ones.

Doses are determined from measured foil activations corrected for radioactive decay. Additional information required, mostly prior to the exposure, is some idea of the type of neutron spectrum involved, the neutron energy ranges covered by the different foils, and the average RBEs for them; obviously, the applicability and accuracy of such prior information affects the precision of dose determination.

As with all measuring devices, there are drawbacks. All of the fission foils have very short half-lives, which suggests not only the need for promptness in counting after activation but also the possibility of saturation and consequent uncertainty in dose measurement for relatively long-lasting exposures. These considerations do not apply to the other materials suggested. Because of its very low $(MPC)_a$, plutonium-239 would necessarily need to be stringently confined—as would be uranium-235 but to a smaller extent. There is little choice between gold or indium on technical grounds. Boron-10 is rather rare and expensive, as is neptunium-237.

In another approach to fast-neutron detection and measurement, neutrons have been slowed down by a hydrogen-rich material, such as paraffin, before their impact on an appropriate detector, foil or otherwise. The more energetic the neutrons, the greater the thickness of such material they will penetrate before being slowed to a specified energy reflected by their actuation of the given detector.

Since combined gamma and neutron doses in the range of 100 to 500 mrad, as originating in a fission reaction, may easily be detected by the thermal activation of indium alone, this substance is now generally included as a component of most personal dosimeters or similar devices used by personnel working in operations in which criticality accidents may occur.

FILM METERS

Principles

Meters using photographic film are probably the most widely used devices for general personnel monitoring today, since not only X-rays and

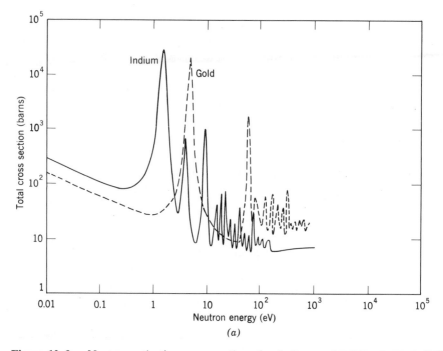

Figure 12–9a   Neutron-activation cross sections for indium and gold.

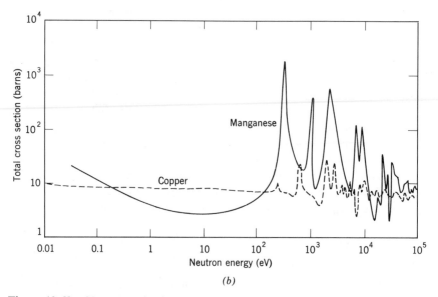

Figure 12–9b   Neutron-activation cross sections for copper and manganese.

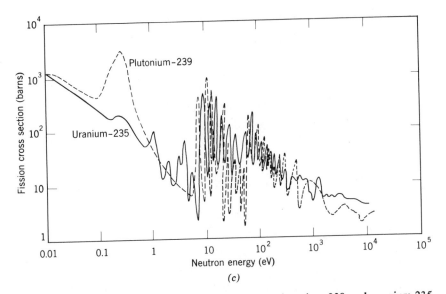

Figure 12–9c Neutron-activation cross sections for plutonium-239 and uranium-235.

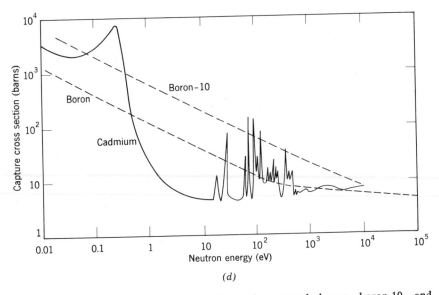

Figure 12–9d Neutron-capture cross-sections for natural boron, boron-10, and cadmium.

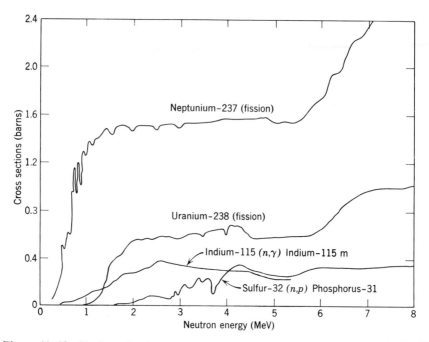

Figure 12–10   Nuclear threshold reactions.

visible light but also essentially all forms of radiation of interest in radiation protection will cause blackening of the photographic film on which they impinge. In practice, film of the approximate dimensions of dental film is placed in a case that is worn by the individual being monitored. To avoid exposure to light the film is not removed from its paper covers until time for development; the paper is an effective shield for alphas, although betas, gammas, and neutrons readily penetrate it and thus expose the film.

Film is generally energy sensitive to electromagnetic wave radiation in such direction that X-rays and low energy gamma photons produce more blackening, comparatively, than the same energy carried by higher energy gammas. Figure 12–11 indicates this effect qualitatively. However, cadmium shielding reduces the energy range in which such enhanced blackening occurs so that it is only at energies below about 30 kV for shielded film that this effect becomes important.

Response to Various Radiations

Film is generally more sensitive to betas than to gammas. However, it is impracticable to attempt an evaluation of film blackening by betas as a

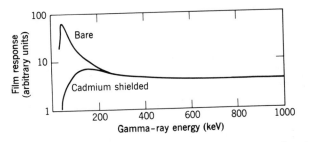

Figure 12–11 Typical film response to various gamma-ray energies (total energy constant).

function of their energies. This is principally because the betas from a given reaction are not emitted at a definite measurable energy—as is the case with other forms of radiation—but in a wide range of energies as described. Similarly, it is not possible to calibrate film in terms of beta energy, but such calibration is possible with respect to a given beta emitter, such as "old" uranium or strontium-90. Thus a given beta exposure, or film blackening from betas, cannot be quantitatively evaluated except in terms of the comparative calibration described. Practically measurement of most beta exposures is considered to be semiquantitative only.

Figure 12–12 gives typical curves showing film blackening as a function of beta dose and of gamma exposures to unshielded film and that shielded by 0.15 cm. of cadmium. Various emulsions have different quantitative responses to betas or gammas, and most monitoring devices thus use more than one film to cover various appropriate ranges.

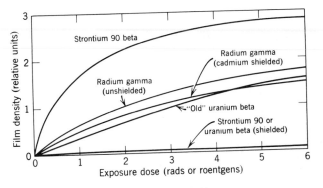

Figure 12–12 Dose-density relation for typical film.

It is the author's experience that the minimum gamma exposure that can be detected with reasonable precision with current films is on the order of 30 mrad, although claims, which may very well be true, are made for 10 mrad sensitivities for some films and conditions of use.

It is generally assumed that film blackening is independent of the way in which the exposure is made. A given exposure to betas and another to gammas will blacken the film to a degree that is independent of whether the exposure was first to beta and then to gamma radiation, first to gamma and then to beta radiation, or is the result of continuing exposure to a mixture of the two types of radiation.

Film is also differentially blackened by neutrons (perhaps more accurately, the secondary ionizing products of neutron collisions) as a function of their energy. Such effects are similar to those observed for gamma rays, but they are even more significant in that much greater blackening results from exposure to thermal and low energy neutrons than their total energy release would indicate. Thus, these responses are not proportional to the relative biological effectiveness of various neutron energies. Some measurements of neutron exposures are made by counting neutron tracks in a special neutron-track emulsion. In this method the ionization produced by the neutron in being slowed down and perhaps stopped causes an indicated exposure along its path.

Use of Shields

Shields of various materials and dimensions are incorporated into the film meter case in attempts to determine both intensities and types of radiation; for example, in addition to its use in reducing the energy dependence of film for electromagnetic waves, cadmium is also of value as a shield to aid in distinguishing between mixed beta and gamma radiation since about 0.15 cm of cadmium will almost completely shield the film from betas but will reduce gamma intensity only slightly. Thus for measurements in beta-gamma fields part of the film is unshielded and responds to both betas and gammas whereas the shielded part responds only to gammas. Although the gamma intensity is also reduced by this shield, this effect is usually so small as to be negligible.

Various filters have also been introduced for the avowed purpose of indicating the approximate energy of incident electromagnetic radiation, particularly to indicate whether or not the exposure is due to gamma or to X-rays, especially X-rays in the widely used 10 to 60 kV range. Although such separation should in principle be possible and some success has attended efforts to evaluate doses from a range of gamma and X-ray energies where the types of incident radiation have been specified, it still appears to be rather impracticable to determine the overall dose resulting from an unidentified mixture of electromagnetic radiations.

Mixed Gamma-Neutron Detection

Since both neutrons and beta-gammas blacken a film, film dosimetry should be useful for mixed neutron-and-gamma fields such as will occur at a criticality accident. However, such usage is significantly affected by the fact that neutrons will activate most materials, particularly metals, making them radioactive and thus generally emitting beta-gammas. Hence, even after the cessation of neutron exposure itself, as by the cessation of a nuclear reaction, any radioactivity induced in the film meter components or other nearby materials will continue to irradiate the badge film and thus indicate a dose that has not been received; cadmium effects have been mentioned. Film blackening is also caused by the incident gammas themselves.

Some of the difficulties in separating the neutron and gamma doses from a mixed exposure are apparent from the above discussions. It is unfortunate, but to date there appears to be no reasonably accurate method of film dosimetry for mixed neutron-gamma exposures. However, a knowledge of the incident neutron dose should be available from activation of the metal components of a given badge, and it should then be possible to determine from prior calibration the blackening due to the neutron dose; the remaining blackening would thus be considered due to gammas.

PROBLEMS WITH PERSONNEL MONITORING DEVICES

There are problems in the use of all types of personnel monitoring devices in addition to those caused by actual exposure of the device under conditions in which the individual does not himself receive the same exposure. This would be the case if, for example, a film meter or any detecting device is left overnight in a beta-gamma field and thus indicates an exposure that obviously did not occur since the meter was not being worn at the time.

Most of the principal technical sources of error in the use of film tend to indicate an exposure which did not occur and are thus "conservative." One of these is the effect of age; as film ages it blackens slightly, and, since low levels of exposure also give very slight blackening, apparent exposures on the order of several hundred millirads may easily be obtained merely by leaving a badge in almost any place for an extended period of time. Heat is another environmental factor that effects the apparent exposure. If film is heated, it becomes fogged and thus indicates an exposure; in fact radiant heat from nearby heated metals can cause localized film blackening. In addition to the direct effects of heat on film, heating of the metal components of a film badge may also have an effect on the film that cannot always be precisely determined. Similarly, chemical fogging can occur in environments with a wide range of non-radioactive contaminants. On the other hand, if film is not processed shortly after an exposure, fading may begin so that the film will not read exposures that are as high as have actually occurred. However, this effect is normally less important than excess blackening.

Electroscope- or ion-chamber-type devices have their own special problems of providing consistently accurate data, primarily because of charge leakage, which will obviously indicate a nonexistent exposure. Insulator problems due to failure or to the insulator's becoming dirty or contaminated with various materials are of particular importance, although some leakage also occurs over a comparatively extended period. These devices—particularly the dosimeters and pocket chambers used in personnel monitoring—may also be damaged by being dropped or otherwise mishandled.

Particular problems in dosimetry with phosphate glass rods result from the possibility of grease getting on the glass either from handling or otherwise and from the existence of checks or other spots that give the appearance of exposure. The effect of a radiation "memory" has already been mentioned.

Heating of thermoluminescent dosimeters (TLD-units) may reduce the dose indication.

The chemical dosimeter, especially at low levels, is subject to incorrect dose indications as a result of age, heat, light, and other environmental factors. Therefore the method has not appeared to be highly successful when used for exposures at the comparatively low levels of interest in personnel exposure considerations. These devices have not as yet appeared to be sufficiently dependable and the methods of reading have not been found to be sufficiently reliable and generally applicable for the method itself to be considered as available for use—certainly not to the extent of other available systems.

In view of all of the unfortunate effects mentioned for the various devices it is rather remarkable that surprisingly consistent results can be obtained in radiation dosimetry.

CALIBRATION OF PERSONNEL MONITORING DEVICES

All monitoring devices require rather careful and known prior calibrations for usefulness. Normally the scales used with dosimeters and pocket chambers are calibrated for radium gamma energies, although 200 kV X-rays are used in precision work. Although only slight differences result from calibrations by a wide range of X-ray and gamma energies, the effect should be recognized.

The radioactive substances that are generally used for gamma calibrations of film meters are cobalt-60 or a radium salt, usually radium chloride. For beta calibrations the calibration source is usually a plate of "old" uranium metal, the betas being those of the UX–1 and UX–2 daughters, which are in equilibrium. Strontium-90 is also useful as a calibration source.

In practice, film is calibrated for gammas and betas by actual exposure to a range of known doses of films from each batch used, and the construction of a density-dose curve. Although such calibration with a single source is usually good for a wide range of gamma energies, this is not true for betas since film blackening depends strongly on beta energy and each source gives a range of energies. In principle, film may be calibrated for neutron exposures, but the facts that all neutron-sensitive film is also gamma sensitive and that there are very few instances where neutrons may be found but not associated with gammas complicate this calibration.

Thus, although appropriately shielded film can be used to distinguish between beta and gamma radiation, and there are indications that such shielding can also distinguish between the high energy gamma radiation and the similar low-energy X-rays, the necessary development, simple calibration, and use of any device for distinguishing between neutron and gamma radiation when both occur simultaneously has not been developed; this comment accepts the fact that neutron exposure may be determinable from activated metals and that lithium-fluoride dosimeters can be prepared to show differential neutron-gamma sensitivities. However, it is thought that further careful study and evaluation could perhaps resolve this particular dilemma. In fact, comparative studies of the effectiveness of various criticality monitoring systems are currently underway among several AEC installations.

PERSONNEL MONITORING FOR ALPHA RADIATION

No attention has been paid to alpha radiation in the above discussion of personnel monitoring. There are two reasons for this. One is the fact that as long as the alpha-emitting materials are outside an individual's body there is no possibility of injury. The second is that none of the devices which have been described can have windows that are thin enough for alpha particles to penetrate. However, there are alpha-sensitive area monitoring devices—including the Samson and similar instruments using ion chambers, as well as others that use scintillation counters. These devices can be used to determine the amount of alpha-emitting material on an individual's body or those portions of his environment with which he comes in contact. In addition, the possibility of internal exposure from these sources may be obtained by air samples, and other environmental monitors can also indicate the potentiality of exposure from airborne radioactive materials as well as from those occurring in water or foodstuffs.

ENVIRONMENTAL MONITORING OF RADIATION FIELDS

The basic principles of devices used for environmental beta-gamma monitoring have been described, a meter using a Geiger-Müller tube (the so-called G-M meter) being the most frequently used device for fairly low level fields. The usual G-M meter is calibrated in three scales to measure radiation fields from even less than background radiation, which is about 0.02 to 0.05 mR/hr in most locations, to fields of about 20 mR/hr; the maxima of the three ranges in a typical instrument are approximately 0.2, 2, and 20 mR/hr, although some devices have added higher scales. However, because G-M meters lack energy discrimination as outlined previously, the dose rates indicated are indicative only and cannot be considered to be precise. The Geiger-Müller tube itself is usually so designed that it may be shielded and thus respond only to gammas, or it may be left unshielded and thus respond to both gammas and betas of sufficient energy to penetrate the chamber walls. Typical instruments are shown in Figure 12–6.

Another difficulty with the use of Geiger-Müller-tube instruments is that in a very high field the detectors become overloaded and do not indicate an exposure. This characteristic has been a serious consideration in evaluations of their usage for monitoring high level radiation such as might result from a possible critical reaction.

For higher fields, instruments that use ion-chambers are probably the most commonly used devices, and they may be used for alpha detection as well as for beta-gamma measurements. These devices usually include shields by which discrimination between alpha, beta, and gamma exposures is possible. For alpha measurements the screens of the chambers are of rubber hydrochloride and are thus thin enough for the alpha particles to penetrate them and be detected. With a shield of a somewhat thicker material, on the order of that of paper, the alpha particles can be eliminated, and the devices will detect beta-gamma radiation. With a thicker shield, perhaps a millimeter or so of aluminum, most beta particles will be removed, and the devices can be used for gamma detection only. It has already been mentioned that calibration is always needed for determining the meter response to the various types of radiation since a simple differential measurement is not always adequate for the data required.

Ratemeters that use scintillation detectors are also coming into wide use for environmental monitoring, although they may not be as stable as would be desired.

Dose measuring instruments of the electroscope type have been described; although these were widely used originally, they are not now much in use for measuring fields but are used to some extent to measure the total dose received at a point. They have always been most useful for precision work.

It is perhaps unnecessary to comment that the monitoring devices described above are adequately capable of measuring radiation fields of concern or interest, whether from specific sources or from material deposited on various exposed surfaces.

AIR AND WATER MONITORING

The normal method for measuring the concentration of radioisotopes in the air (and thus evaluating the possibility of internal exposure from air contamination) is to pass a measured quantity of air through a given collection medium, such as filter paper, and then to read the activity of the material collected on the filter paper itself.

In practice, the air is usually forced through the collection medium, filter paper being the most common one, by an air sampler that is essentially an air-moving device; in fact some of the early samplers were merely converted vacuum cleaners. Today two types of devices are commonly used. One is a direct blower of the vacuum cleaner type, and the other is a constant displacement pump, which is literally a small vacuum-pump-type device that moves air through the paper at a measured rate. Automatic devices have been developed whereby the air sample is passed through a collection paper and then, a measured time later, is automatically counted and recorded.

By use of these devices a prompt indication of the amount of beta-gamma active material in the air can be obtained. However, special problems are involved in determining the alpha activity in the air. One of these is the fact that the normal air background generally contains a certain number of radon and thoron daughters which may adhere to dust or other solids and thus be stopped by the paper; because of their short half-lives they are usually probably in equilibrium with their parents. Unfortunately, this background not only varies widely but may also be several times the permissible exposure levels for most alpha emitters of interest and must thus decay significantly before even reasonably accurate measurements may be made. However, because of the very short half-lives of these radioisotopes, as indicated by Tables 3–1 and 3–2, adequate decay will occur within some 4 or 5 hr, and the counting of alpha air activity must normally be delayed this long, even in automatic counting, for reasonably accurate results to be obtained. Such measurements thus really provide after-the-fact data.

During collection radioactive particles can penetrate into the detection medium, such as the filter paper, with the result that the activity will be partially shielded and not all that is collected will be counted; this is naturally of particular importance in alpha-detection. The possible effects of this penetration, as well as other factors of interest, should be carefully evaluated for use of these and similar devices.

For determining beta-gamma activity in water, it is possible to pass water under meters and continuously record the radiation level. For alpha emitters, however, this method has not been found to be practical, although some efforts are now being made to use certain types of scintillation counters for this purpose. In any event, what are probably the most precise determinations of the concentration of a radioisotope in water are made by the normal radioanalytical techniques of evaporation, concentration of the radioisotope, and then counting the residues.

Relatively precise measurements of air activity are usually obtained by ashing the filter paper and counting the activity in the residue; obviously care must be taken that none of the radioisotope is lost in the ashing procedure. In all of these types of analyses, however, the results are practically always in the microrange and the normal precautions and error probabilities inherent in any radioanalysis by microanalytical techniques also apply to all of these water and air analyses. Information concerning analytical techniques and devices may be found in the literature.

The overall accuracy of any measurement technique is generally limited not so much by analytical techniques as by the efficiency of the collector element used and the adequacy of sampling techniques; for example, if a filter paper collects only about 90 percent of the air contaminant, no precision of analysis can give an accurate value unless the collection factor is known. Similarly, sampling and mixing techniques can introduce analytical uncertainties. Radioactive gases will pass through filter paper and be undetected except as the molecules collect on other particles large enough to be captured by the filter paper or other unit.

OTHER ENVIRONMENTAL EXPOSURES

As has been mentioned, other possibilities of exposure from the environment, particularly for alpha-emitting materials, lie in the possibility of such materials being stirred up from work surfaces or being brushed off clothing and thus being inhaled. In general normal radiation detection methods (including the types of instruments described above) can be used to monitor clothing and to monitor surfaces, although in some cases it has been found desirable to use special types and shapes of probes.

It should be reemphasized that all environmental monitoring of air, water, and particularly surfaces reflects only second and third order possibilities of actual internal radiation exposures since the material must actually enter the body to be a problem. This is true even for beta-gamma emitters, although environmental surveys of beta-gamma fields can indicate a direct potential exposure in the sense that an individual's body is in the same radiation field that his survey meter encounters.

DETERMINATION OF INTERNAL DEPOSITS

Bioassay techniques are usually employed in attempts to determine, usually indirectly, what types and amounts of materials are in the body. The most commonly used techniques are analyses of the excreta, it being hoped that information on the rate of removal of a radioisotope may be related to the material remaining deposited. However, because of the variability of individuals and the elimination process itself, simple relationships have not been observed. Thus comparatively little importance is paid to quantitative results today, although they are useful for indication of potential exposures. The methods used are equally useful for evaluation of deposits of alpha, beta, or gamma emitters.

Fecal analysis is usually used to ascertain the rate of removal for recently ingested materials and may thus indicate possible effects on the gastrointestinal tract. Similarly, it is now thought that much of the initial removal of inhaled materials may also occur through the feces. Breath samples or wipe samples from an individual's mouth or nose (called nose swabs) have also been used to indicate internal-exposure probabilities; this technique has been particularly useful for radon or radium—as in the early radium dial-painter cases. On the other hand, inhaled and absorbed materials are almost invariably removed through the urine, except as noted above, and this is also the principal elimination route for radio-isotopes that have been deposited in the bone or other organ. Urinalysis thus indicates the removal time rate for most materials. With many materials of interest, such as uranium, the radioisotope is normally excreted at almost a constant rate regardless of any variation in the amount of urine produced.

Microanalytical techniques with their inherent difficulties are normally necessary in all of these excreta evaluations.

For beta-gamma emitters—and indirectly for alpha emitters that have beta-gamma active daughters—a technique of attempting to locate deposits of radioactive substances with a survey-type meter (or laboratory type modification thereof) have been extended rather successfully to the so-called total body counter. The method bears some slight resemblance to that used in the usually well-publicized searches for radium "needles" lost by hospitals. In this case the individual is placed in a well-shielded chamber in a sitting or reclining position. He is then surrounded by radiation detection instruments, an instrument may be moved about him in a pre-arranged pattern, or he may be rotated in front of such an instrument. In any event, multiple readings are made at various positions around his body. From information on the location and densities of various organs and parts of the body, deposit sites and amounts may be located. Current spectrum-analysis devices also make it possible to identify the deposited radioisotope. Many facilities now use such body-scanning techniques routinely.

A modification of such methods, and one that has had some success, is the attempt to locate bits of contaminated metal or other radioactive materials that have penetrated an individual's skin as a result of wounds caused, for example, by chemical explosions during regular operations. In this method special probes for measuring radiation inside cuts or other wounds have been developed and described. Analyses of potential ex-posures of these types have also resulted in the attempted removal of such materials by excision. The radiation from both the material removed and that remaining is then determined, and the potential exposure is evaluated. Although this does give an indication of the possibility of certain types of exposure, there are obviously some disadvantages to the method and there are many cases in which employees who have been involved in such accidents have refused to have these materials excised.

REFERENCES and SUGGESTED READINGS

Price, William J., *Nuclear Radiation Detection,* McGraw-Hill, New York, 1964.

Hine, Gerald J. and Gordon L. Brownell, *Radiation Dosimetry,* Academic, New York, 1956.

Sharpe, J., *Nuclear Radiation Detectors,* Wiley, New York, 1955.

Blatz, Hanson, *Radiation Hygiene Handbook,* McGraw-Hill, New York, 1959.

Neutron Dosimetry, Vol. II, IAEA, 1963.

Snell, Arthur H., *Nuclear Instruments and Their Uses,* Wiley, New York, 1962.

Whyte, G. N., *Principles of Radiation Dosimetry,* Wiley, New York, 1959.

Hurst, G. S. and R. H. Ritchie, editors, *Radiation Accidents: Dosimetric Aspects of Neutron and Gamma-Ray Exposures,* ORNL–2748 (Part A), USAEC, Washington, D.C. (Nov. 16, 1959)

Ellis, W. P., L. H. Sipe, A. F. Becker, and H. F. Henry, *Calibration of the Multiplant Security Badge Meter for Application at the ORGDP,* KSA–242. Union Carbide Nuclear Co., Oak Ridge, Tenn. (1961)

The Vinca Dosimetry Experiment, IAEA, 1962.

Proceedings of the First Symposium on the Nuclear Accident Dosimetry Program: Techniques and Uses, Edgerton, Germeshausen and Grier, Santa Barbara, Cal., Oct. 5, 1960.

Attix, Frank H., ed., *Luminescence Dosimetry* (Proceedings of International Conference on Luminescence Dosimetry, Stanford University, Stanford, California, June 21-23, 1965). Published April 1967, USAEC Division of Technical Information. Available as CONF-650637, Clearinghouse for Federal Scientific and Technical Information, National Bureau of Standards, U. S. Dept. Commerce, Springfield, Virginia 22151

Stehn, J. R., M.D. Goldberg, B. A. Magurno, and R. Wiener-Chasman, *Neutron Cross Sections,* BNL-325, Second Edition, Supplement No. 2, Vols. 1, 2a-c, 3, Brookhaven National Laboratory, 1964-1966.

QUESTIONS

1. What is a major problem in neutron dose measurement?
2. Why is a film badge ineffective for alpha particle measurements?
3. Under what conditions is it anticipated that it will be necessary to measure mixed neutron and gamma doses?
4. On what physical phenomena do most radiation measurements depend?
5. Explain how pocket chambers and direct-reading dosimeters are constructed and how they function.
6. What are the meanings of such terms as "saturation plateau," "proportional region," "Geiger plateau," and "tube geometry"?
7. What are some of the ways in which neutrons are detected?
8. What are some of the problems in determining neutron fluxes?
9. What problems are involved in determining beta radiation by the film technique?
10. What are some of the inherent defects in film monitoring and the electroscope- or ion chamber-type detectors?
11. What techniques are used in environmental monitoring?

12. Describe the meaning of the term "counting geometry" and relate it to the various types of detectors.
13. Discuss briefly any problems inherent in, or limits of, the various types of detectors.
14. What are some of the means of measuring neutron exposure? Of discriminating between various energies of neutrons? Include the materials used for shields and the approximate threshold levels of some of the so-called threshold detectors.
15. Why is alpha monitoring important when the range of alpha particles is so short? What are some of the methods of alpha monitoring?
16. Give the principles involved in pulse-height analysis.
17. What are the limitations of the Geiger-Müller tube?

CHAPTER THIRTEEN

Principles of Personnel Monitoring

JUSTIFICATION FOR PERSONNEL MONITORING

The basic justification for personnel monitoring is an attempt to measure the actual dose which an individual may have received. Such measurements are made directly by personnel devices and determined indirectly from field measurements. The basic principles of radiation detection and monitoring devices are discussed in the preceding chapter, which also reviewed some of the problems and possible sources of error in the various dose determining methods. Available devices are capable of measuring both the exposure rate and the total dose produced by external beta-gamma radiation at levels of interest or concern, and, with proper calibration and knowledge of potential exposure conditions, reasonable confidence may be placed in the results of such monitoring. Actual experience is really the only good source of information on the actions necessary in a given situation.

FILM AS A GENERAL MONITORING MEDIUM

The most widely used device for personnel monitoring of penetrating radiation (beta, gamma, X-ray, and neutron) is the film meter using appropriate film and employing components designed for various types of monitoring. Routine dose monitoring by film is required by law in some nations, and in the United States the U.S. Public Health Service has sponsored a film badge services performance and testing procedure through the National Sanitation Foundation. The initial testing of commercial services was inaugurated in 1966; most of the services provided by the U.S. Government agencies, particularly the Atomic Energy Commission and its contractors, however, did not participate in this initial effort. These tests are designed to evaluate performance in determining doses in exposures to gamma, X-ray, gamma plus X-ray, beta, beta plus gamma, and neutron radiations.

In general the film meters currently in use in the United States employ dental-sized film as their radiation-sensitive unit, usually with film developed for the specific type of monitoring desired. Most of the film meters used for beta-gamma monitoring employ a single packet containing two or three individual films of different sensitivities, thus making possible a wide range of dose evaluations. In a widely used three-film packet, one is calibrated for doses in the range of about 30 mrad to about 5 rad, a less sensitive one reflects exposures in the 3 to 25 rad range, and the least sensitive covers the 20 to 100 rad range. These are the ranges for the respective films in which (as indicated in Figure 12–12) the blackening is nearly proportional to the exposure. However, the range of each film may also be extended by careful calibration at both ends. In two-film packets, which are also widely used, the intermediate range unit is generally omitted, the maximum dose measurable may be somewhat different, and a possible smaller overlap may need to be accepted. Development of specific film emulsions will affect the usable dose ranges.

By certain procedures in the development and other processing of the film it is possible to extend these dose ranges—although this requires special preparation, and the action must be taken during processing. One particular modification has been the use of film coated on both sides. The range is thus extended by removal of the blackened emulsion on one side. There have also been cases of high accidental exposures wherein attempts to determine the dose have been made by reading not the density of the film but the actual amount of silver in the developed film. This requires careful microanalytical techniques.

FILM MONITORING FOR NEUTRONS

Some films, principally those with thick emulsions, are sometimes advertised as being particularly neutron sensitive and insensitive to gammas. However, it is the author's experience that such films are not particularly useful in neutron-gamma fields since gammas will markedly blacken the film, generally more than neutrons, for a given dose. These films also reportedly fade more quickly than do other films if they are not promptly developed; the possibility of fading should be evaluated for any particular film.

The use of neutron track emulsions requires rather difficult, time-consuming, and consequently expensive techniques of analysis since the basic premise of the method is a determination of the number and length of the neutron tracks that individually cause film blackening. The difficulty of counting tracks obviously depends on the eye of the person who is counting, and this may introduce some problems.

METER CASE CONSTRUCTION

Since betas are much more strongly attenuated than gammas, even the simplest beta-gamma measurements involve density determinations both of the blackening of the part of the film left essentially unshielded (for beta plus gamma) and of that of another part of the film that is shielded by the meter case and thus responds only to gammas (slightly attenuated). Thus the materials of construction of a meter case or badge, including the shields incorporated therein, are themselves nuclearly important. Most materials used in badge cases—such as plastic, aluminum, or even stainless steel—have little effect on film blackening by beta-gamma radiations other than that caused by their mass densities alone. However, when it is desired to use shields of one kind or another for various purposes, extreme care should be used in their selection, and experimental studies should be made of results obtained under typical conditions of exposure; for example, lead is sometimes used as a backing shield to indicate the direction of radiation entry. However, both because of the scatter of high energy gammas and the secondary emission from beta capture, it will actually provide a significant blackening and thus an apparent exposure for radiation incident on the back of the badge.

The difficulties of using metal shields in mixed neutron-gamma fields have been described. Thus, although cadmium is very effective for separating gamma exposures from betas in beta-gamma exposures, the cadmium itself is activated by neutrons. In exposures that also involve neutrons it will thus actually amplify the apparent gamma exposure because of its beta-gamma activation; at high levels, such exposure may continue after cessation of neutron activation. As a matter of fact, cadmium was one of the first materials used to indicate the presence of neutrons because the film blackening was "intensified" by the neutron exposure of the film.

Many other metals, aluminum being an important one, have also been placed in badges for various purposes, primarily as shields, and some of these may also be neutron sensitive. In particular, indium is now almost universally provided for individuals working where criticality accidents are possible, and in many cases the film meter case has been a convenient location. Other neutron-sensitive materials have also been placed in badges for possible exposure measurements.

FILM CALIBRATION

It should be emphasized that no measurement of exposure by film can be any more accurate than its calibration, typical curves of which were shown in Figure 12–12. Similar curves are, of course, available from the film manufacturers and will necessarily be made by the user himself to take care of the possible effects of the variables of emulsion, handling, and processing. This will obviously include definite limits of detection for a specific film. Also important is knowledge of the energy of the incident radiation, especially if it is X-radiation in the 10 to 100 kV range in which anomalous blackening results.

The principal film calibration sources are cobalt-60 and salts of radium-226 (with its equilibrium decay products). Cobalt-60 is particularly useful because of its almost monoenergetic gammas; on the other hand, the various gamma energies of radium also give certain advantages, the most important of which is the fact that it thus more nearly duplicates the actual exposures which result from radiation by gammas of differing energies. Although actual doses may be best measured by an absolute device such as an electrometer or calibrated R-Meter (short for Roentgen-Meter), it is possible to calculate the doses at various points from the amounts of radium or cobalt used, the distance, etc.

An approximate determination of the gamma exposure rate $I\gamma$ in mR/hr. at a distance S, in meters, from a point gamma source in air may be obtained from the relation

$$I\gamma = 0.5 \; n \; EC/S^2,$$

where C is the disintegration rate in millicuries, n is the number of photons emitted per disintegration, and E is the gamma energy in MeV. Figure 13–1 indicates the approximate radiation levels, in mR/hr., at a distance of 1 m as a function of gamma energy in MeV, assuming only one photon per disintegration; these values are approximately 1.23 and 0.84 for cobalt-60 and radium-226, respectively. For the latter source there is also some capture X-ray emission, which is negligible at distances of 1 m but has some effect at distances of a few centimeters.

Routine calibration exposures are usually made by placing test badges at specified distances from a known and calibrated source, exposing all

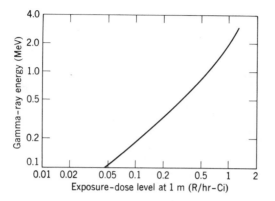

Figure 13–1 Relation of dose level and gamma-ray energy.

of them for a predetermined time to give definite doses, developing them appropriately, and then using a sensitive densitometer to read the density of the blackening as a function of the dose received. This is normally done for each film-use period. Probably the most important of the variables involved in calibration is that of film processing itself; in fact, it is impracticable to attempt a major film monitoring program unless the temperature of the processing solutions, the time in each step of the processing, and the makeup of the solutions may themselves be closely controlled. Film variation from manufacturing batch to batch also can introduce differences.

POSSIBLE INACCURACIES IN FILM USE

Among the difficulties inherent in the use of film for radiation monitoring are the effects of heat and age, which have already been mentioned. In addition there is the possibility that radioactive dust or similar matter getting on the film will produce an intense localized exposure. If this occurs, the film will usually have a spotty or mottled appearance. In these cases the dose is determined from the density of the least blackened portion of the film since it is assumed that the greater blackening elsewhere is not actually related to the individual's exposure but represents contamination or other effect on the specific badge involved.

Otherwise unidentified density variations are usually handled this way, since, other than faulty processing or film defect, almost any irregularities will cause increased blackening. Primarily because of this possibility, the regions for reading exposure should not be too small, since it is otherwise entirely possible that inaccurate readings will result from the entire region's being involved in an irregularity.

The use period for film depends on the possible inaccuracies that may be permitted. The apparent exposures of an unexposed film as a result of heat, age, chemicals, and other environmental effects, as well as the fact that the latent image on an exposed film may fade before reading, have been mentioned. However, there has never been any indication of fading of a developed film image. In any event it appears that badges used for a period of about one month will show no appreciable changes in blackening caused by age alone under normal environmental conditions. However, if the use period is extended to as long as three months, an indicated exposure of as much as 100 mrad of beta-gamma radiation may be anticipated. Such an apparent exposure does not reflect an actual exposure of the individual wearing the badge.

USE OF POCKET CHAMBERS AND DOSIMETERS

Pocket chambers and direct reading dosimeters have been used for measuring gamma exposures and, with special cases, have also been made sensitive to betas and neutrons, as already mentioned.

Most of the instruments currently available are remarkably stable. However, such stability needs to be tested periodically by giving the dosimeter or pocket chamber a charge and determining its rate of leakage; any detectable leakage for the period equivalent to that proposed for use will obviously indicate possible trouble since leakage will simulate an apparent exposure. For actual monitoring with these devices it is customary to use two of them, placed side by side. Where the two indicate different exposures the lower indicated exposure is the one considered to be valid.

These devices are normally used for single jobs or short-term exposures when it is important to know promptly what an individual's exposure is or if his exposure is exceeding a certain amount. In some cases it has been customary to use these devices in conjunction with film meters. When the exposure indicated by these devices reaches a certain level or becomes unknown because the apparent exposure exceeds its scale, the film is developed to determine what the potential exposure actually was. Since the individual being monitored normally carries such devices in his pocket, the shielding effect of cloth or any other potentially shielding material should be considered in evaluating an indicated exposure.

OTHER PERSONNEL MONITORING DEVICES

For reasons that have been already mentioned, chemical dosimeters and phosphate glass are at present used almost exclusively for special types of monitoring. The instability of the former as well as difficulty in its calibration has made its use as a personnel-monitoring device questionable. Although the glass gives rather good results for its range of usefulness, its radiation "memory," as well as the fact that grease or chipping from rough handling can cause a false-positive "exposure," must be considered in its use.

Thermoluminescent dosimeters are rapidly coming into use for beta, gamma, and neutron monitoring. Their excellent characteristics indicate future applicability for a wide variety of uses; in fact, adoption of this device at one AEC installation as the basic device may presage further extension of its use.

One hitherto unmentioned factor in determining the type of personnel monitoring device to be used is legal in nature. In brief, it depends on whether or not an alleged exposure (or lack thereof) must be supported by a permanently fixed dose indicator. Thus, a developed film badge or an exposed and maintained photo luminescent device can provide such a record. Pocket chambers, TLD units, and other reusable devices do not. On the other hand, it is also considered that adequate and well-maintained report records may be as tamperproof and valid as a piece of exposed film itself. In any event, this may be a consideration in selection of a monitoring device, and "permanent record" devices are required, by law, in certain European nations.

MONITORING BY DOSE-RATE-AND-TIME METHOD

Another method of personnel monitoring for beta-gamma radiation involves the so-called dose rate and time method. In this the potential exposure is determined from a measurement of the dose rate in the radiation field of interest and a knowledge of the time that the individual is in the field. This method is generally used to specify work times for certain short-term jobs at relatively high exposure levels, especially where the field is strongly localized and it is impracticable to shield the source and still perform the necessary work. The exposure limitation imposed will of course depend on the importance of the work being done and any dose history of the individual concerned. Actual doses are also usually measured by either the film meter or dosimeter.

Whereas this method obviously does not provide the same precision of exposure determination as would be given by a film badge or other method of direct measurement, it can be used rather well for certain cases.

This general method of personnel monitoring has been used in several other ways. One involves estimates of average low-level exposures over a long period of time for various individuals who work under conditions, as in an individual room, where there is a continuous low-level exposure and their work normally keeps them out of regions with higher radiation levels. Generally, but not necessarily in all cases, it is assumed that the maximum exposure rate in the work location is the average; this obviously gives conservative results. The possibility of exposures, even for short periods, to higher levels of radiation must be recognized and included in exposure records, which must necessarily be kept in detail. This method is not automatically excluded for cases in which the occurrence of somewhat higher than normal exposures for very short periods may also be routinely anticipated. However, caution is obviously indicated since there have been cases where employees working in areas of varied potential radiation exposures have stayed in the region of the higher exposures for a longer time than anticipated because of a comparatively slight change in their work, and a much higher than expected dose resulted.

This dose-rate and time method is actually the one used to limit exposures to the general public or to other individuals. Such usage involves a control boundary that people generally do not pass without special authorization. The radiation level at this boundary is such that it is known that no one remaining outside will get more than a given exposure from the source concerned, which may be an entire operating plant or facility. Thus for people who do not enter this area it can be assumed that at all times they receive a dose no greater than that which they would have received if they had remained continuously at the barrier for the given time. This results in overstating their potential exposures, a practice generally considered to be desirable.

This is also the general method used in attempts to evaluate internal exposures from information on radioisotope concentrations in air or water.

Another special case in which this type of monitoring has been used is in estimates of the exposure of individuals who have been involved in accidental exposures but did not have appropriate monitoring devices at the time. Such cases have involved individuals who have entered unwittingly into regions of comparatively high radiation levels and have received unknown exposures or have been involved in certain types of material releases. For dose determination, attempts have been made to reconstruct the radiation fields to which the individual was exposed, the time for which he was exposed, and other details of the incident.

A modification of this type of personnel monitoring is that used when some members of a party are supplied monitoring devices but others are not, as may sometimes be the case, for example, for plant tours by visitors. Should an accident resulting in radiation exposure occur, the dosage only of the monitored individuals, probably the guides, would be identified. However, it would be assumed that the overall doses of the others in the party would be approximately the same as those who were monitored. It would be hoped and anticipated that this type of monitoring would rarely, if ever, be required to specify actual exposures.

ALPHA MONITORING

Since all alpha exposures are internal in character, there is no simple method of monitoring personnel by such elementary means as placing a single device on an individual. The methods currently used for monitoring potential alpha exposures depend on environmental checks (which are described more fully subsequently), bioassay determinations of excreta, and possible measurements of the beta-gamma daughters of alpha emitters deposited inside the individual's body. The possibility of taking nose swabs has been suggested as have been breath analyses; these are used primarily for evaluation of radon exposures and in addition to the general techniques that have been mentioned. However, it may be repeated that no direct measurements can be made of exposure to alpha emitters.

NEUTRON MONITORING

The characteristics of various devices for personnel monitoring of neutron exposures have been already described. Basically, they fall into the two general types of film and of metal foils and other materials that can be activated and counted. In using film the complications that result from concurrent exposures to beta-gamma radiation have been mentioned, as have been the problems concerning film in general.

Of the materials mentioned as having appropriate thresholds for high energy neutrons only indium and sulfur may be feasibly enclosed in a small compass and are cheap enough for possible use in personnel monitoring. Copper and manganese for intermediate-energy neutrons and either indium or gold used as described in Chapter 12 for thermal and low-energy neutrons may also be used. However, about the only use of such foils for personnel monitoring occurs in conjunction with possible criticality emergencies, which are discussed in Chapter 18.

LOCALIZED EXPOSURES

Since personnel monitoring results are interpreted as showing total body exposures, indicated doses will be affected by methods of using available monitoring devices. Thus a significant factor of uncertainty in obtaining accurate exposure information stems from the fact that essentially all types of direct personnel monitoring other than that by dose rate and time methods involve a single device worn or carried at one position on the body. People usually attach film meters to a pocket or a shirt collar, and wear pocket chambers in a breastpocket; on the other hand, a fair proportion of persons who use these devices carry them in any one of their clothing pockets. Accordingly, although these devices may rather accurately reflect the actual exposure dose at the point of use, the same may not always be said for exposure to any other part of the individual's body or to his total body.

Beta exposures of the hands have been mentioned. Similar highly localized exposures, which are not detected by a single device, could also be caused by beta beams issuing from opened ends of pipes in which beta-active materials have accumulated. A laboratory technician working in front of a hood in which radioactive materials are used may carry his meter in such a position that it is at least partially shielded from the source. It may also happen that the meter is in a highly localized beam, and the indicated exposure may thus be much greater than that which was actually received as a total body dose. In some cases more than one meter has been used because of these factors. In particular, it is frequently customary to use a small meter for hand monitoring, incorporating it either into a ring or into a wrist badge. Film is frequently used in this manner, and the dimensions of current TLD-units make them particularly applicable for such use.

EFFECT OF ORIENTATION

In addition to uncertainties of localized exposures the use of a single monitoring device also introduces orientation effects caused by the human body's shielding capability. The orientation effect may thus be expressed as the ratio of the actual dose and the measured dose, normally an exposure measurement.

It is recognized that an individual's radiation exposure will not be from a single point (that is, the individual will not always be either facing the source nor will he have his back to it) and his exposure will thus generally be essentially nondirectional. However, practically, this is not taken into account, and the reading of the device used is accepted as the dose received. Thus it may be of interest to give some indication of the magnitude of this effect.

The general effect of the relative orientation of an individual in a field has been shown by some simple tests with a phantom. As a result it has been concluded that, if an individual is wearing a film badge on his chest and is facing away from a gamma source, the shielding by his body will cause the resulting reading to be only about 20 to 35 percent of the exposure which would be indicated if he faced the source. Similarly, if the orientation of the field regularly changes to average out the orientation effects, the indicated exposure will be about 60 to 70 percent of that shown if he were to face the source all of the time. A similar effect should be shown by dosimeters and other devices. Neutrons show somewhat the same effect, and beta particles are completely absorbed in the body thickness.

In general the orientation effect is of importance only for comparatively high accidental and short-term exposures (such as would be expected from a criticality accident, as noted in Chapter 18) or an accidental entry into a high field. Long-term orientation effects should probably also be considered when the work is done in a specific position with respect to a source, such as would be the case for routine work with radioisotopes in a hood.

INCORRECT READINGS

For detectors with a strong energy-dependent response, obvious difficulties may be caused by any changes in the energy of the incident radiation. An illustration of what might occur is the fact that, if neutrons are reflected from water or other moderator, the thermal component of the resultant radiation may be picked up by the detector, and the resultant reading would indicate a higher dose than actually received. However, this particular consideration is probably not of too great importance except possibly in the case of an emergency. Similar effects are noted with beta-gamma radiation, but the overall effect is so small as to be almost negligible. As described previously, film calibrated with cobalt-60 gammas will be very inaccurate for X-rays with energies in the range of 30 kV.

In general, monitoring devices remain at the operating facility at all times. Even under this condition there are an adequate number of possibilities of inadvertent incorrect readings, not to mention deliberate exposures of the device only by the organization clown to "see what will happen" or as a practical joke. However, film meters also frequently double as security or identification devices and are thus frequently worn outside the plant areas. In this case the possibility of spurious readings becomes much greater—all the way from apparent exposures being caused by dental X-rays, to those caused by leaving the badge over a weekend in the full glare of the sun coming through an automobile window, or even to the badge being dropped in water during ablutions.

MONITORING FOR INTERNAL DEPOSITS

An individual can receive no exposure from internal deposits, regardless of environmental conditions, unless radioactive material actually enters his body. As already mentioned, the general techniques for identifying such deposits or intake of material include total- or partial-body counting, and analyses of urine, feces, nose swabs, and expired air. Such evaluations are unfortunately, and perhaps mistakenly in many cases, not considered as giving adequate quantitative data because of variations in people, differences in time between material intake and excretion or other measurements, and various other factors.

On a long term basis the probability of identifying many actual deposits depends on the removal rate, and this rate is variable, depending upon the material concerned. Some materials, such as radium and plutonium, once deposited, are apparently so slowly removed biologically as to be considered almost permanently fixed; the same is generally true of pulmonary deposits of inhaled solubles. In general the elimination of an actual deposit from a given organ is independent of the intake method. However, only a small fraction of the intake of any material is actually deposited, the remainder being transiently in the body for a comparatively short time before being eliminated, generally in the first few hours or days. Thus, probably the greatest possibility of detecting intake of materials occurs with measurements made shortly thereafter; for example, urinalyses should be made about 4 to 24 hr after uranium inhalation for the greatest excretion rate to be observed, and fecal analyses, which have been described as well correlating with total body evaluations for many radionuclides, should be made during the first few days following exposure or intake.

It may be emphasized at this point that although environmental studies, as will be described later, are important indications of possible internal exposure, measurable excretion always accompanies a sufficiently large internal deposit even though actual long term deposits above the maximum permissible body burden for some materials may be necessary to give excreta concentrations above the limit of detection. However, even in these cases the transient component of the material taken in can usually be readily detected.

CONCLUSION

The difficulties of making direct determinations of potential internal deposits of radioactive materials have been described, and it has been noted that indirect methods of evaluating such possibilities include environmental measurements. The next chapter discusses this monitoring method and some of the problems inherent therein.

It is true that there are difficulties in accurate personnel monitoring. However, it is just as true that in most cases they are self-compensating, particularly over an extended period of time; that they usually indicate higher doses than those actually received; and that most of the sources of inaccuracies can be taken into account by comparatively simple techniques and evaluations. In addition, instruments are available to measure exposures much below permissible limits of exposure, and these limits are themselves set sufficiently low so that there appears to be little, if any, possibility of a severe exposure's not being accompanied by a clear indication of a possible problem, provided the available personnel monitoring devices are properly used.

REFERENCES and SUGGESTED READINGS

Radiological Health Handbook, U.S. Public Health Service, U.S. Department of Health, Education, and Welfare, Washington, D.C. (1960)

Ellis, W. P., L. H. Sipe, A. F. Becker, and H. F. Henry, *Calibration of the Multiplant Security Badge Meter for Application at the ORGDP,* KSA–242, Union Carbide Nuclear Co., Oak Ridge, Tenn. (1961)

"Kodak Personal Monitoring Films," *Kodak Scientific and Technical Data,* Eastman Kodak Co., Rochester, N.Y.

"DuPont Dosimeter Film," E. I. DuPont de Nemours and Co., Wilmington, Delaware. National Bureau of Standards Handbooks 42, 57, and 75.

Blatz, Hanson, editor, *Radiation Hygiene Handbook,* McGraw-Hill, New York, 1959.

Standards of Performance for Film Badge Services, U.S. Public Health Service Publication No. 999–RH–20, U.S. Department of Health, Education, and Welfare, Washington, D.C., 1966, NBS Handbook 57 (See Appendix I).

Cusimano, J. P., and F. V. Cipperley, Personnel Dosimetry Using Thermoluminescent Dosimeters, *Health Physics,* **14,** 339-344 (1968).

QUESTIONS

1. Why is more than one film generally used in film badge monitoring?
2. How do heat and age effect film exposure?
3. What effect does orientation have on exposure to film?
4. Why is internal monitoring still largely in an experimental and theoretical state?
5. What are the two basic types of neutron-monitoring devices?
6. Describe the dose-rate and time method of monitoring personnel for beta-gamma exposure.
7. Why is personnel monitoring necessary?
8. What respective advantages do cobalt-60 and radium-226 salts have in calibrating film for personnel monitoring?
9. What are some of the problems with film detection?
10. Under what conditions are pocket chambers and direct reading dosimeters used?
11. Enumerate some ways in which incorrect readings occur with various types of monitoring devices.
12. How are internal deposits monitored?

CHAPTER FOURTEEN

Principles of Environmental Monitoring

PURPOSE

The purpose of environmental monitoring is to determine the possibility of an individual receiving radiation exposures as a result of his work activities. Thus it generally involves measurements of such factors as the levels of penetrating radiation in work locations; contamination of the air, water, or food by radioactive materials; and the presence of such materials on clothing, work surfaces, and other surfaces available for contact.

MEASUREMENTS OF BETA-GAMMA FIELDS

Beta-gamma radiation fields may be readily monitored by any of the various types of rate-meter instruments discussed in Chapter 12, with the Geiger-Müller meter probably being the most widely used device because of its sensitivity. In any case, the gamma and beta components of the existing radiation field are determined from measurements made with the detector tube shielded and unshielded.

These measurements obviously indicate an actual potential exposure only if an individual remains at the point of measurement, and the results of such measurements must be coupled with an estimate of exposure time as previously described to give a quantitative estimate of potential exposure. The source of a beta-gamma field may be a reactor, an encapsulated source, beta-gamma emitters on work surfaces, and so on.

Although the result of time and exposure-rate dose determinations should be the same as that shown by a personnel monitor which the individual uses, this is not generally the case. Personnel monitors normally are worn in a single position and thus may not be in the same place as the field measurement. Furthermore, they are subject to shielding by the individual's body and their energy and intensity sensitivities may differ from those of the environmental monitor; for example, the energy sensitivity of film meters is not identical to that of any other type of meter for various types and intensities of radiation. The discrepancy between direct dose measurements and those obtained from time-intensity calculations would probably be especially significant with a Geiger-Müller-tube meter because radiation energies are generally not a major factor in its response. However, the same conclusion would also be drawn for ionization chambers and other types of meters. Hence, as for all monitoring, careful attention should be given to the conditions of calibration, recalling that any standardization is usually based on a given radiation energy or group of energies and that not only are the responses of various types of devices energy dependent but also that such dependence varies among these devices.

Although these calibration problems are generally of comparatively minor importance for a wide range of gamma energies, major problems that can be very important are involved in beta calibrations. As a matter of fact, most measurements of beta radiation are considered to be qualitative rather than quantitative since reasonably accurate determinations mean not only rather careful calibration of the device itself but also some knowledge of the maximum beta energies involved. However, it may be reiterated that it is very simple to detect the existence of a beta-gamma field and to get some indication of its magnitude.

SOURCE LOCATION

Environmental measurements can also show how the beta-gamma radiation fields change with time, direction, or location of the monitoring position. If for some reason radiation fields are changing as a result of an increase or decrease in the amount of the available radioactive material that forms the source, this will also be indicated by this type of monitoring, and studies of the relative intensities of a field as a function of position may make it possible to locate the source of the radiation if this is unknown.

A convenient method of determining the direction of a source is for an individual to use his body as a shield. To do this, he holds the sensitive element of the meter close to his body and turns through 360°, observing the indicated field. His body will be between the source and his meter when the indicated radiation field is a minimum. Of course, such a method may not be feasible for an extended source which literally surrounds an individual.

For more specific location of an unknown source—especially a large one such as may result from a criticality accident or from radioactive materials being released and perhaps accumulating elsewhere—a somewhat complicated method may be necessary. Triangulation and an evaluation of the position of isodose lines are two applicable methods. In triangulation, the direction of the source from several external points is found. Then, using a map, the intersection of the lines drawn from these points in the indicated direction of the source should intersect near its location. In the second method, points with the same measured field are located on a map and connected by pencilled lines; the approximate source location is the center of the resulting curve. Similar treatment of groups of points that represent different dose levels may improve the accuracy of locating a source by this method.

Such measurements necessarily require an evaluation of the effects of shielding, particularly if such shielding is variable and rather localized. One would expect that the field measured immediately outside a door to a room containing a source would be higher than that on the other side of a concrete block wall surrounding the same room, even though the points of measurement are the same distance from the source itself.

AIR CONTAMINATION AS A PROBLEM

Radioactive materials dispersed in the air of a work location may not only be a possible source of exposure themselves but may make it difficult to identify the field; in particular, if the particles collect on the detecting unit, thus contaminating it, a relatively constant reading may result regardless of the actual field or the meter's position. Air monitoring is useful not only for beta-gamma emitters but also and particularly for alpha emitters, for which air contamination may be the most significant problem.

If there is any possibility of radioactive materials getting into the air, the importance of air monitoring as a means of indicating possible internal exposure can hardly be overestimated since breathing is not only an involuntary action but it also provides a very convenient and simple way for many materials to enter and be distributed in the body, especially if they are soluble in the body fluids. For most alpha emitters of interest, such as uranium and plutonium, control of inhalation is of particular interest in that breathing is manyfold as dangerous, even by several orders of magnitude, as ingestion of the same amount of material. This is indicated by the figures in Appendix II.

AIR MONITORING TECHNIQUES
Basic

In current methods of air monitoring a measured quantity of air is generally passed through a collector, such as a piece of filter paper, and the activity collected is then measured. The average concentration of the radioactive materials in the air is then readily determined from this total activity and the air volume as calculated from the flow rate and the total sampling time. Obviously, radioactive gases may not be measured by this method except as the molecules become attached to dust particles that are collected.

The placement of the sampler itself and interpretation of the data are subjects of considerable discussion; for example, the air is usually drawn continuously through the sampler, and the collecting unit does not thus have the in-and-out motion of an individual's breathing procedure. Of perhaps greater significance, however, is the fact that usually the concentration of the radioactive materials in the air is extremely variable throughout a room, depending as it may on air currents, possible deposits of materials in the room, and almost any other consideration one would wish to include.

Air sampling is usually best employed when the operations involved are continuing and relatively constant, since abrupt and indeterminate changes in the air activity are difficult to evaluate by convenient sampling techniques, especially if the activity involved may be significant and relatively uncontrolled.

Three general methods of air monitoring are currently in use. It is recognized that each of these has peculiar advantages and disadvantages, and that the actual methods used at a given facility may be modifications, combinations, and extensions of these respective sampling methods.

Breathing-Zone Sampling

The first method is known as breathing-zone sampling. In this the sample collector is placed as close as feasible to the nose (or face) of the individual being sampled during a typical operation or period of work, and the collection rate may or may not be at his approximate breathing rate. In a modification of this method the sampler is placed as close as feasible to the actual working location of the individual and the air is monitored while the individual is actually working at that location. This method is considered to be particularly applicable to evaluation of the potential exposure of a person who works relatively continuously at a given location in which the air activity remains relatively constant or varies at a reasonably predictable rate. It has been typical of industrial hygiene work for a long time. Not only does it thus have considerable justification from a theoretical viewpoint but it also is rather highly regarded by many people in the field.

The breathing-zone sampling method is frequently used to establish working times at a given location on the supposition that the air activity which results from a given operation will remain at a measured average level and a person will work at this operation a specific length of time. His average exposure on this job is then considered to be the product of his work time and the average air activity. Such evaluations are thus part of a method for determining the average air exposures of individuals working in a given location on a definite group of jobs. In such an evaluation the overall average air activity an individual breathes over a given period is obtained by measuring the activity at each job, multiplying this by the time he spends at that job, and then summing this up over the entire work period. This may be expressed by the relation

$$C_A = \frac{\sum\limits_{k=1}^{k=n}(Ct)_k}{\sum(t)_k}, \quad \text{or} \quad C_A = \frac{C_1 t_1 + C_2 t_2 + C_3 t_3 + \cdots + C_n t_n}{t_1 + t_2 + t_3 + \cdots + t_n},$$

where $C_1, C_2, \cdots C_n$ represent the air concentrations at the various jobs; $t_1, t_2, \cdots t_n$ are the respective periods the individual works at these jobs; and $\sum(t)_k = T$ is the total work period. Obviously C_A will be kept below the MPC for the radioisotope concerned. If the exposures involve different radioisotopes, the Cs are really fractions of the various MPCs as described previously, and C_A is thus the average fraction of an overall MPC breathed.

This averaging method may obviously include work in locations of comparatively high activity for short periods of time coupled with longer periods of work in locations of lower activity. This action inherently faces a possibility that an operation itself may change markedly and unbeknownst to the individual, with a possible result that the average air concentration actually breathed will also vary from that assumed. An individual's specified work time on a given job may also change as a result, for example, of changes in other operations not at all related to that involved. It is thus apparent that dependence on a method such as this (whereby average weighted exposures are determined from the product of breathing-zone samples and work time) requires continual and careful review of work conditions to be sure they do not change unexpectedly and correspondingly cause higher exposures than are expected.

A hypothetical situation may illustrate this problem. It is assumed that a certain task requires six different operations, only one of which is performed under conditions of comparatively high level air activity. Normally the man's average weighted exposure is well below permissible levels, provided his daily work at the comparatively high level location is no more than 30 min each day. For a period of 1 week, however, increased output from other plant activities, combined with illness and vacation of his co-workers, make it desirable for him to work at the high level operation for the entire 8 hr shift for several days. If in the absence of a clear understanding of exposure potentials this assignment were made, he would then obviously receive a much higher exposure than that on which his work load was predicated. Although this might not produce a dangerous exposure, it might present an undesirable increase in the total dose received.

A similar difference in exposure could occur if one operation normally involves materials of a certain activity, but, because of an experimental or developmental project with higher activity materials, the same operation is done with the much higher activity substances. Under these conditions exposures much greater than those anticipated could occur. Of course, these illustrations merely point out that it is always necessary to have careful evaluation of work conditions, especially when it is not feasible to monitor the operation continuously.

The breathing-zone sampling method also has the obvious disadvantage that it cannot be used while the individual is himself working except at some inconvenience to him. Moreover, since it may also require someone to operate the sampler itself, it may become prohibitively difficult and expensive, particularly in relation to the benefits gained. In addition, unless the sampler is rather continuously used (and this is not normally the case) unexpected changes in the concentration of the air breathed will not be detected.

In some cases a continuous sampler is placed very near the individual's work location, and the collector head may thus be near the individual's face. Even under this condition, however, this type of sample may not truly indicate the activity in the air actually breathed by the individual unless the contaminated zone is fairly large; for example, the author and co-workers in investigating the possible inhalation from clothing contaminated by uranium found a twofold difference in the air activity indicated by material collected by a person's respirator and that collected by a sample head strapped to the individual at about eye level and 6 to 8 in. from his nose.

Continuous Area Monitoring

The second method is identified as continuous area monitoring wherein the air is continuously collected, usually on a shift-length basis, by a sampler placed at a given position in a room or work area. Coverage may be improved by use of several samplers distributed about the work location. In some cases, several sampler heads placed throughout a location and connected to a common vacuum header will provide this multiple coverage.

It is generally assumed that a change in the concentration of radioactive material in the air, even if produced at a single point in the room, will be reflected by a proportionate change in the concentration at the position of the air sampler. Accordingly, an estimate of the average activity in the room or, with proper point-to-point calibration, the average air activity at any given point of concern may, in principle at least, be evaluated. This method is particularly useful with individuals who move about in a rather large work area and are accordingly not specifically exposed to any one source or at any one location.

The continuous sampling method has the disadvantage that a sampler, or sampler head, is not usually at the specific location where a person is, with the result that changes in the air concentration which the individual breathes may not be accurately reflected. This is especially true if the operation near the individual causes the activity or if the change results from a process leak occurring near him. It does have the particular advantage however, that changes in the general air activity will be indicated and, at best or worst, can thus indicate the need either for personnel protection or for additional attention to the sampling procedure. It may very well indicate rather promptly the existence of a process leak or even the fact that materials of higher activity than normal are being processed.

Spot Sampling

The third general monitoring method, a rather versatile one, is area spot sampling. In this method the sampler is used for successive short periods, normally much less than shift length, at one or more locations in a room or work location to get an indication of what might be expected in routine operation in that location. Thus it may be used on an irregular schedule to get a reasonable indication of the average long-term air activity in a given location, it can determine if the type of sampling technique employed in a particular operation should be changed or even if any sampling is needed, or it may indicate whether or not more radioactive material than anticipated has gotten into the air as a result of process failures or other minor emergencies. These are also the techniques usually employed in emergency monitoring.

The short-term area sampling method has the obvious disadvantage that it does not indicate an average activity for a given room or other location as does a continuous monitor. Furthermore, it does not indicate either an individual's exposure or the average activity he may encounter in a routine operation. Thus, it is not intended to replace either of the other methods when there is need for information on a continuing basis. However, as a periodic monitoring technique it has the advantage of being a relatively economical method for insuring that operations or facilities in which the air activity is normally small, perhaps negligible, will not continue for an extended period without being located. It can also provide back-up data concerning the reliability of any other method used or the validity of the various assumptions made to justify the conclusions otherwise reached, perhaps from data obtained by the other methods. As a matter of fact, the breathing-zone sampling method as usually employed may actually be considered to be a special adaptation of this technique.

In addition to the use described above, spot sampling is obviously peculiarly applicable to pilot-plant and operation startup monitoring, emergency monitoring, and monitoring of special short-term operations, especially if comparatively high air contamination levels are anticipated.

FILTERING

Filters placed in a building's air distribution system between an individual's work location and a source of air activity can reduce air contamination to a small fraction of that present otherwise. Such personal respiratory filtering devices as respirators, contained air units, or gas masks can be used to limit the amount of the air contaminant actually inhaled. Obviously, if the fraction of the incident air activity that penetrates a filter is known, a permissible concentration in the air that is thus filtered may be determined.

WATER AND FOOD MONITORING

The problem of potential environmental exposure from ingestion is basically different from that of inhalation, since in principle a person can control the materials that he ingests and consequently his internal radiation exposure from such intake. However, it is possible to ingest radioactive material from various sources, and the monitoring of these is also part of a radiation protection program. Of these potential sources of contamination, water is probably the most widespread one, although food may also frequently be a consideration, and there are various other sources of ingestion; for example, a person may ingest radioactive material from cigarettes or food that he handles while his hands are contaminated.

The possibility of intake by ingestion is generally monitored by attempts to measure the activity of a given sample of water or food. For water this may include the use of a continuous monitor that routinely and continually measures its activity at some point before its use point.

Beta-gamma activities in water are comparatively readily obtained by normal counting techniques with Geiger-Müller counters, using several tubes arranged as a bank. However, it is much more difficult to make similar determinations for alpha emitters, and no continuous monitor using detectors placed above the water surface has been developed. For one thing, the water vapor itself ordinarily causes trouble with the screen of the ion chamber because the short path length of the alpha particles makes it necessary for the screen to be near the liquid surface. Attempts have also been made whereby the water is allowed to pass across scintillation crystal surfaces to be counted. However, partly because of the very low activities that are normally involved as well as the self-shielding of the water, these methods have not been completely successful.

Radioactivity in water is also determined by the normal microanalytical methods of evaporating the water to dryness, this assuming that the material that is dissolved or otherwise included in the water will not also evaporate. The activity is thus concentrated, and the self-absorption of the water itself is markedly reduced or eliminated. However, although this method has the disadvantage that it is an after-the-fact determination of contamination levels, it can be useful in controlling exposure where the water can be held in storage after sampling and before its distribution or consumption.

Monitoring procedures for food are rather indefinite. Even though analytical methods are adequate, there are extreme variabilities in other factors, both as they refer to individual habits and to concentration in food materials. Probably sparked by the recent fallout scares, continuous and intermittent monitoring of various foods has been undertaken, with milk having received probably the greatest attention.

NEED FOR OTHER MONITORING

General

Measurements of such secondary possibilities of radioisotope intake as the radioactive contamination of various surfaces—particularly those of the work location or equipment, hands or other exposed portions of the body, and clothing—are also of value in an overall radiation protection program. Such measurements can provide information on the overall efficiency of the control measures used and may also give some indication of possible exposures; for example, an individual may contaminate food if he handles it while his hands are covered with radioactive materials, and material dusting off his clothing may produce some highly localized air contamination even in locations in which the overall air activity is itself negligible. Furthermore, widespread transferable surface contamination in a work location may be directly related to a more or less general air contamination since such a deposit may indicate (a) the existence of a prior air contamination or (b) that fans or other work activities are stirring radioactive material from the surface itself.

Work-Surface Contamination

Although contamination of surfaces by beta-gamma emitters may be a problem both as direct external exposure and as they affect the probability of internal exposures, the only potential problem for alpha emitters concerns possible body intake of these radioactive materials. It is very probable that the external radiation from a beta-gamma emitting surface contaminant is a greater problem to an individual working in the contaminated location than is the potential hazard of its becoming airborne or otherwise posing an internal exposure hazard. This conclusion, which includes consideration of ease of detection on surfaces and the comparatively high permissible limits for most beta-gamma emitters as air and water contaminants, is certainly true in comparison with alpha emitters. Hence the most serious surface contamination problem, which may result in either ingestion or inhalation, involves alpha-emitting materials. Control procedures that are adequate for alpha emitters of a given activity will be at least adequate for beta-gamma emitters of about the same activity.

Although it is possible for material on work surfaces to enter an individual's body through wounds or by absorption, this probability is comparatively small for most radioisotopes and is relatively independent of surface conditions other than the possibility of splintering, etc. Moreover, since the conditions leading to such intake are relatively amenable to control, attention is usually necessary only to the possibility of surface material being inhaled or ingested, with inhalation probably being a greater potential problem than ingestion. For radioactive materials contaminating a surface to present an inhalation hazard, they obviously cannot be fixed to that surface—and the conditions of wind, air movement, and particle size must also be such that they can readily become airborne and thus be inhaled. Both the work activities of the employees concerned and the work operation itself, including the existence of fans in a work location, can obviously affect this possibility. Such simple maintenance activities as disassembly of equipment may result in an uncovering of concealed material. In fact routine maintenance activities such as buffing, sanding, and welding are particularly effective in making surface contamination airborne. Similarly a running motor can produce breezes that stir up radioactive material on adjacent surfaces, and an individual's vigorous activity may shake such material from his clothing.

Hand and Clothing Contamination

Radioactive materials from contaminated surfaces may get on a person's hands or he may contaminate his hands directly from materials he is handling. In general, the problem of hand contamination is more one of potential ingestion than of inhalation, although an individual with contaminated hands may also contaminate the cigarettes he smokes. Skin contamination, other than that of hands, is normally a considerably smaller problem except as a possible external beta-gamma exposure.

An individual's clothing is perhaps the most intimate surface, other than his skin, that can become contaminated. Omitting special beta-shielding items such as leather gloves, however, the principal protection afforded by clothing against any sort of contaminating material is that it may be removed and cleaned, thus eliminating a possible continuing problem. There is no particular type of clothing which provides significantly better protection against alpha emitters than does any other.

It does not seem likely that clothing can release radioactive material in such a way that it will become an ingestion hazard. However, it would be expected that, if clothing were covered with contaminating dust, the individual's normal movements would result in some of this dust being shaken off to become airborne and perhaps then inhaled. Thus, the inhalation problem from contaminated clothing is probably a short-term one unless the individual is continually recontaminating his clothing. In this case the principal problem is contamination control of air or other source rather than that of clothing. If clothing is not being continuously recontaminated, the initially deposited radioactive material is either released to contaminate the air, in which case the amount that becomes available for release becomes progressively smaller, or it is not being released because it is too tightly bound to the clothing. In the latter case there is obviously little problem to the wearer, since the radioactive material cannot be an inhalation problem if it does not become airborne. However, it may be a consideration in laundry or other cleaning operations if a clothing contaminant is dusty and readily removable (obviously, there is a much smaller problem if it is not readily removable). Considerations of waste disposal resulting from cleaning activities should not be neglected.

MONITORING METHODS

Problems in Measuring Surface Contamination

Measurements of surface contamination should include not only general work areas but also the specific work locations of an employee; for example, an individual who works regularly at a single bench would have an exposure potential that is tied rather closely to this particular location. On the other hand, a person whose work takes him to various locations would have an exposure potential which is more consistent with the contamination levels in these several locations, or associated with the several jobs, rather than with any one of them; perhaps his potential exposure would be related to an average for the entire area. The contamination of surfaces of machinery with which an employee is working may have a significant effect on his potential exposure—especially if he is performing such operations as buffing, grinding, or welding, all of which have high possibilities of producing air contamination. Under such conditions simple measurements might not too accurately indicate the actual exposure potential.

Alpha-emitting materials inside a pipe or other closed system are generally no problem to an individual, the same also being the case generally for beta emitters, except at system openings. However, such a situation changes and a possibility of exposure is presented if it becomes necessary to do work that requires the system to be opened.

Monitoring Techniques

Methods of direct air monitoring for alpha and beta-gamma contamination regardless of the source of material have been described, as have been the usual beta-gamma monitoring systems. For hand contamination evaluations a simple large probe the size of a hand and with an appropriate detector for the type of radiation considered possible or probable, including alpha emitters, may be used successfully.

Beta-gamma monitoring of surfaces requires merely the use of beta-gamma meters used in the same manner as for radiation field determinations. However, the basic method of estimating any problem caused by alpha emitters on surfaces, including clothing, requires placing the alpha detector as nearly as possible in direct contact with the surface being measured. Because of the short range of alpha particles and their easy absorption the overall geometry and consequent detection efficiency for curved surfaces is not as good as that for flat surfaces. The roughness of the surface, its porosity, or the presence of liquids or other potentially shielding material also affect the accuracy of measurement.

In general, direct measurements of this type indicate a combination of both fixed and removable contamination, although a measure of removable surface activity will probably most nearly indicate the possibility of surface material becoming an air hazard. In the most common method used routinely for this type of measurement a pad of paper is rubbed across the surface being monitored, and the activity thus wiped off the surface is then counted either by a portable meter or by laboratory counting equipment as convenience, importance, or other factor may dictate. Ordinarily paper towels or similar cheap and readily available papers are as generally effective for this removal and counting as are filter papers or others. Although the method of wiping the surface may not be of extreme significance, it is important that, whatever the method employed, it be consistently used and that some effort be made to calibrate it so that a measured wipe activity may be related to the actual removable surface contamination. Comparatively reproducible results can be obtained by folding a 4-in. square piece of paper into about four thicknesses and then rubbing it carefully and fairly thoroughly over an area of approximately 100 cm². The alpha activity on the pad is then counted by a survey meter or perhaps eventually in a laboratory counter. Although both the vigor of rubbing and the individual's ability to estimate an area of 100 cm² are factors of concern, the most important factor is that the method used be both consistent and approximately calibrated.

Calibration of such methods is rather difficult, and the confidence limits may be very wide, especially if a variety of situations and surfaces are averaged. Thus, insofar as practicable, attention should be directed to the conditions actually anticipated in various operations and facilities, and efforts directed to interpretation of potential measurements should be made; for example, it may very well be desirable to establish a relationship between a wipe (or direct surface) measurement on a given surface and the amount of transferable material thereon. As a part of a possible technique for making such a determination a given amount of contaminating material in a form anticipated in practice is spread on the surface or surface type concerned and a wipe test is made in the usual manner. This measurement is then compared with the activity measured and also with that estimated from the amount of material used. The remaining material is then taken up by an acid wash or other appropriate agent until no measurable wipe activity remains. The total amount of material accumulated by these techniques is then determined analytically and compared with the amount known to have been deposited. If essentially all of the material originally deposited has been collected, the ratio of the wipe measurements to the total material on the surface may, with some confidence, be accepted as a calibration constant. Actually surface measurements are at best qualitative only, and such calibration techniques are usually indicative only.

Since different people may have different monitoring techniques, they should obtain somewhat different results. However, if comparable results are expected from measurements by these various individuals, it is obviously important either that the measurements be all made in essentially the same way or that a proportionality constant among the individuals involved be established. In any event it is important that, whatever method is used, it be (a) consistent, (b) as nearly well calibrated as possible, and (c) applicable to the type of monitoring required.

CORRELATIONS BETWEEN ENVIRONMENTAL MONITORING AND EXPOSURE DOSE

Rather good correlations have been obtained between the average magnitude of external radiation fields and resultant personnel exposures, particularly of groups of individuals working therein. On the other hand, measurements of the contamination of air, water, or food (which are probably representative of the most direct methods by which radioactive material may be taken into the body) have correlated very poorly with even the best indications of the possible internal deposit.

Since an individual is himself a "collection agency" for the air and other environmental factors resulting in radioisotope intake, it would appear that excretory data should reflect the environment. However, data thus obtained are at best qualitative only. To the best of the author's knowledge no facility has developed a useful quantitative correlation, except with such comparatively wide statistical confidence limits as to be almost meaningless, between the air activities, as measured by current techniques, and the radioactivities of materials excreted either by an individual or by groups of individuals. This has been especially true for the alpha-emitting radioisotopes of interest for which such correlations would be of special value.

Similarly, not too much correlation has been found in operating facilities between the average air contamination as measured by current methods and the average surface contamination, also as measured by current techniques, even though reasonably good surface-air correlations have been noted under laboratory-type investigations. In this the results are similar to those obtained in attempts to relate excreta measurements to air contamination, even for individuals whose work activities and locations are relatively constant. On the other hand, reasonable correlations between air contamination and excreta have been observed in controlled animal experiments. However, the author has noted rather qualitatively that the confidence limit for correlation between average surface contamination and excretory measurements for groups of individuals is somewhat higher than that for either a similar surface-air correlation or an air-excretion relation. Such observations, however, may require considerable verification and are mentioned only as an indication of evaluations that may be undertaken and problems which may be encountered.

GENERAL OBSERVATIONS

Some of the methods for evaluating exposures and potential exposures, especially internal ones, have been described above. In general all of these methods, as well as others that may suggest themselves as being peculiarly appropriate to a given operation, are usually necessary and desirable, and any operating group should itself make arrangements to evaluate the meaning of its various measurements, regardless of how obtained, and their interrelations.

Some of the difficulties in meeting applicable criteria have been mentioned, together with the fact that all too little effort has been made to study them completely and thoroughly. In fact, there is today no good indication of what excretion rates mean in terms of body deposits, air activity levels, or surface contamination levels. It is also rather surprising that the correlation between an individual's excretion rates and the overall average surface contamination levels of his work location appears to be at least as good as is the similar correlation between excretion rate and measurements of breathing-zone air samples or other direct measurements of air activity. A great deal needs to be done in this field in order to make the readings actually meaningful.

One problem in making such evaluations for general applicability has been that each major operating facility has essentially its own method of making samples and then of reading and analyzing the data. Thus, it has been almost impossible for the experience of one facility to be translated into considerations of another because of these differences in monitoring techniques. Far be it from the author to recommend the standardization or universal consistency of such techniques, because there are major differences in all of the factors that go into making the determinations, but the need for calibration of one technique against another would obviously be advantageous—as would be a thorough knowledge and publication by the various facilities of the calibration, significance, and meaning of their particular methods, criteria, and measurements.

REFERENCES and SELECTED READINGS

Blatz, Hanson, editor, *Radiation Hygiene Handbook,* McGraw-Hill, New York, 1959, National Bureau of Standards Handbooks 48 and 51 (See Appendix I).

QUESTIONS

1. Distinguish breathing-zone sampling, continuous air monitoring, and spot sampling. Under what conditions should each of these techniques be used?
2. What is done to account for monitoring both fixed and removable contamination?
3. What relations have been established between excreta and air contamination?
4. What factors does environmental monitoring measure?
5. How can a person locate an unknown source?
6. Why is air monitoring so important?
7. What are the advantages and disadvantages of continuous air monitoring?
8. What could cause some of the poor correlations derived from measurements of air, water, and food contamination with the possible internal deposits?
9. What are some of the problems involved in making data from different facilities interchangeable?
10. Why is general surface monitoring not sufficient for an environmental program?
11. What are the problems of filtering to reduce radioisotope concentrations in air?

CHAPTER FIFTEEN

Fundamentals of Radiation Intensity Reduction

PRINCIPLES OF CONTROL

An individual's exposure to penetrating radiation may be limited by control of the radiation intensity and thus the dose rate at his work position. It has been noted that the radiation emitted by a given source, and thus the radiation intensity at any point produced thereby, decreases with time in the process of radioactive decay. The radiation intensity at a given location may also be reduced by increasing the distance between the location and the source or by interposing radiation-absorbing material, or shielding, between the location and source.

EFFECT OF DISTANCE

Radiation sources emit radiant energy isotropically, or in all directions, and the radiation intensity is defined as the radiant energy that passes through a unit area per unit time. Quantitatively it may be shown that the intensity I (in consistent units of ergs/cm²-sec) of the radiation at a distance R (in cm), from a small source emitting energy at a given rate Q (in ergs/sec), is given by the simple relation

$$I = gQ/4\ \pi R^2,$$

where g is a constant of proportionality that is dimensionless in the consistent units of the above illustration. If g is independent of the distance, which is the case if there is no absorption in the medium, the intensity is

263

obviously inversely proportional to the square of the distance from the source. Hence, the intensities I_1 and I_2 at differing distances R_1 and R_2 from the same source are respectively given by

$$I_1 = gQ/4\pi R_1^2 \quad \text{and} \quad I_2 = gQ/4\pi R_2^2.$$

Accordingly

$$I_1/I_2 = R_2^2/R_1^2, \quad \text{or} \quad I_1 R_1^2 = I_2 R_2^2.$$

Thus, if $R_2 = 2R_1$, then $I_2 = 0.25 I_1$. This relation reflects the fact that the flux density of the radiation, whether waves or particles, becomes smaller with increasing distance since the same total energy is spread over successively larger areas.

The above relation assumes that the source is very small, essentially a point. Although most sources are finite and this is thus not strictly true, the above relation will give errors that are small enough to be negligible if the distance R is large compared with the source dimensions. This is a frequently encountered condition involving penetrating radiation, particularly for specific sources. If the source is very large compared with the distance—as might be the case if a gamma-emitting radioisotope is spread over the wall of a building and measurements are made close thereto—then the inverse-square distance relationship would have no meaning, and any intensity reduction would depend on factors other than distance. For smaller finite sources the intensity falloff caused by distance would depend on the ratio of the distance to the source size and would be less than the inverse square but approaching it for large ratios. Although it is only in a vacuum (and in the vacuum of open space) that the inverse-square distance relation holds precisely for small sources, it must, in practice, always be used in combination with absorption and other factors relating to intensity reduction, as is discussed subsequently.

GENERAL EFFECT OF ATTENUATION

The intensity reduction that results from absorption of energy in the medium in which the radiation travels is known as attenuation. All materials attenuate radiation, the amount of such attenuation depending on the types of radiation concerned and the characteristics of the medium itself. For comparatively short distances of travel, air produces negligible attenuation for neutrons and gammas, and, to a less extent, for betas.

When particles pass through matter they lose energy as a result of absorption, reduction in velocity, or being scattered out of a finite beam.

Electromagnetic waves may lose energy similarly, except that there is no reduction in velocity. An analogous condition may be a reduction in frequency, as observed in the Compton effect. The overall attenuation effect of a given medium on the radiation concerned thus depends on the respective effects of these various factors.

One very important difference between attenuation of particles and that of electromagnetic radiation is the observation that the particles have a definite range, or maximum distance of travel, in a medium, whereas there is no comparable condition for gammas and X-rays. This effect apparently results from the fact that, as previously mentioned, the particles may be slowed down as a result of impacts until they have only thermal velocities and are thus readily captured or, at least, have no further preferred direction of travel.

Because a neutron is actually radioactive, changing to a proton with emission of a 0.78 MeV beta particle with a half-life of about 13 min, the neutron also has a limited range, or a maximum distance of travel even in vacuo. No comparable process affects other radiations of principal interest.

ALPHA PARTICLE ATTENUATION

Alpha particles, which are slowed as they ionize or excite the atoms of the medium they are traversing, have a range in air of about 4 to 5 cm for particles that are of principal interest from radiation protection consideration. Figure 15–1 shows the range of alpha particles in air as a

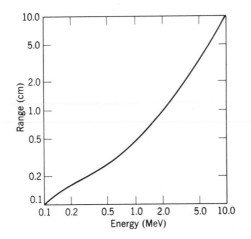

Figure 15–1   Range of alpha particles in air. (From Lapp, Andrews, *Nuclear Radiation Physics,* Prentice-Hall, p. 117.)

function of their energy. Almost any thickness of other materials, such as a sheet of paper or the dead cells of the skin, is greater than the alpha range therein. Hence no further attention is given herein to alpha shielding.

BETA PARTICLE ATTENUATION

Betas lose energy by elastic and inelastic impact with atoms of the medium through which they travel. As a result of the former process they lose only a small amount of energy per collision because of their small mass compared with the atomic mass, and they may be scattered out of a beam very much as a light ball striking a heavier one loses some energy but is also deflected. Such impacts are actually the interaction of the electron's charge with the electric field produced by an atom's planetary electrons. As a result of scatter, betas also enter the geometric shadow of a shield and are reflected from surfaces.

Inelastic impact results in either excitation of atoms or their ionization, with consequent emission of electromagnetic radiation in the optical range. This process is particularly important for intermediate-energy electrons. Higher energy electrons also lose energy by their deceleration in the electric field of the atom, with the emission of electromagnetic waves. Such radiation is called *bremsstrahlung,* and its production is essentially the same as that by which the continuous X-ray spectrum is produced. Betas and X-rays (or gammas) have different absorption and attenuation characteristics. Thus bremsstrahlung introduces not only the possibility of radiation in the geometrical shadow of a shield but also radiation problems that are different from those of the incident betas.

Since the energy loss by the electron at impact and consequent X-ray (or bremsstrahlung) frequency are proportional to the square of the atomic number of the material concerned as well as the square of the electron energy, it is apparent that the higher the atomic number of the material concerned, the more energetic become the X-rays produced by a given beta source. This is the reason that beta shielding normally uses materials of low atomic number, such as plastics or aluminum.

The fact that betas from a single reaction are not emitted at a single energy but in a range of energies extending from essentially zero to a maximum (as shown by Figure 3–1) has been mentioned. Hence the effective attenuation of betas in a medium is the overall result of the different attenuations of these various energies. Figure 15–2 indicates a typical beta-absorption curve, and Figure 15–3 is a plot of range as a

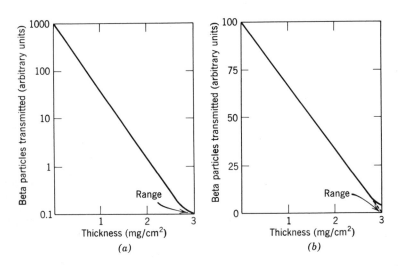

Figure 15–2 Beta-particle absorptive curves: (a)—typical emission spectrum with maximum energy same as in (b); (b)—monoenergetic beta particles. (Adapted from Kaplan, I., *Nuclear Physics,* Mc-Graw-Hill, p. 348.)

function of energy, with the values shown being based on data obtained with aluminum. Since the most penetrating betas are those of the highest energy, the range of betas in a given medium is normally considered to be that of the most energetic ones, which is given approximately by the relation

$$\rho R = 0.54 E_m - 0.13(1 - e^{-3.2E_m}),$$

where E_m is the maximum energy of the beta particles emitted, in MeV; R is the beta range in centimeters; and ρ is the density of the medium in gm/cm³. For beta energies higher than about 0.8 MeV, the term $e^{-3.2E_m}$ may be neglected. Betas have significant ranges in air; for example, the range is about 1 m for 0.5 MeV energies and about 10 m for 3 MeV particles.

Betas are so efficiently attenuated that the inverse-square-distance relation applies for only a very short distance, and, although the intensity of a monoenergetic electron beam decreases essentially exponentially for much of its range in matter (as described in the next section for neutrons), the overall effect is masked by the multiple energies of the betas in a given beam.

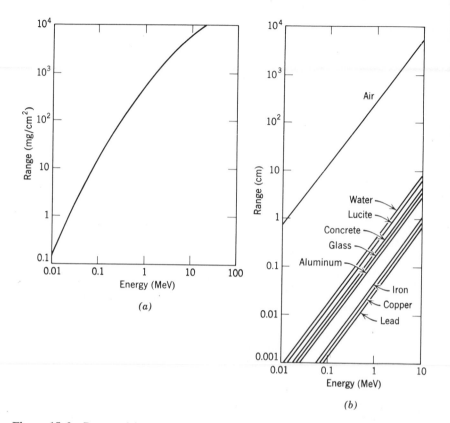

Figure 15–3 Range of beta particles. (*a*)—Range in terms of area density (based on experiments with aluminum). (From Kaplan, I., *Nuclear Physics*, p. 353.) (*b*)— Range in various materials. (From *Radiological Health Handbook*, p. 155.)

ATTENUATION OF NEUTRONS

For neutrons the absorption effects are somewhat more complicated, although the energy loss basically occurs from slowing down as a result both of elastic (scattering) collisions with atomic nuclei and of inelastic collisions with these nuclei, which may result in capture and possibly the emission of other radiations as well as fission. Overall neutron absorption thus depends on the energy of the neutrons and the nuclear characteristics of the attenuating materials. Important in these nuclear characteristics is the cross section of the nucleus for the appropriate reaction. These considerations have been reviewed in a preceding chapter, and some cross sections were shown in Figures 12–9 and 12–10.

It may be shown that a constant fraction of the monoenergetic neutrons incident on a given thickness of material will be removed. Accordingly the relation for such removal is

$$\frac{dn}{dx} = -kn,$$

where n is the number of neutrons, dn is the change in this number as it travels through a material thickness of dx, and k is a proportionality constant. As shown previously the solution of this equation is

$$n = n_0 e^{-kx},$$

where x is the thickness of the material, n_0 is the number of neutrons incident at a point $x = 0$, and n is the number of neutrons transmitted through the material. Obviously the number of neutrons removed is

$$N = n_0 - n = n_0(1 - e^{-kx}).$$

The fact that neutrons apparently interact only with the atomic nuclei has been stated. Hence the number of nuclear events that occur as a beam of neutrons traverses a given thickness of an element depends on the number of atoms present, the probability of a neutron impact on an atom, and the number of neutrons available. The number of atoms, S, available in a volume V may be obtained from the density of the material, ρ; Avogadro's number, A; and the atomic mass of the nucleus, M. Let m be the mass of a volume V. Therefore $\rho = m/V$. Now $(m/M) = (S/A)$ since the mass of any material is proportional to the number of atoms of which it is composed and there are A stems in a mass M. Accordingly

$$S = mA/M = \rho V A/M.$$

The ratio $\rho A/M$ is actually the particle density h, expressed in particles/cm^3. Now assume that a neutron flux of n neutrons/cm^2-sec is incident on a cubic volume dV with a surface area of x^2 and a thickness of dx. Thus $dV = x^2 dx$, and the total rate of neutron impingement on this surface is nx^2 neutrons/sec.

Since each of the S atoms in a volume dV has a microscopic cross-sectional area σ, the atoms in this volume thus have a total (macroscopic) cross-sectional area of $S\sigma$. A neutron entering dV must thus pass within this area to interact with the atomic nuclei of this volume. This assumes that no nucleus shields another one, which is very nearly the case if the equivalent nuclear radius r (where $\pi r^2 = \sigma$) is small compared with interatomic distances and the volume thickness dx is so chosen that $S\sigma$ is small compared with x^2. The probability P that a neutron entering the volume dV will be captured or otherwise affected is thus the ratio of this area of interaction to the total area of incidence. Hence

$$P = S\sigma/x^2 = \rho dV A\sigma/Mx^2 = \rho A\sigma x^2 dx/Mx^2 = \rho A\sigma dx/M = h\sigma dx.$$

Since the atoms of molecular systems apparently act completely independently in neutron interactions, this probability may be expressed as $P = \Sigma(h\sigma)dx$, where each value of $h\sigma$ is that for each of the respective elements. Obviously $\Sigma(h)$ would be the total particle density.

The probability P is also equal to dn/n, the fraction of the total incident neutron flux n which interacts with the atomic nuclei in passage through the thickness dx. Since the flux change dn represents a removal from the beam, its value is thus given by the relation

$$dn = -(\rho A\sigma/M)n\ dx,$$

or, for several atoms

$$dn = -n\Sigma(h\sigma)dx.$$

Thus it may be seen that the constant k given in preceding equations has been determined so that the fraction of incident neutrons penetrating a thickness x is given by

$$n/n_0 = e^{-(\rho A\sigma/M)x},$$

or, more generally,

$$n/n_0 = e^{-\Sigma(h\sigma)x},$$

and the fraction removed from the beam in this penetration is

$$N/n_0 = 1 - e^{-(\rho A\sigma/M)x},$$

or, more generally,

$$N/n_0 = 1 - e^{-\Sigma(h\sigma)x}.$$

If σ in the above relations is the scattering (elastic impact) cross section, the above relations refer respectively to the fractions of neutrons that are not scattered or scattered. Similar fractions for other interactions (as noted in Chapter 3) may also be determined. If σ is the total cross section, these fractions refer to all neutron interactions.

ELASTIC SCATTERING OF FAST NEUTRONS

For fast neutrons elastic scattering is the principal method by which energy is lost. For a head-on collision it may be shown that f_0, the fractional energy lost by a neutron of mass m colliding with an atomic nucleus of mass M (assumed as being initially at rest), is given by the relation

$$f_0 = 4mM/(m + M)^2.$$

It may readily be seen that the more nearly equal m and M are, the greater is the fractional energy loss on collision; for example, essentially all of the energy is lost if the collision is with a proton (hydrogen nucleus), whose mass is very nearly the same as that of the neutron. Similarly, for carbon with an atomic mass of 12 the fraction is about 28.4 percent, and for a heavy nucleus, such as lead with an atomic mass of 207, it is about 1.9 percent.

The above conclusions are based on an assumption of a head-on collision, which is possible regardless of the dimensions of either the neutron or the atomic nucleus; in other words, both could be point masses. However, the probability of impact is a function of the nuclear cross section σ, and for elastic impacts the neutron may be considered to be a mass point and the nucleus a perfectly elastic sphere with a cross-sectional area σ. Accordingly, for any impact other than a head-on one the neutron will be deflected through an angle θ, much as is the case with billiard ball impact.

From elastic-impact relations—assuming the mass M is very much greater than that of the neutron (and is thus little moved by the impact)—it may be shown that E_2, the final energy of the neutron, is a function of θ and is related to the initial energy E_1 by the relation

$$E_2 = E_1 \frac{M^2 + m^2 + 2Mm \cos \theta}{(M + m)^2}.$$

Similarly it may be shown that the fractional energy lost by the neutron on impact, f, is related to that for a head-on collision, f_0, by the relation

$$f = f_0 \frac{1 - \cos \theta}{2} = f_0 \sin^2 \frac{\theta}{2},$$

where θ is the angle of neutron deflection. Obviously, $\theta = 0$ when no collision occurs. Essentially all values of θ will be observed for a large number of these interactions. Hence, since the average value of $\sin^2 \left(\frac{\theta}{2} \right) = 0.5$, the average fractional energy lost per collision is

$$f_a = 0.5 f_0 = 2mM/(m + M)^2.$$

It should be pointed out again that the above refers only to monoenergetic neutrons and that the cross section σ varies with energy. Thus, whereas the average fractional energy lost per collision may remain the same for all impacts, the probability of a collision will change as the energy of the neutron changes. The fact that essentially all collisions of high energy neutrons with the atomic nuclei of most elements are elastic also means that the capture and inelastic impact cross sections at these energies are so small as to be negligible. Since at lower energies this is no longer true, the probability increases that an impact will result in a total "loss" of the neutron, with a consequent transfer of its energy into some other form as described previously.

NEUTRON SHIELDING APPLICATIONS

Nearly all materials have comparatively large capture cross sections for thermal or low energy neutrons. Hence, once neutrons are slowed to thermal velocities, which means that their energies assume the same Maxwellian distribution as that of the atoms with which they are associated, they will eventually be captured and so may be said to have reached their range. In general neutrons are captured long before there is a significant probability of decay, and in shielding considerations it is generally acceptable to ignore this possibility.

From the above it is obvious that the range of neutrons in a given material depends on the masses and cross sections of the atomic nuclei in this material, and these cross sections are not particularly predictable from other atomic constants. Hence neutron shielding specifications depend basically on experimental information concerning the cross sections of various materials since little success has been encountered to date in efforts to determine these values theoretically.

Probably the best neutron shielding material is hydrogen, both because its nuclear mass is nearly equal to that of the neutron and because of its comparatively large capture cross section. Figure 15–4, which is adapted from neutron "age" curves, indicates the approximate range of neutrons in water as a function of their initial energy. Actually the distances are those for slowing down to indium resonance (about 1.44 eV), but the additional distance required for neutrons to slow to thermal energies is small. Figure 15–5 shows the approximate energy distribution from a fission source that is probably of most interest for radiation protection considerations.

As with all particles, neutron attenuation for distances that are short compared with their range apparently follows an exponential relation. Similarly, in media such as air, in which the attenuation is slight and the

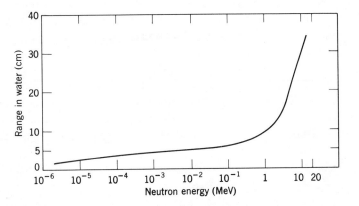

Figure 15–4 Range of neutrons in water. Range is taken as approximately the distance required for neutrons to slow down to indium-resonance energy of about 1.4 eV. (Values are adapted from neutron "age" calculations.) (From K-1486, p.5.)

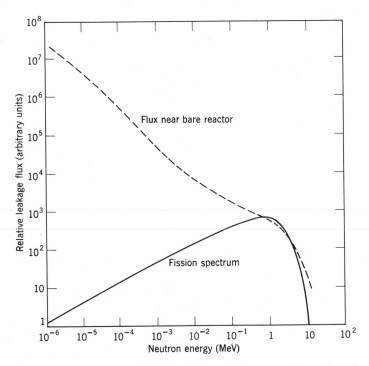

Figure 15–5 Typical neutron-energy distributions near source. (From K-1486, p. 23.)

range is long, the intensity decreases by the inverse-square-distance relation. Hence the neutron flux n (in particles/cm²-sec) at a distance R from a neutron source N (in particles/sec) is given approximately by the simple relation

$$n = gNe^{-kR}/4\pi R^2,$$

where g and k are constants, the former relating to the source and the latter to the nuclear characteristics of the intervening medium. Such relations are particularly of interest for nuclear weapon analysis, and Figure 15–6 shows the dependence of the flux on the distance from a specific nuclear weapon. For various sources which emit different neutron energy spectra the curve will also vary somewhat.

In most cases where neutron shielding is specifically used, the shield thickness is small compared with distance from the source. Thus the distances from the source to the entrance and exit sides of the shield are essentially the same, and so the effect of the inverse-square-distance considerations is negligible. Under these conditions flux attenuation in the

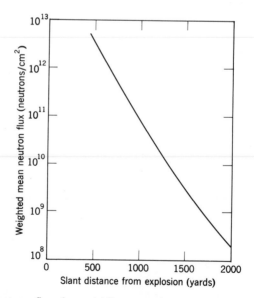

Figure 15–6 Neutron flux from 1-kiloton nuclear weapon. The ordinate is an average neutron flux with the various neutron energies weighted to reflect their RBE effectiveness. The actual slow neutron flux (in the low eV range) is about 10 to 20 times the fast flux (above about 2 MeV). (*From Effects of Nuclear Weapons*, p. 412.)

shield is due only to the shield material and is thus essentially exponential only.

For these comparatively thin shields a useful defining figure is the half-value thickness. This is the thickness of material through which half of the incident radiation is transmitted. Thus for a given material with a linear absorption coefficient k the half-value thickness $X_{1_2} = 0.693/k$. Obviously 25 percent of the incident radiation will be transmitted through two half-value thicknesses, and so on.

The fact that neutrons are also reflected as a result of elastic impacts with heavy nuclei may become of importance for radiation protection considerations. This is true in the case where a neutron shield, although adequate, is comparatively small and individuals are expected to remain in its shadow. In this case neutrons that do not strike the shield may be reflected by air or other materials into the shield shadow. If the shield is not too large, such reflection can result in a flux in the shield shadow on the order of 10 percent of the flux that would be measured in the absence of the shield.

Similarly beta and gamma radiation may result from radiative capture of neutrons. In this case the radiation protection problems are similar to those for these radiations. With respect to the results of reflection and radiative capture, neutron problems are similar to those of the bremsstrahlung described for betas which also produces problems due to various radiation types in shield shadows.

ATTENUATION OF GAMMAS

As is the case with neutrons, the fractional energy lost by a beam of monoenergetic gammas in passing through a given thickness of material is proportional only to the intensity of the incident radiation. The attenuation of the radiation as it passes through the matter is exponential, and the emerging intensity I is related to the incident intensity I_0 by the expression

$$I = I_0 e^{-\mu x},$$

where μ_1 is a constant called the linear absorption coefficient and x is the thickness of the shield.

If a small source emitting energy at a rate Q is placed in an absorbing medium with a linear absorption coefficient μ, the intensity I at a distance R from the source will be given by the relation

$$I = dQe^{-\mu R}/4\pi R^2,$$

where d is a constant that with appropriate choice of units may be unity. It μ is 0, which means that there is no absorption, this relation obviously reduces to the value $I = dQ/4\pi R^2$ as given earlier for a nonabsorptive medium.

The respective intensities at two distances, R_1 and R_2, will be given by

$$I_1 = dQe^{-\mu R_1}/4\pi R_1{}^2 \quad \text{and} \quad I_2 = e^{-\mu R_2}/4\pi R_2{}^2.$$

Hence the relative intensities at these two distances will be given by the ratio

$$I_2/I_1 = e^{-\mu(R_2 - R_1)} (R_1/R_2)^2$$

Since this relation gives only the relative intensities at the two points, its use does not require knowledge of the actual source strength nor is it necessary for the source to be in the same absorptive medium as that between the points. Thus, if a shield of thickness x is placed at a distance R_1 from a source and $x = R_2 - R_1$, the relative intensities on the two sides of the shield will be given by $I_2/I_1 = e^{-\mu x}(R_1/R_2)^2$. Furthermore, if the distance x is small compared with R_1, the ratio R_1/R_2 is nearly unity, and the intensity ratio reduces closely to $I_2/I_1 = e^{-\mu x}$, which is the relation for attenuation in an absorbing medium alone.

If a second shield of thickness y and of different absorptive characteristics as defined by another linear absorption constant, k, is placed adjacent to the first one as shown in Figure 15–7 and its outer edge is at a distance R_3 from the source (where $y = R_3 - R_2$), it may similarly be shown that

$$I_3/I_2 = e^{-ky}(R_2/R_3)^2$$

and that

$$I_3/I_1 = (I_3/I_2)(I_2/I_1) = e^{-(\mu x + ky)}(R_1/R_3)^2.$$

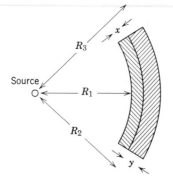

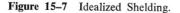

Figure 15–7   Idealized Shelding.

Again, if the total shield thickness, $x + y$, is small compared with R_1, the intensity relation depends only on absorption in the two media, and the intensity ratio becomes simply

$$I_3/I_1 = {}^{(\mu x + ky)}.$$

The overall relation involving both attenuation and distance is obviously applicable to evaluating intensity reductions by use of distance and shields in situations such as that indicated by Figure 15–8. It should particularly be noted that the two distances chosen, R_1 and R_2, need not be those from the source to the shield edges nor, if more than one shield is used, need the shields be adjacent to each other. However, it should also be noted that the distance relationship depends on the distance-to-source size ratio as already described.

A quantity that is frequently used in identifying the shielding characteristics of some material is its half-value thickness, which is defined as the thickness of material through which half of the incident radiation is transmitted. Similarly the tenth-value thickness is that through which only 10 percent of the incident radiation is transmitted. If μ is the linear absorption coefficient of a material, its half-value thickness is given by $0.693/\mu$, and its tenth-value thickness by $2.303/\mu$. Obviously, only 25 percent of the incident radiation will be transmitted through two half-value thicknesses and only 1 percent through two tenth-value thicknesses. The obvious similarity in the relations used for determining the half-value thickness of an absorbing medium from its absorption coefficient and the half-life of a radioisotope from its decay constant is due to the fact that both radioactive decay and radiation absorption are described

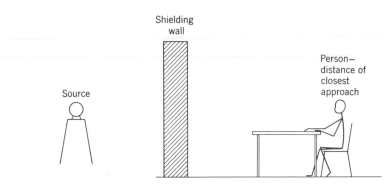

Figure 15–8 Typical shielding configuration.

by exponential functions. Table 15–1 lists half-value thicknesses for several materials.

TABLE 15–1

HALF-VALUE THICKNESSES FOR VARIOUS MATERIALS

Gamma Energy (MeV)	0.1	0.2	0.5	1.0	2.0	5.0
Material	Half-Value Thickness (cm)					
Aluminum	1.60	2.14	3.05	4.17	5.92	9.11
Iron	0.26	0.64	1.07	1.49	2.09	2.84
Copper	0.18	0.53	0.95	1.33	1.86	2.47
Lead	0.012	0.068	0.42	0.90	1.34	1.44
Lead [a]	0.024	0.050	0.31	0.80	1.20	
Water	4.14	5.10	7.17	9.82	14.05	23.02
Air [b]	35.5	43.6	61.9	84.5	120.5	195.8
Concrete [c]	1.75	2.38	3.40	4.65	6.60	10.28
Concrete [a]	1.75	2.54	3.26	4.51	6.12	

[a] These are half-value layers for heavily filtered constant-potential X-rays.
[b] For air, distances are given in meters and inverse-square-distance effects are neglected.
[c] Average concrete density is 2.35 g/cm.[3]

The constant μ depends on both the frequency of the incident radiation and the properties of the attenuating material. Important in the gamma-attenuating characteristic of any material is its overall average density ρ. Hence μ is sometimes expressed in terms of a mass absorption coefficient μ_m to which it is related by the factor

$$\mu = \rho\mu_m.$$

From this the attenuation relation given above becomes

$$I = I_0 e^{-} e^{\mu_m x}.$$

Obviously the factor ρx has the units of gm./cm^2; this is sometimes called the area density of the shield. If this area density is represented by X, the above expression is

$$I = I_0 e^{-\mu_m X}.$$

Figure 15–9 shows values of μ_m as a function of gamma photon energy for several materials. To a reasonable degree of approximation for a

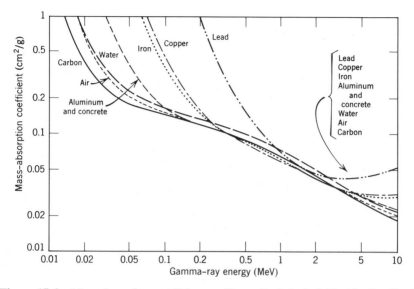

Figure 15–9 Mass-absorption coefficients. (From *Radiological Health Handbook*, pp. 144–150.)

wide range of gamma energies, the mass absorption coefficients for a variety of materials are nearly a constant for a given incident gamma-ray frequency. Therefore the above form of the attenuation relationship is particularly useful in computing an approximate attenuation for a shield composed of a mixture of several materials since only the gross area density of the shield itself must be determined. Similarly for a succession of individual shields of various materials an effective area density may be obtained for the entire shield by merely adding the respective individual area densities. In this case the above relation would become

$$I = I_0 \exp[-\mu_m \Sigma(X)] = I_0 \exp[-\mu_m(X_1 + X_2 + \cdots + X_n)].$$

Since μ_m is not actually a constant, a more correct value of the above expression would be of the form

$$I = I_0 e^{-\Sigma(\mu_m X)}.$$

It may also be seen that the respective thicknesses of different materials necessary to produce a given attenuation are approximately inversely proportional to their respective gross densities.

GAMMA ABSORPTION

Three processes apparently account for the absorption of gammas. For those of low energy (below about 1 MeV) the most important of these is the photoelectric effect whereby part or all of the photon energy is absorbed in ejecting a bound planetary electron from an atomic orbit. For intermediate energy photons (in the range of 1 to 5 MeV) Compton scattering is probably of the greatest significance. In this interaction the photon is scattered by a loosely bound electron, losing some of its energy in the process. Since the photon velocity is that of light and is unchanged by the interaction, this energy loss is reflected by the scattered radiation having a slightly lower frequency than the incident radiation. A series of these scattering collisions will reduce the radiation frequency to the point at which other absorption methods, primarily photoelectric absorption, become important.

For high energy photons (generally above about 5 MeV but necessarily more than 1.02 MeV) pair production is important. As a photon passes near a heavy nucleus it splits into an electron and a positron, each possessing half the photon energy. At least 1.02 MeV of the photon energy are necessary for the energy transformation into the particle masses; the remainder appears as kinetic energy, evenly divided between the particles. The electron loses energy as does any beta particle. The positron very shortly disappears by collision with another electron. As a result two other photons are produced, each of which has an energy that corresponds to half the sum of the energies of the colliding positron and electron, including their masses. If the positron loses no energy before collision and if the electron with which it collides has only thermal energy, each of the two photons which result will have energies that are somewhat above 25 percent of the energy of the initial photon, this being the lower limit of such photon energies. In no event, however, can the energy of these photons be less than the 0.51 MeV mass energy of the electron or positron.

In addition to the above, low energy photons are also scattered elastically, with negligible degradation of energy. Hence they, as well as those suffering Compton scattering and those resulting from pair production, may be scattered out of the incident beam, as was the case with neutrons. Under these conditions the attenuation is what is called narrow beam attenuation. However, if the shield is large compared with its distance from the detector or the source is large enough compared with its distance from the shield, these photons that are scattered out of the direct beam

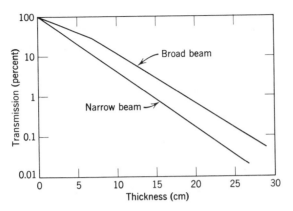

Figure 15–10 Broad- and narrow-beam attenuation. The transmissions given are for cobalt-60 gamma rays through concrete. (From *Radiation Hygiene,* p. 83.)

are replaced by others that are scattered into its path, and the resulting attenuation under these conditions is called broad beam attenuation. Figure 15–10 indicates the difference in absorption coefficients for broad beam and narrow beam attenuation.

The effect of reflection in the shadow of a shield, or skyshine, must also be considered for gammas. However, except for pair production, the secondary radiation is always electromagnetic waves and does not give a different radiation type as is the case for betas and neutrons.

LOW BACKGROUND MEASUREMENTS

For precision measurements of low-level radiation sources it is necessary to provide a counting facility that is shielded to reduce as much as feasible the effects of background radiation. In one such facility, which has made precision measurements of gamma-ray spectra with a 3 in.-diameter x 3 in.-thick cylindrical sodium iodide (NaI) crystal, the basic shielding of the 32 x 32 x 32 in. enclosure containing the crystal is provided by 4 in. of lead. However, the lead emits lower energy gammas in attenuating the incident high-energy radiation and is thus itself lined with 0.030 in. of cadmium which appropriately shields out this secondary emission. Since the cadmium in turn also emits gammas of still lower energy, a second liner of about 0.015 in. of copper is provided. Only the crystal need be shielded; the auxiliary electronic circuits and other counting and recording devices may be conveniently placed outside the enclosure.

Since it is necessary in total-body counting to measure very low radiation levels, such shielding is also provided for these devices even though wall areas several meters square are involved.

SELF-SHIELDING

Self-shielding is a phenomenon of considerable interest, particularly where large masses of radioactive materials of low specific activities are concerned. These low specific activities could either be characteristic of the radioisotopes themselves or they could result from the mixing of radioisotopes with inert materials, such as would be the case for dilute solutions, ores, or mixtures of solids.

As its name implies, the material of the source attenuates its own emitted radiation. Thus the radiation level produced by a given amount of a radioisotope located at a significant depth inside a source is smaller than that produced by the same amount of the radioisotope on the surface. In fact the depth may be sufficient for the contribution of this part of the source to the total radiation emitted to be negligible. This effect is particularly important for beta emitters; for example, the beta radiation from a block of "old" uranium several centimeters thick is at a maximum value and is thus unchanged for any greater thickness of material. It is thus obvious that there are limitations to the accuracy of estimates of the amount of a radioisotope available as determined from measures of the radiation emitted.

Self-shielding is also of concern for gamma sources, especially low activity ones. For high activity sources the radiation level of a source sufficiently large for self-shielding to be a factor may be so high that maximum precautions are required.

REFERENCES and SUGGESTED READINGS

Radiological Health Handbook, U.S. Public Health Service, U.S. Department of Health, Education, and Welfare, Washington, D.C., 1960.

Blatz, Hanson, editor, *Radiation Hygiene Handbook,* McGraw-Hill, New York, 1959.

Goldstein, Herbert, *The Attenuation of Gamma Rays and Neutrons in Reactor Shields,* USAEC, Washington, D.C. (1957)

National Bureau of Standards Handbooks 63, 73, and 76. (See Appendix I.)

Raleigh, Henry D., *Radiation Shields and Shielding,* TID–3547. U.S.A.E.C., OTS, Washington, D.C., Feb. 1960.

Brucer, Marshall, *Marble Used as a Radiation Shield.* Marble Institute of America, Mt. Vernon, N.Y. (1954)

Moteff, John, *Miscellaneous Data for Shielding Calculations,* APEX-176 Atomic Products Division, General Electric Corporation, U.S.A.E.C., Washington, D.C., Dec. 1, 1954.

Blizard, E. P. and J. M. Miller, *Radiation Attenuation Characteristics of Structural Concrete,* ORNL–2193, USAEC, Washington, D.C., August 29, 1958.

Spielberg, David and Arthur Dunear, *Dose Attenuation by Soils and Concrete for Broad, Parallel-Beam Neutron Sources* AN-108, Associated Nucleonics, Inc., Garden City, N.Y., USAEC, Washington, D.C. (1960)

Leipuns Kii, O. I., B. V. Novozhilov, V. N. Sakharov, *The Propagation of Gamma Quanta in Matter,* Pergamon Press, Ltd., London, (1965) (Translated by Prasenjit Basu; Translation edited by Spinney, K. T., J. Butler, and J. B. Sykes)

Henry, H. F., and J. R Knight, *Neutron Energy Degradation by Water Moderation* K-1486; Union Carbide Nuclear Co., U.S.A.E.C., Washington, D.C., Sept. 27, 1961.

Heath, R. L., *Scintillation Spectrometry, Gamma-Ray Spectrum Catalog,* 2nd ed., Phillips Petroleum Co., Atomic Energy Division, IDO-16880-1, August 1964. Office of Technical Services, U. S. Department of Commerce.

QUESTIONS

1. What is the definition of radiation intensity?
2. What limits the range of a neutron in any particular medium?
3. What is meant by bremsstrahlung and under what conditions does it apply?
4. With what part of the atom do neutrons interact and why?
5. On what does the range of neutrons in a given material depend?
6. How far would you have to be from a source providing a dose rate of 10 rems/hr at 5 m to reduce the intensity to 1 rem/hr? Assume that there is no shielding involved and that the source is a point. Justify your method.
7. What difference is there between attenuation of particles and electromagnetic waves?
8. Explain the processes by which alphas and betas are attenuated.
9. How are neutrons absorbed? Of what importance are their energies and the cross section of the nucleus?
10. Of what value are mass absorption coefficients?
11. If the half-value thickness of an absorber is known, what is a quick way of finding the linear absorption coefficient? The mass absorption coefficient?
12. Discuss narrow and broad beam attenuation.
13. Determine the half-value thickness for iron if its density is 8.3 g/cm^3 and its mass-absorption coefficient is 6.8 cm^2/g.
14. The mass absorption coefficient of X-rays at 0.710 A for nickel is 48.1 cm^2/g. The density of nickel is 8.75 g/cm^3. For what thickness of nickel will the intensity of the X-ray beam be reduced to one-third its original value?
15. A photon of 8 MeV passes a heavy nucleus and splits into an electron and a positron. What is the velocity of the electron after this pair production?
16. If the density of iron is 8.0 g/cm^3 and the maximum energy of a beta beam is 3.6 MeV, what is the range of the beta beam in iron?
17. What are the differences between particle and wave attenuation?
18. What is the fractional energy loss for a neutron in an elastic collision with an oxygen-16 nucleus? With a uranium-238 nucleus?
19. How do elastic and inelastic scattering of neutrons differ?

CHAPTER SIXTEEN

Guides to Practical Protective Measures

BASIC PREMISES

The basic aim of a radiation protection program in an operating facility is to control the radiation exposures of its employees and all other individuals, including members of the general population, who may come into contact with radiation or radioactive materials for which the facility is responsible, to values that are as low as feasible but always below the appropriate permissible limits. As a legal matter, specific laws or other regulations must also be met. Actual protective measures may generally be classified as source protection, direct protection of an individual or group, or combinations of these. The methods used may be technical, administrative, or both, and their application to specific conditions may be obvious.

Basic to any evaluation is the fact that radiation problems can result from the use of radioisotope quantities which are frequently extremely small compared with dangerous quantities of most nonradioactive toxic materials, especially those in widespread use. Thus comparatively extreme measures are frequently needed to be sure that the radioactive materials are adequately confined and that radiation levels are appropriately limited in work locations. A radiation protection program may affect essentially all aspects of an operation involving radioactive materials even though the locations actively concerned may be rather limited.

The specific action in a given operation may involve considerable engineering or administrative ingenuity, it may be limited by concurrent problems related to other types of hazards or activities, and it may need to fit into general administrative patterns. Frequently specific choices between two or more possibilities, either of which would be acceptable, must be made. The final choice may as often be based on administrative convenience, cost, or similar factors as on purely technical considerations.

Although this chapter deals primarily with technical matters, it also includes certain of the administrative procedures of engineering review and radiation monitoring that are at least semitechnical in nature and affect technical considerations. What are probably more nearly nontechnical administrative matters are discussed in Chapter 19.

SOURCE CONTROL

Initial efforts are made to control the source. This normally necessitates confinement of the radioactive material to identified locations and implementation of specific provisions to limit the radiation levels in accessible locations by shielding the source or by using barriers to limit personnel to a minimum distance of normal approach. Shielding and distance limitations are particularly appropriate to beta-gamma and neutron radiation, whereas confinement of material is generally applicable to all types of radioactive materials and is the basic method used for alpha emitters.

Where adequate source control may not be feasible, as is the case in many normal maintenance activities, appropriate protection for the individual concerned can be provided. This may include the use of protective devices of various kinds and administrative actions such as limitation of exposure time. The results of both personnel and area-type monitoring, as well as other appropriate methods, are used to indicate the need for, and consequent success of, the actions taken.

PRACTICAL SHIELDING CONSIDERATIONS

The fundamentals on which practical shielding plans may be based have been reviewed. No particular effort may be necessary for alpha radiation, which is stopped by a minimum of shielding, such as the dead outer skin layer. A coat of paint will reduce the alpha activity from a contaminated surface to a negligible value. Similarly, if an alpha-contaminated surface is covered by a thin layer of water, the activity will be undetectable.

It has been pointed out that the shielding required for beta-gammas depends on the types of radiations involved, their energies, and their intensities. Although most shielding calculations probably give rather desirably conservative results, such determinations should be used only as guides, and the actual effect of a given shield for a given source should be measured by the appropriate meter where possible.

A few centimeters of plastic or water or a few millimeters of aluminum or other metal are adequate to stop completely most betas, and frequently it is necessary to provide only localized beta shielding; this may be the case, for example, when a normally enclosed piece of equipment containing beta emitters is opened for maintenance, inspection, or other purposes.

The criterion of gamma shielding adequacy for a given situation is its reduction of incident radiation to an acceptable limit. The shield thickness required depends on the gamma energies and intensities involved, the gross density of the shield material, and, to a less extent, the material itself. Where visibility through a shield is necessary, special dense glass is available. Where the source may change with time, as may be the case in an industrial operation with differing output schedules, the radiation levels may correspondingly vary. Under these conditions the level at a specific location may be calibrated to indicate the corresponding levels at other locations and thus indicate any necessary changes in shielding or other source control. In addition to direct shielding of the source, it is also necessary to take account of any secondary radiation production, such as that causing bremsstrahlung or resulting in skyshine. Any shielding that is adequate for gammas is normally also adequate for the associated betas.

Neutron shielding is generally necessary only for protection near reactors or specific neutron sources. The effectiveness of a neutron shield depends on the nuclear properties of the materials used. Most substances, including those that are effective in gamma shielding, are generally rather poor neutron shields, even though they may slightly attenuate neutron beams. Since neutron shielding depends on the incident neutrons being slowed down and stopped (or captured), materials which have good neutron shielding properties are those with relatively light nuclei but with high capture and scatter cross sections, particularly if such characteristics apply to a wide range of incident neutron energies.

As indicated previously, most materials have significant cross sections for low energy and thermal neutrons. Thus, since most neutrons originate at comparatively high energies, those of fission averaging about 2 MeV, they need to be slowed down or moderated to facilitate their capture by most materials. Among common elements cadmium and boron are considered to be among the best shields for low energy and thermal neutrons since they have particularly high capture cross sections at these energies and are thus among the more widely used nuclear "poisons." Hydrogen is unexcelled as a moderator, and its capture (or poisoning) cross section is also significant, particularly for low-energy neutrons. Most materials have some neutron-capture capabilities, and some of the rare earth materials—such as vanadium, europium, and gadolinium—have high capture cross sections, particularly for thermal neutrons. Other elements which have good neutron moderation or slowing-down properties are carbon, deuterium, and perhaps lithium. These materials do not have scattering cross sections as large as hydrogen, they are not as efficient in neutron energy reduction per collision, and their capture cross sections are so small as to be negligible.

Hydrogen-rich materials, such as water and paraffin, are considered to be among the more effective neutron shields, particularly if combined with compounds of boron or cadmium. In general some 8 to 12 in. of water or paraffin will moderate and capture essentially all incident neutrons of fission energies, and some 1 to 2 mm of cadmium will capture all thermal neutrons. Iron may also be used to moderate high-energy neutrons to an energy range in which hydrogen becomes effective in moderation and capture.

In practical situations involving neutrons, gamma shielding must also be provided both for the gammas associated with neutron emission and for those produced as a result of radiative neutron capture. Concrete is a good general purpose shield if sufficient space is available. It is fairly massive, cheap, and also contains elements that are useful in neutron shielding. The use of certain additives can increase its effectiveness. Another general purpose shield is formed by a combination of a beta-gamma shield of lead or other dense material and a neutron shield incorporating water or paraffin with possible neutron poisons such as cadmium. In practice this shield can take the form of a general mixture of materials or it may use successive sheets or thicknesses of the two types of material, with a gamma shield being preferably farthest from the source.

It is frequently desirable to provide only partial shielding between a source and the work area, this assuming that people will remain in the shield shadow and not be in the area that is unshielded—or that the shielding will not be "lost" while people are adjacent to a significant source. Such possibilities will be determined administratively and technically. The possibility of skyshine or other reflection when partial shields are used has been mentioned, as has been the possibility that neutron and even gamma reactions with various materials may produce additional radiation which should also be considered in shielding determinations.

EXPOSURE TIME LIMITATIONS

Where it is not feasible to shield a gamma source so that permissible limits will not be exceeded in work locations with indefinite exposure time, a basic control premise is a time limitation. Such limitations usually require rather close measurements, especially if the radiation field involved is such that permissible dose limits will be exceeded in a comparatively short time. It is sometimes necessary to have each of a group of individuals do a certain part of an operation, with the amount of work done dependent on the exposure time permitted. In fact individuals have been scheduled to receive a quarterly or annual permissible dose during a single job, with their subsequent work assignments made to permit no exposure for the remainder of the period involved. For all such work the respective individuals are also provided with personnel monitoring devices such as film meters or dosimeters to show the actual exposure received. These may also indicate the validity of the safety factors used in specifying the time limitation as well as the need for possible future changes in these time limits.

It is obviously feasible to use exposure times and a total anticipated dose to determine a maximum permissible radiation field. This could well affect the amount of gamma shielding required, which in turn may have considerable economic significance; for example, if it is decided to make 5 rads the total exposure for a year of fifty 40-hr weeks, it is obvious that in any locations in which the radiation level can be no greater than 2.5 mrad/hr or 100 mrad/week no shielding or other protection is probably needed, and stringent monitoring of personnel or work activities may not be required. On the other hand, there may be little reason why the entire 5 rads may not be received in a single hour, provided there is no additional exposure the rest of the year, and the radiation field for this work would thus be 5 rads/hr. Ordinarily, however, a dose limit of 3 rads, which is the permissible limit for a quarter, would be imposed for a single exposure, and a limit of 12 rads may also not be impossible. Other time-field figures may similarly be developed, such as 300 mrad/hr for a 10-hr exposure to meet a quarterly limit of 3 rads. For other exposure limits, such as those to extremities, the permissible values would be correspondingly different. It may sometimes be preferable to use a given shield with a comparatively short time limit rather than to use a thicker shield and extend the time. The use of small shields for localized sources has been noted.

CONFINEMENT OF MATERIAL

Sealed Sources

What may be identified as sealed sources range all the way from bits of radioactive metal, such as cobalt that has been irradiated in a reactor, to radioactive powders such as radium compounds that are placed in sealed containers.

For a sealed metal source it is usually necessary only to know of its actual location and that it is adequately shielded, as determined with an appropriate meter. If these sources are normally kept in a shielded enclosure except when used for calibration, irradiation, or other purposes, it is obviously necessary to be sure that the source has been returned to its shielded container before anyone approaches it. This is particularly true for the very large cobalt-60 gamma sources used for various irradiation purposes. Normally there need be comparatively little concern of the spread of radioactive material as a result of a possible failure of the container integrity, since such failure will result in only a comparatively slight transfer of activity to a surface with which the radioisotope comes in contact.

Various accelerators and X-ray machines may also be generally treated as sealed sources of this type since there is usually no radiation problem when they are not in operation, certainly by comparison with that encountered during operation. Precautions are necessary to see that the machine is truly not in operation when this is supposedly the case, in addition to those taken to ensure that no one is inadvertently exposed during normal operation.

If the sealed source is in powder or other finely divided form, the result of container failure becomes much more serious since such failure could result in a wide and possibly dangerous dispersal of the material. Obviously, the degree of hazard is associated with the seriousness of the container leak. If the container breaks and quickly releases all of its contents, the radioactive material may be quickly spread throughout a wide area. Although such dispersal may be easily detected by various radiation monitoring devices, there may be a significant degree of personnel hazard, which will depend on the type of radioactive material involved, its form, and its amount. Similarly, if it is widely dispersed, a significant contamination problem of the type discussed in the next section may result.

Massive releases as described above are not generally anticipated today because of the use of metal capsules, although in the early days when radium was used for therapy and similar activities, the radium "needles" employed were usually first encapsulated in glass and possible breakage was a real concern. However, the possibility of such releases cannot be completely discounted, and precautions should be regularly taken to check the condition of sources, to verify the presence of the radioactive material inside the capsule when it is removed from storage or returned thereto, and to determine if any radioactive material is spread about the source location.

Slow leaks, which may also present a problem, are of two types. One is a slow release of the radioactive material itself, with eventual results similar to those described above. In this case the material released in even a comparatively short time will probably quickly produce environmental radiation levels that are much above permissible limits. The other type of release is that of daughter products. This is exemplified by such materials as radium, which has a gaseous radioactive daughter, radon, that may itself escape and radioactively contaminate the environment. However, because of radon's short half-life, it may be comparatively simply determined if the contamination of the capsule environment is caused by the radium itself or is only a result of the radon. For this determination, the capsule is merely moved from its usual location for a few hours. If the radiation level drops rapidly to negligible levels, the escaped material is probably radon; but if this is not the case, the radium itself will be suspect.

Any leaking radioactive sources should be placed in appropriate sealed containers and returned to the manufacturer or other agency equipped for handling them; it is generally very inadvisable to attempt a source repair.

Contamination Control

Radioactive contamination may be identified as radioactive material in locations where it is undesirable. Contamination problems thus frequently involve radioactive materials escaping from closed containers into the air or into waste or other systems in which their presence is unexpected and people may or may not be directly involved. Moreover, such material may settle on work surfaces or be transferred throughout the plant or even outside it. Although it is not the purpose of this presentation to provide an engineering guide to confinement of radioactive materials, a few of the methods that may be used are indicated. The best general contamination control obviously is the elimination of the material escape routes. This means tight joints, seals, perhaps special covers for valves, and so on. It must be remembered again that most radioisotopes have such large specific activities that only small masses or quantities are necessary to produce significant radiation levels. As a result, normal criteria for confinement as developed for nonradioactive contaminants may very well be completely inadequate for radioisotopes.

As a first step in contamination control, equipment should be made as generally leakproof as feasible. Next, if feasible, the confinement cell or the room in which the radioisotope is kept should be maintained at a lower pressure than outside locations since this will tend to prevent air-borne escape of the radioisotopes in the event of a leak. Such a reduced pressure, especially for large volumes, is usually provided continuously by exhaust fans that have been backed up by filters. In such cases, however, the possibility of filter leakage and deposit of material outside the confined area may be of concern and possibly require regular monitoring. Where the confinement area is a work location itself, this filtered exhaust system may very well remove some material from the air before it can build up to a significant air concentration or be deposited on the floor or other work surfaces.

For handling low activity materials or those with low penetrability, such as alpha emitters and especially in laboratory-type operations, "dry boxes" are frequently used. These are sealed boxes into which rubber gloves are fitted so the material may be handled and still be well confined. Where entry into equipment or into a confinement location is necessary, special jigs and fixtures, possibly including specially designed dry boxes, may be required to provide adequate localized containment.

Since it may be necessary for health or economic reasons to prevent radioactive material from escaping into sanitary sewers or other drains as a result of spills or accidental releases, such drains should be designed to lead to confinement containers from which the radioactive material may later be obtained. Because of the possibility of inadvertent criticality, such treatment of fissionable materials may present real problems.

If small quantities of radioisotopes are released to the atmosphere, and this may even be possible despite the use of filters, tall stacks are usually employed to distribute the material so widely that its concentration on any surface, including particularly areas outside a controlled location, is negligible. This is discussed more fully in Chapter 17.

In the field of contamination control by confinement there are almost limitless possibilities for proper specialized actions and methods. It is sometimes not feasible to prevent some escape of contamination from containers, and this usually results in air activity, surface contamination of the facility, or both. It is generally necessary to remove this contaminating material from the surfaces, and, since the quantities necessary to produce even relatively high radiation levels are sometimes extremely small, they can become embedded on many surfaces to the point at which normal cleansing will not readily remove them.

DECONTAMINATION

The basic methods for removal of contamination are rather obvious, but
their overall efficiency in a given situation may not always be adequate.
Removal can be either mechanical, as by scrubbing; or chemical, as by
dissolving or by formation of a soluble removable compound; or by a
combination of both. Simple vacuuming is the first choice for dusts, and
various cleaning agents, primarily liquid, are widely employed with due
consideration for the chemical properties of the radioactive isotopes being
removed. In fact, certain chemicals are almost specific for removing given
radioisotopes. Water is a relatively good overall decontaminant for many
materials, various acids are good for many others, appropriate caustics
may be used, etc. Solvents that have been fairly widely used for decontamination include the following:

1. Water.
2. Alkaline solutions: trisodium phosphate, sodium carbonate, and
sodium bicarbonate.
3. Acids: hydrochloric, nitric, chromic, and other.
4. Various commercial detergents, including Terpitol.
5. Various wetting agents.
6. Chelating agents: citric acid, ammonium citrate.
7. Organic solvents: kerosene, trichloroethylene, carbon tetrachloride,
and various Freons.
8. Hydrogen peroxide.

DEPENDENCE OF CONTROL METHODS ON LEVEL OF
POSSIBLE EXPOSURE

Although the basic principles of radiation source control as given above
are always applicable, adequate protection in various situations may
require widely different orders of control effort and efficiency; for
example, shielding for beta-active materials or for gamma radiation where
the maximum radiation level is in the subrad-per-hour range obviously
requires significantly different attention than shielding for reactor wastes
or fission products where the radiation level is in the hundreds of rads
per hour. This emphasizes again that control efficiency, both technical
and administrative, should be much higher where a possible radiation
exposure resulting from a failure of control can cause serious injury in
a short time than should be those necessary where the worst result of
such failure may be only a possible minor overexposure as based on some
permissible long-term limit.

The confinement requirements for a given material depend on its specific activity and the relative hazard, as shown by permissible limits, of the emitted radiation. Thus the permissible limit of plutonium is approximately 10^{-6} that of natural uranium, with the result that 1 μg of plutonium will create about the same air and surface contamination problem as will 1 g of natural uranium. This overall factor is the result of the specific activity of the plutonium being some 10^5 times that of the uranium and the internal deposit problems being such that the alpha radiation emitted by plutonium is considered to be some 10 times as hazardous as that emitted by uranium. Similarly, enrichment of uranium in the $^{235}_{92}$U and the $^{234}_{92}$U isotopes, with a resulting specific activity some 100 times as great as that of normal uranium, requires greater confinement control than that of the natural uranium itself.

The so-called "hot cell" frequently used for work on highly active materials normally provides very significant shielding and a high level of material confinement. Since such operations must frequently be viewed, it has sometimes been found necessary to use a periscope or other reflector to maintain the necessary shielding. Materials inside the cell are handled with mechanical hands, and material transfers within the cell, between two such cells, and between the cell and the outside are made by ingenious methods designed to limit potential exposure; as one example, toy trains have been pressed into such service—to the probable pleasure of their users.

The possibility of accidents should also be considered. Thus a nuclear reaction, which is always accompanied by dangerously high radiation fields, may occur at a place in which the normal radiation level is negligible. Similarly, a break or other failure in process equipment may result in high air and other contamination in locations in which such activities are normally negligible.

OTHER CONSIDERATIONS

The general principles of protection outlined above are also applicable to special types of radiation problems such as those involving accelerators or other specialized equipment; for example, protons are amenable to essentially the same types of control as those used for alpha particles, and certainly any controls that are established for beta particles are entirely adequate for positrons. Reactors present a major problem in that the whole gamut of radiations—including neutrons and gamma rays of a wide range of energies as well as protons, alphas, betas, positrons, and even various other high energy particles—is emitted. Protection against all of these is required.

A high degree of ingenuity is frequently necessary to fit considerations of confinement, shielding, and exposure time into the normal type of chemical processing operations, as well as those of the laboratory and the medical facility. This ingenuity applies not only to radiation protection itself but also to the fact that other hazards—such as fire, chemical release, and so on—must concurrently be considered; for example, a special radiation protection criterion which leaves a system susceptible to fire cause or spread may actually defeat its purpose in the event of fire, which could destroy both the shielding and the confinement on which radiation safety may be based.

EQUIPMENT FOR PROTECTING PERSONNEL

In locations or operations in which it is not feasible to shield or confine materials adequately and still continue to operate efficiently or at all, adequate localized protective devices may be provided for the individuals involved. Leather gloves and lead-lined aprons or gloves and similar devices provide a measure of protection from beta- and comparatively low energy gamma- or X-radiation for the part of the body shielded. Any gloves or other clothing can prevent beta-gamma emitters from being directly on an individual's body, and removal of this clothing will prevent these sources from staying near the body for an extended period of time. This is particularly valuable for reducing potential exposure from short-term operations. Small surgical caps can prevent radioactive material from getting into an individual's hair.

Efforts to prevent radioactive materials from getting into an individual's body can range all the way from simple arrangements to monitor clothing and remove dust therefrom to rather elaborate plastic "balloon" coverings that completely enclose an individual. Efforts to prevent body contamination have also included complete taping-up jobs wherein the bottoms of a person's trousers or coveralls are taped to his shoes, gloves are taped to his sleeves, and so on. Although there may be cases in which such vigorous protection is advisable, the potential exposure control provided thereby may be so negligible in comparison with other factors that justification may be difficult. However, it is recognized that this is a matter for local consideration and decision.

If it is desirable to prevent radioactive materials from coming into contact with the skin, a different type of cover or clothing will be used when the material concerned is a dust than will be the case when it is a liquid or solution.

Routine removal of contaminated clothing by an individual leaving the contaminating source may also be of some very limited value in preventing subsequent internal exposure from inhalation or ingestion, although it may help prevent the spread of the radioactive contaminant to other plant areas or outside a plant. Such assistance is the only value of clothing in protection from alpha emitters.

Since alpha emitters are a problem only if they enter an individual's body, respiratory protection is the major personnel action to be taken against a potential problem therefrom and is similarly useful in preventing inhalation of beta-gamma emitters. It has already been pointed out that any methods or devices that provide adequate protection against internal exposure from alpha emitters are also appropriate for beta-gamma emitters.

The specific respiratory protective equipment used in a particular situation will depend on the chemical and physical form of the material involved. Thus a dust filter may be adequate for various radioactive dusts, but it may be entirely inadequate for radioactive material in the form of a gas. Similarly a chemical cartridge filter may be adequate for certain radioisotopes if the chemical used is appropriate to the particular compound involved, but a chemical cartridge may be completely useless against dust. Properly fitted self-contained air masks or oxygen-supplied masks generally provide a very high degree of protection.

An individual may prevent ingestion of radioactive materials by such simple means as some control of his eating, drinking, and smoking habits; fingernail biters or those with other nervous habits may present special problems. The obvious conclusion has already been stated that any wounds should be well bandaged before work in contaminated locations.

A requirement that employees who go between contaminated and uncontaminated locations change clothing and possibly take showers in change houses or rooms provides a rather elaborate method of preventing radioactive materials from being carried from one area to another. As a matter of fact, such arrangements are probably more important for contamination control than for health protection.

It is sometimes observed that rather extreme precautions taken in the name of radiation protection may actually partially obviate their purpose; for example, too extensive shielding and clumsy clothing may so impede the working capabilities of an individual that the increased time he needs to do his job may more than offset the reduction in radiation exposure rate, with the result that the actual exposure on the given job is increased. This may be especially true of work that involves a short term activity in a comparatively high field; for example, if an individual can do a job in 30 min in a 1 rad/hr field and if shielding reduces the field to 0.5 rad/hr but the same job takes 90 min, the total exposure will be greater in the "protected" case than otherwise.

TYPES OF ENGINEERING DESIGN CONSIDERATIONS

In establishing a radiation protection program for an operating facility it is important that as much radiation safety as possible be designed into the facility and its operation at the earliest feasible time in its development. This usually means beginning while a project, a facility alteration, or even a major operation is still in its initial planning stage, even at the time of the engineer's "gleam in the eye". It is during these initial considerations that those aspects of radiation safety that are most readily amenable to engineering control may be best considered and incorporated into the design.

Of principle concern to overall design are the types and amounts of radioactive materials or radiation sources for which the facility or operation is planned, the actual radiation levels or activities anticipated in various parts of the operation under normal conditions, and the possibilities and results of emergencies or other abnormal operations.

Among specific considerations given would be the necessity for shielding, ventilation, automatic controls, and special equipment such as dry boxes. Also reviewed would be such items as (a) the effect of operating efficiencies other than those assumed, (b) the possibility that remotely controlled operation will be required, and (c) the need for special methods or materials to ensure maintenance of the integrity of the various systems to an even greater extent than might be normally anticipated.

An important concern will be the possible problems involved in routine maintenance. These may include the need for special equipment or shielding, the handling of contaminated tools and fixtures, the storage or disposal of contaminated scrap and waste, and even an evaluation of possible time limits and manpower requirements for work in a given facility or on a probable job.

Planning to meet abnormal operations will obviously include the possibility of more material being held up in certain parts of equipment than has been specified for normal operation, the breaking of equipment lines, loss of container integrity, shielding loss, equipment and power failure, etc. Detailed plans should also be made to handle possible maintenance after abnormal occurrences or accidents; this will include the provision of devices commensurate with the potential hazard foreseen and the establishment of adequate control procedures. Insofar as possible, the principal controls of utilities and operations should be at positions which would most probably not be involved in a release of radioactive materials or other failure of radiation source control.

The fissionable uranium-235 and plutonium-239 both emit alphas with which no significant penetrating radiation is associated. However, in the event of an accidental fission reaction the beta-gamma radiation would become extremely important. Thus the possible high level radiation that would result from a nuclear accident in equipment or facilities in which the radiation levels are normally extremely low will require evaluation.

REVIEW DURING CONSTRUCTION

After design comes construction, although a pilot plant may be first operated to provide appropriate data for the final design and construction; this is especially true if major technical information is needed. Pilot-plant problems in general are similar to those of operation, though usually on a smaller scale, and experimental changes may be anticipated; the health-physics concerns will be essentially the same.

Periodic studies during construction should be made to ensure that any special health-physics consideration is not omitted, that apparently minor on-the-spot changes will not affect evaluations already made, and that some factors of more or less concern from a radiation protection standpoint may not have been inadvertently missed during the design review. For example, a change in the materials used for a wall separating two cells or items of equipment may obviate the shielding supposedly provided, or a room which is to provide containment of radioisotopes in case of a leak may have its air tightness violated by normal plumbing or piping lines passing through the walls; this latter condition might not have been readily apparent at the time of design review. During these reviews hitherto unnoted items that would make maintenance or operation difficult or could cause a greater exposure of personnel than would otherwise be necessary may be detected and readily corrected with only slight changes. Similar correction might be much more costly, inconvenient, and difficult after construction is completed.

INITIAL STARTUP

Beginning with initial startup operation, a radiation monitoring and evaluation program should be established. Startup review should be rather detailed because it is at this time that the adequacy of containment provisions can be tested, the distribution of radioactive materials can be measured, or the extent to which actual operations necessitate changes in health-physics preparations can be evaluated; for example, if a chemical reduction efficiency of 80 percent has been assumed in shielding evaluations for a certain process, the discovery that the actual operating efficiency is 99 percent with a resulting decrease in the self-shielding provided by the nonradioactive material may indicate the necessity for additional shielding. Conversely, an observation that an initial estimate of radiation hazard was overly pessimistic might permit an operational change leading to increased efficiency. Similarly the problems inherent in various maintenance operations, both those that are foreseen and those later found to be necessary, may be evaluated to establish definitely the adequacy of their radiation safety. In particular, the times required for certain types of maintenance operations and the radiation exposures that might be expected therefrom may be evaluated.

All told, initial operation is a rather important period to a radiation control program, requiring close attention to see that the operation is in accord with design, that the assumptions made (or changed with experience) are valid, that under normal operations it is relatively simple to ensure that radiation safety is maintained, and that adequate preparation is made for abnormal operations. After this shakedown operation, with the plant being as radiation safe as can reasonably be expected, routine health-physics activities can be established. As will be discussed in succeeding chapters, the actual procedures adopted will depend on such nontechnical factors as the plant organizational structure, insurance requirements, etc.

ROUTINE RADIATION MONITORING

A significant part of any routine radiation protection program is a system of radiation surveys, the frequency and thoroughness of which obviously depends both on the potential hazard and the degree of personnel involvement. These, in turn, depend on the radiation problems of the respective operations as affected by the various factors mentioned previously. Thus the frequency of any survey will depend on such factors as the operation itself, the type and level of radiation encountered, and the rate at which changes could unknowingly develop; for example, if the radiation problem is an encapsulated beta-gamma emitter in a solid form that remains in a relatively stationary position and cannot move as a result of normal operation or even a misoperation, then it may be possible to place appropriate warning signs and to monitor the location only occasionally to see that its radiation problems have not changed. This is especially true where personnel who may have occasion to work in or near this location use personnel monitoring devices. On the other hand, a much more frequent monitoring schedule may be necessary for locations, operations, or facilities in which there can be sudden and perhaps unexpected changes; even the use of a beta-gamma source in powder, rather than solid, form may require additional consideration.

It is generally assumed that in locations in which people regularly work, changes that can seriously endanger anyone will not occur without this becoming known either through alarms, automatic or otherwise, or by other methods. Thus a continuous air monitor with alarm may readily indicate a leak or a break in process piping long before a hazard could occur, or appropriate alarms could indicate unexpectedly high radiation levels resulting from shielding failure, criticality, etc. These alarms could either give notification of misoperations, which can be taken care of in due time, or they may be danger signals with the urgency depending on the process and the potential hazard indicated. If dependence is placed on alarms, surveys will obviously include evaluation of the continuing integrity of the alarm systems. Alarms and indicators could very well also be advantageous from an operational standpoint, and conversely, operational alarms and indicators may provide independent data of value to radiation protection activities.

Since there may be possibilities of radiation or contamination occurring in locations in which it is generally unexpected, it is usually desirable on occasion to monitor so-called cold areas, such as cafeterias or office space, where in the normal course of events radiation is not expected. Such monitoring is usually required on occasion regardless of the basic evaluation methods employed.

EFFECT OF ADMINISTRATION ON MONITORING

Monitoring criteria, including determination of an appropriate survey frequency, are usually inextricably involved in administrative decisions. Thus they are affected not only by changes in the design, construction, or operation of a facility but also by the way in which these are authorized and instigated. Also pertinent is the corresponding availability of significant data on such changes, particularly those of radiation protection concern. Similarly, employee knowledge of the radiation protection aspects of his job, personnel job assignments, use of personnel protective equipment, and other general administrative actions may affect monitoring criteria on a continuing basis.

The use of air monitors, high level radiation monitors, bioassay results, medical findings, and the availability of results from related types of review (such as safety or fire protection audits) can also affect a radiation monitoring program even as do the survey methods used and the way by which information obtained is reviewed to indicate the need for action and the required frequency of monitoring. This is discussed in more detail in Chapter 19, as are types of monitoring organizations.

Special monitoring considerations generally apply to locations in which changes are essentially a routine part of operations. These are typified by various research and development laboratory activities. In these activities, especially if they may involve significant quantities of radioactive materials or if high levels of radiation are possible, it is probably important that rather frequent audits be made and that the operating personnel be particularly well versed in the effect of changes on radiation exposure probabilities.

Special monitoring is also required for shipping, primarily because of the regulatory requirements of the Department of Transportation and the IAEA or foreign nations to which shipments are made. Since such requirements usually require that all items be checked, the monitoring usually involves the adequacy of the routine checking being provided, even though a special staff group itself provides the routine checks.

SPECIAL PROBLEMS

In addition to routine operations as noted, the health-physics organization should also be prepared to handle certain special problems, including emergencies. These may be of special plant interest or they may have widespread applicability. They may involve particular facilities or activities for long term operation, or they may involve only short term operations. In some cases decisions concerning the possibility and extent of exposure in a particular type of operation may play an important part in determining its feasibility.

Special problems frequently pose difficulties that are not clearly foreseen and for which technical skill and know-how are required for successful completion; for example, information available on a given project may have been obtained from actual experimental work but with such a small amount of material that the true hazard involved in handling the amounts of material and resulting radiation in the actual plant operations presents an entirely different and much more serious problem that may not be initially appreciated. There are many ways in which such a problem can be solved, depending on the actual conditions, but in all cases it is assumed that any solutions will be on the safe side despite the various emphases which may be involved.

CONCLUSIONS

The above indicates the types of factors that should be taken into consideration in devising adequate protection for individuals with the actual methods to be used under a given set of conditions depending on the activities concerned. It cannot be emphasized too highly that determination of the most effective methods of protection should depend basically on experience, including engineering studies, pilot plant operations, and routine surveys, with subsequent analyses of any problems thus indicated.

REFERENCES and SUGGESTED READINGS

Various of the National Bureau of Standards Handbooks, particularly Nos. 23, 27, 42, 48, 50, 51, 60, 63, 66, 73, and 76. (See Appendix I for complete list.)

Glasser, Otto, Edith H. Quimby, Lauriston S. Taylor, J. L. Weatherwax, and Russell H. Morgan, *Physical Foundations of Radiology,* 3rd ed., Harper and Row, New York, 1961.

Braestrup, Carl B., and Harold O. Wyckoff, *Radiation Protection,* Thomas, Springfield, Ill., 1958.

Blatz, Hanson, editor, *Radiation Hygiene Handbook,* McGraw-Hill, New York, 1959.

Various publications by the U.S. Atomic Energy Commission, the International Atomic Energy Authority, the United Kingdom Atomic Energy Authority, and those prepared by some industrial establishments for their own use can give helpful information.

Jefferson, Sidney, *Massive Radiation Techniques,* Newnes, London, 1964, (distributed by Wiley).

RCA Service Company, *Atomic Radiation,* 1958; Part II, 1960.

Safe Handling of Radioisotopes, IAEA, 1962.

"Health Physics Addendum", *Safe Handling of Radioisotopes,* IAEA, 1960.

"Medical Addendum", *Safe Handling of Radioisotopes,* IAEA, 1960.

Basic Safety Standards for Radiation Protection, IAEA, 1962.

QUESTIONS

1. What type of radiation must be guarded against most stringently? Why?
2. Protons are similar to what other basic type of particle from a protection standpoint?
3. What function can clothing have in protecting an individual?
4. What function can a pilot plant serve toward plant design?
5. What devices can be used to protect personnel directly? Is there a possibility that these defeat their purposes?
6. Why must a competent health physicist be involved in the building of a new facility from its very inception?
7. Why are routine radiation surveys so vital?
8. Comment on the uses and limitations of personnel protective equipment.
9. Enumerate some of the engineering considerations in radiation protection.
10. In what general ways can exposure be controlled?

CHAPTER SEVENTEEN

Offsite Monitoring and Surveys

NEED FOR OFFSITE MONITORING

Although the principal problems of radiation control occur where radiation sources or radioactive materials are used, the possibility that radioactive materials may be released into the surrounding environment must also be considered—even though to date there has been little evidence that any long term problems have resulted either from routine releases or from the few comparatively serious emergencies which have released radioactive materials into offsite locations. In particular, there have been few cases of external or internal exposures above permissible limits for the general population, even though such limits are only 10 percent of the corresponding values for in-plant use, and no recognized injuries or long term damage have resulted from such events. However, despite these limited health hazards, adequate information concerning radiation conditions in offsite locations becomes almost an imperative for a facility known to handle radioactive materials, such activities being peculiarly sensitive to legal and public relations considerations as represented by possible allegations of injury and damage. Almost any undesirable condition in the areas surrounding such a plant is all too frequently considered to be a result of the plant operations and thus subject to suit or compensation, even though the same thing may have happened many times before the plant even came to its present location. Irresponsible "scare" statements concerning possible results of radioactive material releases or other incidents obviously accentuate the possibility of such allegations. In any event, probably the most effective defense has been the availability of appropriate monitoring data.

WASTE DISPOSAL

Although it is assumed that the plant design itself takes into account the need for limiting and controlling the release of activity to the environment, both routinely and as a result of a credible accident, a major problem for many facilities is the disposal of radioactive wastes. The sheer volume of many wastes and their increase with time, their frequently extremely high activities, and the fact that they may often involve off-plant individuals and locations can make such disposal a significant problem for certain types of facilities, particularly reactor operations.

It is beyond the scope of this book to discuss waste disposal in any detail. Many of the bibliographical references treat the problem at some length, and its various aspects are strongly subject to governmental regulation. In general, specific criteria for release to the environment are not always clearly specified, but it is ordinarily considered that such release is permissible if it does not increase the activities of the air or water above their respective breathing and ingestion limits at locations to which access by the general public is possible. This normally does not include locations within a plant boundary, and such sites may actually be far from this boundary.

A plant is considered to lose control over released materials at the point they leave the plant boundary even though actual physical control may be lost earlier; for example, a plant can exercise little, if any, control over radioactive gases that have left a stack, but technical "control" may itself be considered to include the recognized dilution of the air between the stack and the plant boundary. Similarly, although actual control over a liquid release may be lost at the point the liquid emerges from a pipe into a stream that subsequently leaves the plant boundary, potential dilution in this stream may usually also be considered to be part of the technical control. Obviously further air or water dilution may take place between the plant boundary and public access locations, and the effect of reduced flow because of drought may require evaluation and control.

If there is a possibility of air contamination by plant effluents, the meteorological conditions of the location and the proposed plant operations should be carefully evaluated. As a major method of control a stack appropriate to the hazard may need to be provided for their release. Filters, scrubbers, and other methods of removing residual air activity may be required in the line leading to such a stack, and it should be realized that a break in such a filter or a failure of the scrubber may lead to the release of an unexpected amount of material.

Similarly, efforts should be made to limit the release of liquids that do or may contain radioisotopes, particularly those with long half-lives. The possibility that lines used to transfer radioisotopes may leak and thus release materials to the environment, perhaps through a waste system that is normally used for nonradioactive materials, should be frequently checked. Dilute solutions of radioisotopes can be evaporated to reduce the volume of solution to a value which may be readily held, and hold tanks or other vessels may be used to permit the radioisotopes to decay to acceptably low activities before final release. This last is particularly useful for radioisotopes with short half-lives.

The general methods used for disposal of liquids usually fall into one or more of the following categories, with due acceptance of the importance of the activities of the materials concerned, their radioactive half-lives, and the rigidity of control necessary to limit the possible involvement of the general population:

1. Release to sanitary sewers or storm drains only for small quantities meeting release specifications.

2. Release in natural craters or earth fissures. The problems of contaminating underground water or the eventual appearance of the radioactive material in streams or other locations at distances remote from the disposal site will always require geological evaluation (currently not done).

3. Storage, either directly or in containers, in relatively impermeable underground caves or similar locations; abandoned salt mines have been suggested for this purpose.

4. Disposal in the ocean, usually in tightly sealed containers.

5. Chemical reactions, including precipitation, that produce relatively insoluble solid compounds or other bonds with the radioisotope. The resulting solids may then be appropriately stored above or below ground. Ceramics have been especially proposed for this purpose, as has been metal smelting, and resulting materials have been buried in opened trenches.

The release of radioisotopes into streams has caused some concern over their possible concentration in the stream bottom and in the vegetation or the animal life, such as algae or fishes. A particular problem in release to streams has been the fact that the natural radioactivity of some of the western streams is normally near or even above the permissible limit.

The health-physics problems of direct interest are obvious. These may involve in-plant or off-site situations, including those of different states or even those of international concern. Thus knowledge of radiation levels and the amount of activity, in curies, that has been released should be obtained at the point at which plant control is lost. This may be obtained by continuous monitoring or by continual sampling and eventual analysis of the air and water released at the last controlled location, spot sampling as necessary, or other appropriate methods. These activities should be accompanied by actual environmental surveys, both routine and spot, made at appropriate points perhaps distant from the plant site.

IMPORTANCE OF CHANGE IN ENVIRONMENTAL RADIATION CONDITIONS

Almost independent of any evaluation of actual material releases, however, is a determination of any overall effect on the environment, especially on a long term basis. This obviously points to the importance of information concerning any change in the radioactivity of the environment as a function of the plant and its existence, whether such change results from routine releases or from an accident uncontrollably releasing large quantities of radioactive materials. Of significant importance in such evaluations is a preoperational or even preconstruction evaluation of the background radiation at the plant site, including that of the air, water, soil, vegetation, and even possibly animal life.

BACKGROUND DATA

Background radiation data for watercourses should be taken from a mile or so upstream of the operating facility to a distance of several miles downstream. Particular attention should be given to a watercourse that drains the plant area and flows into a river or other source of potable water for downstream towns. Due concern should be given to the flow pattern of the water in a river. Samples should be taken at more than one position in the river at the general positions selected, especially if they are at any significant distance from the plant, and they should be obtained at various times during the year, especially if the water level changes significantly as a result, for example, of a spring runoff or late-summer dry spell. Not to be neglected are similar background measurements of any tributaries of the main stream, particularly those into which wastes may be released, either planned or not. These evaluations should include normal waste, such as storm drainage, with the sampling locations being very carefully identified. In all cases sampling should include both the stream bottom and the water itself.

Some evaluation of both normal and abnormal radioactivity in the air should also be obtained, it being recognized that there is always a variable alpha-active background due to radon and thoron, which can be eliminated by time delay in counting. Routine air sampling techniques as described previously for in-plant monitoring are adequate. The samples should be obtained at different altitudes up to about 6 ft, or even higher in some cases, and under various conditions of wind velocity, direction, and other meteorological conditions insofar as possible.

Similar background radiation information should be obtained from soil measurements and from analyses of various native grasses and trees of the region. It is recognized that the radiation levels which must be distinguished in these environmental surveys are usually so low as to require careful and tedious microanalytical techniques. At the least, the data obtained should indicate a level of radiation above which activity is not observed. It should be a cause of no surprise that marked differences in radioactivity at various locations and in differing types of soil and vegetation will be obtained.

Such geological considerations as those of water movement through rock strata should be considered since this factor can have an important bearing on waste disposal and on the location of such items as burial pits for contaminated equipment or other items. One of the difficulties of monitoring is the fact that sometimes ground water will travel rather long and tortuous courses. Thus soluble radioisotopes put into the ground at one point can emerge at distances very remote from the holding point or other source itself. It has sometimes been found necessary as a result of this type of problem to seal the burial pits or locations planned for disposal or even to change the locations themselves.

ROUTINE SAMPLING

Based on all of the information obtained in preoperational and perhaps operational evaluations, certain locations may be selected as eventual places for sampling during routine operations. Once these environmental monitoring locations have been selected a regular schedule for a continuing program should be established. The extent of the program may actually be dictated by the regulations of governmental units as well as by possible health problems. The schedule selected, as well as the emphasis placed, will depend on the items to be monitored, anticipated results of routine activities, and the estimated results of more serious occurrences. If they are feasible and warranted by the importance of the problem, continuous monitoring systems for air and water should be established.

Where there are possibilities of major releases or air contamination, a wide-spread grid of detectors placed at different elevations as well as over different geographical locations and terrains has been proposed; results of the continuous monitoring are telemetered to a central location and recorded.

PROBLEMS OF MONITORING

In environmental monitoring it is particularly important to realize the inherent inaccuracy of any single measurement for indicating an overall condition. Thus it is almost invariably necessary to treat offsite environmental monitoring results by statistical means, with error limits if possible, because of the inherent variability of the analytical system and the sampling techniques themselves; for example, the content of a given radioisotope in the leaves of a botanical specimen will usually differ markedly from that in the stem, flower, or other component. Similarly the age of a particular type of growth could have an effect; new leaves on a tree may have a significantly different content of a radioisotope than do older ones. Significant differences in the indicated radioactivity of different soils, and even of different samples of one soil, should be anticipated. If plans are made to sample animals or other fauna, differences in the potential deposit sites of various radioisotopes should be recognized. With the possible exception of fishes and other water or mud dwellers, animal monitoring is not often used because of the formidable difficulties not only in sampling and analysis but also in the interpretation of results.

It goes without saying that any change in environmental contamination should be carefully and closely checked to determine if a possible unexpected problem noted is the result of plant activities. Such checks are generally of importance not only from a health-physics viewpoint but also, and of perhaps greater importance in some cases, for operational purposes. In fact, as in the plant itself, the first intimation that all is not well in a process operation has often been the discovery of radioactive materials leaking into the environment.

EMERGENCIES

In addition to the general problem types already noted, each emergency brings its own particular problems and monitoring considerations, some of which will be discussed in Chapter 18 for certain specific types of emergencies. Obviously the results of known material releases or abnormal operations should be evaluated as soon and as completely as possible. This will include some knowledge of such items as the direction of the wind at various heights if the release involves gaseous radioisotopes and the flow rate in a stream into which liquid radioisotopes have been released.

Control problems should not be underestimated since some material releases, as illustrated by the Windscale incident in England in 1957, can certainly require a tremendous amount of offsite monitoring. Although tests of nuclear weapons are obviously not emergencies, their results, at least insofar as atmospheric conditions are concerned, can give an indication of the maximum problems that might be encountered in emergencies.

MONITORING FOR SHIPMENT

A particular type of offsite monitoring but one which is really done at the plant as a routine part of plant activities is that for materials being shipped elsewhere. Included are shipments not only of radioactive materials but also of other supplies or equipment that may have been contaminated in processing operations. Standards which such shipments must meet have been almost universally established by the various countries that may handle them. Such standards are probably more than safe from the health protection standpoint since they may reflect political considerations as well as limitations imposed for protection of certain types of equipment or other items. A classic example of the latter aspect is the need for protecting photographic film from the unwanted blackening, or "exposure," caused by radioactive material or items contaminated with such materials. The bibliographical references give some of the current shipping regulations; however, it is recognized that these can change frequently and apparently arbitrarily.

In general, monitoring for shipment is similar to that for other uses, although the limits specified may be different, and, unless the future use of the equipment or other items is known, greater consideration is usually given to the possibility that radioactive material is held up inside a container; for example, a steel pipe can contain beta-active material that may produce no problem at all for plant use and may never be discovered by outer surface monitoring; such material may also be lodged in the coils of a motor, and it is frequently almost impossible to determine that this is the case from external monitoring procedures. As may be expected, most "hidden" deposits of alpha-emitting materials are almost impossible to discover by normal monitoring techniques. Accordingly, regulations for uncontrolled shipment usually provide for a determination of the possibility of hidden materials which are not otherwise observable, this including the plant history of the item concerned in addition to obvious evaluation techniques. Similar techniques and criteria are also frequently used as a part of contamination control procedures inside a plant to limit the locations in which radioactive materials may be found.

As is the case for in-plant conditions and all types of control programs, a decision must usually be made on whether shipping requirements will be met only by continuous and routine monitoring of all items or if routine checks on a periodic (but unscheduled) basis, supplemented by special monitoring at a time of unexpected conditions, will be adequate. Use of this latter method will depend strongly on the levels that will be expected, the types of materials involved, and the shipping requirements themselves.

SCRAP AND SURPLUS SHIPMENTS

Special shipment problems result from the release of surplus materials and of scrap or other discarded items. Surplus materials, particularly equipment items, are those that will probably be reused as originally intended, and there is probably little problem in transferring equipment, for example, a motor, from one facility handling radioactive isotopes to a similar facility. Such transfers are frequently made between AEC-contractor facilities where the hidden radioactivity may introduce no particular problems beyond those of information concerning the existence of such material and its acceptability to the recipient; if common-carriers are used in shipment, the conditions to be met will differ from those where such transfer is by contractor-controlled vehicles. On the other hand, any general release to the public must assume service in noncontaminated operations, and the possibility that traces of radioisotopes may remain in the equipment might make undesirable its transfer for broader usage.

Release of scrap usually recognizes that the material (particularly metal) will be melted down and subsequently intermingled with similar materials elsewhere so that its eventual location and use are completely unknown. One approach which is frequently used is a specification that the contaminated scrap can form no more than a certain small fraction of the total material of its type being released. Even this carefully controlled release of scrap has introduced some problems in industries such as those of film manufacturing, which can tolerate essentially no radioactivity at all. It has accordingly been customary for such industries to specify that virgin materials only are permitted in manufacture of items of their interest.

SHIPMENT OF RADIOISOTOPES

Shipment of radioisotopes, both nationally and internationally, has been reasonably well standardized by the appropriate regulatory agencies. For interstate shipments in the United States the Department of Transportation has recently taken over such responsibilities as had previously been vested in the Interstate Commerce Commission, the Post Office, the Federal Aviation Agency, and the Coast Guard. Local governments, notably New York City, have also added some regulations, and international shipments obviously must meet the requirements of the nations to which and through which they are sent; these generally conform to suggestions of the International Atomic Energy Authority, although this is not always the case. The recent Department of Transportation regulations have attempted to bring United States requirements more nearly in line with those promulgated by the IAEA.

Details of shipping requirements are obviously beyond the scope of this book, but appropriate bibliographical references are given. Regulatory limitations usually fall in one or more of the following general categories:

1. Radiation levels at the surface of an individual container or shipping unit such as a boxcar; these also usually include maximum permissible levels at specified distances from these surfaces.

2. Quantities of radioisotopes shipped in individual containers or in a single shipment; in some cases these refer to actual quantities of specific radioisotopes, although these may also be expressed as a multiple of a "radiation unit" for other shipments. Such units are now identified in terms of 7 Transport Groups into which the various radioisotopes are divided.

3. Considerations of the degree of control to be exercised by the carrier during shipment. Identified as the Transport Index, this was originally developed from criticality considerations to identify the amount of fissionable material placed in a single container and thus limit the number of such containers which may be assembled. The index has been expanded also to reflect a maximum dose rate near a single radioisotope container.

4. Types of materials which may be shipped or stored together, the quantities that may be so handled, and sometimes minimum distances of separation between types of materials.

5. Degree of hazard of respective materials. This may also include an exemption specification.

6. Extent of radioactive contamination of shipping containers or of carrier equipment and terminal facilities.

7. Ability of containers to maintain their integrity during shipment and handling; this also includes maintenance of shielding integrity where necessary. In general, containers must meet rather rigorous tests which include those for corner drop, penetration, compression, water spray (except for certain materials such as metals) with a subsequent free-drop. In many cases specific containers have been approved for specific types

of shipments; details of some of these are also available in the references, although most prior approvals given by the Bureau of Explosives are privately obtained and kept on file.

Shipments are also identified by appropriate labeling, some illustrations of which are given in Figure 17–1; label colors are also of significance in indicating the type of radiation involved. The current labels are similar to those of the IAEA. Obviously, radioisotope shipments can require a significant amount of paperwork.

The requirements for shipment of radioactive materials introduce no types of actions that have not been already accepted for industrial shipment of nonradioactive materials such as toxic chemicals, flammable and explosive substances, corrosive materials, and materials which are individually relatively inert but can be hazardous on contact. However, the special "hazard" associated in the lay mind with radiation has frequently produced exaggerated responses to comparatively minor incidents. Apparent leakage from a container marked as carrying radioactive material or a similar incident involving several such containers, especially if it occurs at the terminal facility of a common carrier, has invariably resulted in a relatively tremendous uproar and a consequent exhaustive investigation by the shipper, the carrier, and a regulatory agency itself, even though the incident has proven to be a false alarm.

A very cogent consideration in many shipments is the radiation hazard that would be caused by a major accident such as a serious fire, a truck wreck, the derailment of a train, and so on. Although containers of radioisotopes and fissionable materials are designed to maintain their integrity in certain severe conditions of handling (and have presumably been tested), an accident obviously produces unpredictable conditions, the handling of which is not amenable to hard-and-fast rules. However, the basic principles and methods used are those already described for taking care of radioactive materials anywhere. Accidents in over-the road transport rarely occur where appropriate monitoring and control equipment and facilities are available. Thus, plans for offsite monitoring should include considerations of emergency conditions; this includes appropriate warning, identification, and emergency contact signs, as well as specifications limiting actions by bystanders.

Of special importance in handling offsite emergencies is the desirability of not unduly alarming the populace in the surrounding areas. Unfortunately in many accidents requiring offsite monitoring the initial health-physics approach has too often seriously overemphasized the potential hazard of the incident and has thus actually created a perhaps undue and

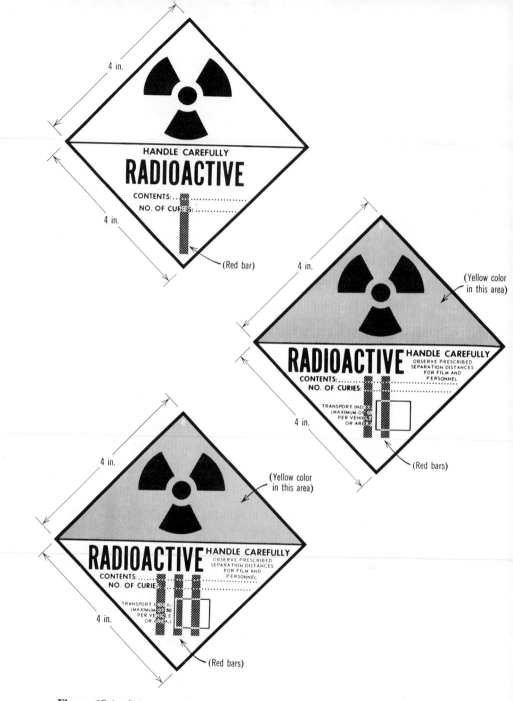

Figure 17-1 Shipping Labels for Radioactive Materials. (I) Radioactive white–I label. For use on each package not exceeding 0.5 millirem per hour at any point on the external surface of the package. (II) Radioactive yellow–II label. For use on each package (a) not exceeding 10 millirem per hour at any point on the external surface of the package and not exceeding 0.5 millirem per hour at 3 feet from the external surface of the package; or (b) for which the transport index does not exceed 0.5 at any time during transportation. (III) Radioactive yellow–III label. For use on each package for which (a) either of the limits for Radioactive yellow–II labeling is exceeded; or (b) certain types of shipments of radioactive or fissile materials. (Note: Certain types of radioactive devices, some specified small quantities of radio-active materials, and certain shipments of low specific activity materials are exempt from labeling.)

unjustified mental problem among the populace. It should be apparent that such hysteria should be avoided. By no means should the above be interpreted as saying that there should be no concern at an incident or that any resulting hazard should be understated, but it does point out that the potential hazard of an incident should not be unduly or extravagently overemphasized.

INTERSTATE AND INTERNATIONAL ASPECTS OF OFFSITE ACTIVITIES

Waste Disposal in Streams and Seas

Because air currents and river flow are no respecters of political boundaries, international relations as well as relationships between different states, or even different locations within the same state, may present special offsite control problems. The problem itself is not new since rivers have long been polluted with wastes that are at least equally as hazardous as radioactive materials. Similarly, disputes over pollution are not limited to radiation control.

With respect to stream movements it is possible that the release of waste materials at one site can "use up" the permissible limit for a given river so that a site further downstream cannot put additional radioactive wastes into this river without the permissible limits being exceeded, unless, of course, the originally released radioisotopes are deposited between the two points; subsequent release of these deposits may then present a problem.

For an international river some agreement will obviously be necessary concerning the respective amounts of radioactive materials to be released by the various nations involved. The nuclear program of one or the other of the countries may be severely hampered by the different limits that could be required for their respective wastes if agreement has been reached on overall contamination limits and these are not to be exceeded at any point in the river. Possible differences in the permissible limits as adopted by the individual nations would compound the problem. There may also be some ethical concern at downstream facilities being required to use water which is more highly radioactive than that upstream even though the degree of radioactivity in the water may not actually be harmful except as a possibility which is incorporated into the stated permissible limits.

In the case of atomic energy installations located near an international border, wastes deposited in the ground may enter the other country by travel through rock formations as has already been described.

Waste disposal in the ocean is also of international concern, since the release of radioactive waste materials into the ocean can mean their eventual transfer to the shores of other, and perhaps far distant, nations. No particular problem has been introduced by such radioactive waste disposal in the ocean to date, although there have been some international complaints, principally by nations without nuclear capabilities, concerning such waste disposal by others—even though the waste was "well canned" and thus probably of negligible hazard. In general such complaints have been based on political implications arising from possibly extreme estimates of possible ocean contamination rather than on direct indications of any measurable additional radioactivity from the deposit reaching the complaining nation or state. Various methods of preventing the spread of these wastes after deposit into the ocean have been suggested and used. Since these usually involve confinement of the waste in drums, present attempts are probably feasible only if the volume of the waste is small.

It is with problems of this type, among others, that the International Atomic Energy Authority has administratively concerned itself, and the International Commission on Radiation Protection, as well as other international groups, has provided useful information for the necessary evaluations.

Similarly, there have been interesting interstate complaints inside the United States concerning river usage, this resulting on occasion in studies of river usage agreements.

Materials Shipments

Another phase of atomic cooperation and possible consistency obviously involves shipping specifications. Again, this is nothing new, since international transportation has been a continuous necessity and frequently a problem, extending not only to language and customs differences but even to the use of railroads with different track gages, which has necessitated trans-shipment at international borders.

In addition to the problems of shipment of radioactive materials there are similar problems involving the international shipment of fissionable materials. Attempts are continuing, principally under the sponsorship of the International Atomic Energy Authority, to promulgate consistent regulations for the international scene.

Interstate shipments have presented problems even though current regulations are generally considered adequate. Perhaps particularly serious has been the promulgation of rather stringent local regulations that have frequently been less difficult to meet technically than administratively. No attempt is made herein to evaluate the validity of these regulations and transport activities either as to their adequacy or to the need for more stringent regulations in the United States or elsewhere.

Weapons Testing

Of particular concern to the international aspects of nuclear activities is contamination of the air and consequent fallout in one country as a result of nuclear activities in another. To date man has been unable to control the course of the winds. Thus, if radioactive materials become airborne, they will obviously be essentially uncontrolled at the point of release. However, with the exception of weapons testing, significant international problems of air contamination have not been involved in the peaceful uses of atomic energy to date, even for the accidents that have occurred. On the other hand, the fallout from weapons testing has been both a major political football and an ethical problem even though it may be of comparatively little concern from a technical viewpoint. Similar considerations are obviously inherent in actual nuclear war that would result in fallout on a neutral or nonbelligerent nation. This possibility has also produced considerable discussion and activity.

Among the measurably undesirable effects of weapons testing have been personnel exposures at certain specific locations in some of the western states, the comparatively high radiation exposures resulting from certain tests in the Pacific, and the results of localized fallout in areas in which photographic film was being manufactured. These have been rather well publicized and will be discussed in a subsequent chapter together with the overall effect of nuclear weapons.

REFERENCES and SUGGESTED READINGS

Handbook of Federal Regulations Applying to Transportation of Radioactive Materials, AEC, Washington, D.C., 1958 and 1966.

Mawson, C. A., Processing of Radioactive Wastes, IAEA, 1961.

Krasin, A. K., and B. A. Semenov, Operating Experience with Nuclear Power Stations, IAEA, 1961.

Farmer, F. R., The Packaging, Transport and Related Handling of Radioactive Materials, IAEA, 1961.

Radioactive Waste Disposal into the Sea, Safety Series No. 5, IAEA, 1961.

Regulations for the Safe Transport of Radioactive Materials, Safety Series No. 6, IAEA, 1961.

"Notes on Certain Aspects of the Regulations", Regulations for the Safe Transport of Radioactive Materials, Safety Series No. 7, IAEA, 1961.

Glueckauf, E., editor, *Atomic Energy Waste, Its Nature, Use and Disposal,* Interscience, New York, 1961.

Blatz, Hanson, editor, *Radiation Hygiene Handbook,* McGraw-Hill, New York, 1959.

National Bureau of Standards Handbooks 48, 49, 53, 58, 59, 65, 69, and 73. (See Appendix I for complete list.)

Agent T. C. George's Tariff No. 19, Publishing Interstate Commerce Commission Regulations, August 5, 1966 (Including Subsequently Issued Supplements).

Federal Register, Volume 33, Number 194, Department of Transportation Hazardous Materials Regulations Board, Part II, Washington, D.C., Oct. 4, 1968.

QUESTIONS

1. What factors may affect the level of concentration of radioactive material in a river?
2. Does the vegetation surrounding a plant absorb much radiation?
3. Why must shipping products be monitored so closely?
4. When is a release of waste material permissible in the environment?
5. What five general methods are used to dispose of liquids containing radioisotopes?
6. How would you go about determining whether or not a change in environmental radiation occurred after a plant has been established?
7. What facts must be known as quickly as possible in the event of an emergency?
8. What action should be taken if a common carrier hauling radioactive material happens to be in an accident of some sort?
9. How major is the problem of radiation contamination in the overall stream pollution problem of the United States?
10. Does fallout from nuclear testing seem to be a major hazard or is it principally a political hazard?
11. What are the limitations of each suggested method of waste disposal?
12. Categorize the general controls of shipping radioactive materials.
13. Do surplus and scrap shipments require special control approaches?
14. List some of the international considerations of offsite monitoring.

CHAPTER EIGHTEEN

Plant Emergencies

GENERAL COMMENTS

Other than accidents related to weapon tests, which are treated elsewhere, emergencies of health-physics concern fall into the following rather general classifications:

1. Criticality accidents which, in general, involve all aspects of radiation emergencies and thus will be specifically treated.

2. Releases of liquids, gases, or finely divided dusts into the environment.

3. Lack of shielding or failures thereof, this including sudden reductions in the amount of shielding surrounding a radioisotope.

4. Alleged incidents that are subsequently shown to have little or no significance.

Significant accidents have fortunately been extremely few in number as compared to their possibilities, and it is obviously beyond the scope of this book to detail many specific ones. However, it may be instructive to describe a few more or less typical accidents, with special attention to criticality incidents which have generally not only caused the most serious exposures but have also required the most detailed health-physics analyses.

Carelessness in familiar routine activities is obviously a potent accident cause. Thus it is perhaps not surprising that exposures resulting in actual injuries have most frequently occurred in routine operations in which comparatively minor operational changes were not recognized as introducing significant injury potentials and any administrative procedures designed to see that employees are aware of the changed conditions failed. Such results may, and frequently do, point to similar lapses in the activities of the radiation protection staff groups. Actually it is probably a high tribute to all of these groups that serious accidents have been so few. In particular the author knows of no accidents that have occurred in special operations in which specific radiation protection problems were identified and plans to meet these problems were established and followed as part of the operation.

SHIELDING FAILURES

What is identified as a shielding failure can result from the loss of a shield, as would be the case if a water shield around an operating neutron source were to be ruptured and the water escaped. Similarly, an explosive internal rupture involving a containment unit in a reactor or processing system could not only result in destruction of the shield but could also permit fission products or other highly radioactive materials to enter inadequately shielded locations and produce high levels of radiation. Although there have been such leaks, the integrity of the basic shielding was not seriously compromised, and the principal accompanying hazard was that of internal exposure, mainly from inhalation of an escaping gas or finely divided particles.

Another type of shielding failure has resulted from individuals unknowingly (or willfully) either bypassing existing shielding, working with more active materials than those for which their shielding was designed (this possibly resulting from unexpected events elsewhere), or even working where the shielding construction left unexpected "holes."

Other than the problems encountered in criticality incidents and the results of weapons tests (principally highly localized deposits of fallout materials), by far the most common type of shielding failure noted has been the result of individuals handling materials that were much more active than anticipated. Most of these have resulted in beta burns of the hands, although such incidents have also resulted from the use of X-rays or radioisotopes when the shields normally employed were deliberately bypassed for reasons of efficiency or when higher radiation levels or intensities were used than those for which the shielding was planned. There have also been incidents where ion-accelerators (including high-voltage X-ray machines) have been left on, but this fact was unknown by technicians or maintenance people working nearby without benefit of the designed shielding which might even have been removed. In general all such incidents are peculiar to the facilities involved, and the corrective measures are rather obvious, including particularly the use of radiation detection and alarm devices along with proper administrative actions. Some of the exposures received in these incidents in recent years have been dangerously high, but this has not been the usual case.

However, it may be illuminating to review briefly an incident that occurred in one plant, because it is one of the few with serious injury potentials, which were fortunately not realized. In this instance an individual remembered leaving a hand tool on the floor of a nearby cubicle, which was one of a group of similar units. He used a master key to open a cubicle, which was unfortunately not the one in which he had left his wrench but an adjacent one wherein some very active material made the radiation levels dangerously high. Almost as soon as he got well inside this room the employee realized he was in the wrong place and left immediately. He was wearing a pocket chamber, which indicated an exposure above its maximum detection limit of 200 mrad. Hence, as was normal, the film meter he also wore routinely was sent for processing as an irregular. However, because of changes in supervision, vacation, and the fact that the incident occurred at night and on a weekend, the processing was delayed, and it was thus several days before it was established that despite his short stay he had received an estimated exposure of 60 rems. Obviously a longer stay could have had much more serious results.

In this particular case the exposure causes were primarily administrative, and the corrective actions indicated are rather obvious. There should not have been available a single master key to cubicles which were designed for intermittent storage of highly radioactive materials. It may have been desirable that two locks be specified for cubicles containing highly active material, clear labeling of the cubicle contents should have been placed on the door, and employees should have been aware of the possibility of changes in cubicle contents. Other possible actions could also be suggested.

MATERIAL RELEASES

A comparatively large number of material releases have occurred, many being the result of reactor failures. Such releases have not usually been accompanied by explosions or other violent incidents; most of them have been of gases, although there have been some liquid releases to watercourses. Although there have been no cases in which sufficiently large amounts of radioactive materials have been released to cause immediate danger to life, there have been releases of sufficient comparatively short half-lived materials for the environment—including air, ground, or stream—to have been contaminated above permissible limits for short periods of time.

The causes of most incidents of this type have been rather commonplace. Filters have occasionally failed, and more radioactive material thus released through the stack than had been expected. Similarly, containment vessels for liquid radioactive waste have ruptured and released these materials into streams. Even such events as the overfilling of exterior storage containers with water during a rain has permitted some of these materials to escape into the stream drainage. Fortunately in such cases to date the stream itself has provided sufficient dilution so that the permissible limits have not been exceeded, and, if this dilution had not been sufficient, the river into which the contaminated streams empty would have provided it.

Windscale Incident

The accident with probably the greatest potential for damage or injury occurred at Windscale, England, in 1957. In this event the reactor was shut down as a result of the uranium metal fuel catching on fire during an annealing operation, and the fire continued for some four days before being quenched by water. A comparatively large area of the countryside was contaminated by the gaseous releases, this resulting in probably the most extensive environmental contamination that has occurred to date other than that from weapon tests.

The highest radiation level encountered during and after the incident at ground level was about 4 mrad/hr, which occurred directly under the gas plume and about 1 mile downwind. The highest radiation level caused by deposited radioactive materials was found to be about 0.2 mrad/hr at a point about 3 miles downwind. However, even after 5 days there were several locations of about 5 sq. miles in area where the radiation level was as high as 0.15 mrad/hr, which is about five times that of normal background radiation.

An exposure to external radiation in the range of 30 to 50 mrad is estimated as the maximum that anyone remaining completely unshielded for an indefinite period would have received as a result of the entire accident, and there is no evidence that anyone received such a dose. This is obviously much below permissible limits.

The greatest exposure potential was considered to be that resulting from iodine-131 in the milk-food cycle, the maximum thyroid dose received by anyone being estimated as about 16 rad. The iodine-131 hazard was considered to be much more of a problem than any direct inhalation of radioactive materials in the air or exposure to direct external radiation, although it was observed that the locations with the highest residual external radiation fields were also those in which the milk had the highest iodine-131 activity. The general usage of milk from contaminated areas was immediately controlled, and all milk with an activity greater than 0.1 μCi/l was discarded. All restrictions on milk usage were not removed for some 42 days after the quenching of the fire, although the area under restriction was steadily reduced in line with measurements during that time.

During the entire period of the release some 12,000 air samples were taken near the plant site and over 1000 in the environment. Among the observations made at this incident was the large demand for individuals with some radiation protection training, which seriously taxed the available technical personnel.

Chemical Explosion

In another incident a chemical explosion was accompanied by a widespread dispersion of plutonium. The incident occurred on an offshift, and it is interesting that the health physicist assigned to the facility on shift was not even aware of the presence therein of plutonium in critical (sufficient to produce a chain reaction) amounts. He also did not trust his alpha-monitoring instrument and did not immediately accept the presence of the alpha contamination he observed. In fact, this result of the incident was actually not established until the cause of the accident was ascertained and the presence of the plutonium was verified by operations personnel. By this time contamination had been wellspread by people walking through the contaminated area, by vehicles being driven through it, and so on. Although only some 15 g of plutonium were apparently "lost," the severely contaminated areas were so widespread that the cost of cleanup was probably in six figures and some of the operations were out of service for more than a month.

In addition to this contamination spread there was also a short-term air contamination problem. Fortunately, no one was injured as a result of the explosion nor was there any indication of any significant personnel problem from subsequent plutonium intake. One interesting facet of control in this case was the fact that, once it was established that plutonium was involved, it became necessary to take adequate precautions to prevent a possible critical accumulation since it was not initially known how much of the fissionable material might have been involved.

CRITICALITY AND REACTOR ACCIDENTS

General Conditions

Criticality accidents involve not only some aspects of essentially all these types of emergencies but also some rather special problems that require special attention. Thus a critical reaction in a unit of normally operating equipment generally means that an item which has hitherto been safe becomes highly radioactive, this giving essentially the effect of a shielding failure. Similarly, a reaction can occur so rapidly that materials may be physically displaced or equipment failure can result in a possible release of radioactive materials; significant material releases have not been observed in criticality accidents to date, however.

The fact that so very few criticality incidents have occurred during the development of nuclear energy for both war and peace is truly remarkable, considering the large quantities of fissionable materials prepared and used under a variety of conditions, as well as the complex and frequently expensive processing activities necessary to produce these materials in useful and usable amounts. Safe handling of fissionable materials has introduced entirely new problems of control, with the result that appropriate procedures and criteria for safe operation had to be developed along with the similar development of the processing and handling techniques themselves. To this must be added the facts (a) that there is no way of measuring an approach to criticality except under extremely limited conditions and (b) that no obvious mechanism such as heat or pressure change is necessary to initiate a reaction. The only requirement for an accident is the accumulation of sufficient material in a favorable geometry. A system that has been subcritical may without warning quickly become critical with the addition of even a small amount of fissionable material. Since criticality is possible with as little as 1 kg of uranium-235 or with some 2.5 kg contained in a 5.5 liter volume (21 cm diameter sphere), the great paucity of accidents becomes even more surprising. Fissionable materials have neither brain nor conscience, and an incident can occur when conditions accidentally become appropriate as readily as when they are planned. In fact, reactor engineers have sometimes felt that the odds were against planned criticality.

Criticality incidents have occurred in processing plants, in reactor auxiliary units, and in criticality or reactor research facilities. In the last category the possibility of criticality incidents has been accepted as a normal hazard of research activity (although this was not initially the case), and appropriate shielding and other protection have been provided the experimenters. Similarly it is usually necessary to shield all reactor-connected equipment and facilities because of fission products, and such shielding is also normally adequate for accidental criticality. For the incidents in other facilities, most individuals were fortuitously absent at the time of the incident and for reasons other than the possibility of criticality, with the result that there have been few injuries or fatalities.

Accidents in Processing Facilities

Some six incidents have occurred under conditions that are directly related to processing activities. Two of the most significant of these occurred in 1958, one at the Y–12 Plant in Oak Ridge and the other at the Los Alamos Scientific Laboratory. Of all of these the Y–12 incident was probably the most representative of the more individualistic aspects of such occurrences in complex industrial operations.

In this incident an employee drained wash water from a pipe, which supposedly contained little uranium, into a 55-gal drum. Unfortunately, a leak and misoperation occurring during prior accountability activities (designed to determine that there had been no unauthorized or accidental diversion of fissionable materials from their normal locations in the processing cycles) had allowed a significant amount of uranium highly enriched in the $^{235}_{92}U$ isotope to accumulate in the pipe, the dimensions of which made a critical reaction therein impossible. This control no longer existed after the solution drained into the drum, and criticality ensued.

The high radiation level that accompanied the critical reaction activated the alarms installed for that purpose, and the individuals in the large building left in accord with emergency plans. There was some grumbling as usual because there previously had been false alarms, but employees did leave the building promptly, and it is thought that at least one employee, in addition to the one actually draining the pipes, owes his life to his prompt evacuation. Meter checks at the building exterior showed that this was no false alarm, and all employees were further evacuated to assembly stations.

A year or so prior to the incident Y–12 had been the third plant in the nation and the world (the first having been the Oak Ridge Gaseous Diffusion Plant), to place indium in all security identification badges as a criticality monitoring device (see Chapter 12), and these badges were promptly checked for activation. All of those who had been close enough to the reaction to have received a dose above some 100 to 200 mrad were quickly picked out, and the eight men receiving doses later estimated to be on the order of 20 to 365 rads were promptly hospitalized; to date all have recovered. The indium also gave the first dose indications, although considerable subsequent experimental effort was undertaken at the Oak Ridge National Laboratory, including exposures of burros to a similar reaction in a criticality research facility to evaluate better the doses and other radiation information of interest.

This is the only incident to date wherein a reaction has occurred in a typical operational facility where the fissionable material was being processed in many types of equipment using a variety of processes, any one of which could logically be the source of the reaction. Hence the source could immediately be no better located than being somewhere in a very large room; as with the equipment, a large number of people could potentially be involved. However, the three most important basic emergency responses to an incident went as planned. First, the alarms responded to the high radiation level; second, all employees promptly left as instructed and went to a location that was known to be remote from the source; third, the activation of the indium in the badges permitted the most seriously exposed individuals to be promptly picked out of the hundreds for whom it could not otherwise have been proven quickly that no exposure, or a low one, occurred. This has probably been the most thoroughly and exhaustively investigated industrial incident in history.

A fatality resulted from the 1958 accident at Los Alamos. An employee had placed a large amount of waste materials, much beyond the normal amount and without the usual prior analysis, into a single 225-gal tank for the purpose of dissolution and digestion. He was standing beside the tank when he turned on a stirrer, the mixing action of which caused the solution to become critical. He succumbed 2 days later, and it was subsequently estimated that his exposure was above 2000 rads. In this instance the equipment was sufficiently isolated so that when the alarms sounded there was no question of the location of the reaction.

An incident at the Idaho Falls Chemical Processing Plant in Arco, Idaho, in 1959 resulted from uranyl nitrate solution being accidentally siphoned from a safe to an unsafe geometry. However, it occurred behind the heavy shielding necessitated by the processing of fission products, and thus there were no personnel problems other than the release of some gaseous fission products which activated some air-monitor alarms. A similar incident also occurred in 1961 in the same plant.

The fifth incident occurred in a tank during cleanup and plutonium recovery in a processing-system hood at the Hanford Works of Richland, Washington, in 1962. The system was not definitely known to be subcritical for some 37 hr after the initial burst, and the sequence of events leading to the accumulation was not positively identified. The alarm system was activated, and all but two of the 22 employees in the room left promptly. Fortunately neither of these entered a high radiation field at any time. Only three individuals received significant exposures, these being in the range of 19 to 110 rems.

A fatality occurred in a private plant in Rhode Island in 1964 when an individual mistakenly poured uranium-235 ($^{235}_{92}$U) solution from a geometrically safe container into a large washtank which was not geometrically safe and received a radiation dose estimated to be over 10,000 rems. It also appears that a second criticality excursion occurred during initial control activities, and two individuals received small exposures. This was the first criticality accident in a private facility operating under AEC license; the other facilities were contractor-operated for the AEC itself.

Significant Reactor Incidents

Two significant criticality incidents have occurred during reactor operation or testing. The first occurred in Yugoslavia in 1958 when a control rod was inadvertently withdrawn too far while experimental personnel were in the room. The exposed individuals were flown to Paris for medical treatment that included bone marrow transplants, and all but one subsequently recovered. Subsequent to the accident, personnel from the United States, the United Kingdom, and other nations used a mock-up experiment to evaluate the radiation exposures actually received by these experimenters. They concluded that the doses were in the range of 200 to 440 rads.

Probably the most violent of criticality incidents to date occurred in the Stationary Low Power Reactor No. 1 (commonly called the SL–1) at Idaho Falls, Idaho, in 1961. In this incident an employee apparently manually withdrew a control rod further than the specified limit, with the result that the core went prompt critical and built up a high steam pressure in the core water. This blew out the shield plugs and other items from the top of the reactor vessel, actually impaling one of the three men present on the ceiling of the building. All three of the men at the accident site were killed by the explosion, although the radiation fields in the immediate work location following the accident were sufficiently high to have eventually proved fatal. The reactor had had a history of trouble, and sometimes the corrective measures taken were apparently rather makeshift. The extremely high radiation fields in the building after the accident made rescue attempts (before it was known that the men were dead) very difficult, as was the case for the resultant control and cleanup efforts. This is probably the only incident in which individuals not directly involved but arriving later for appropriate health-physics actions received sufficiently high exposures that it was considered desirable to limit their future exposures.

One characteristic of all of these industrial and reactor incidents has been the comparatively long time before actions could be taken to insure that the system would remain under control and not release another burst; for example, it was more than 48 hr before the Los Alamos incident was thus "buttoned up." This delay is because it cannot be established that even such slight changes as the approach of an individual close to a barely subcritical system will not reactivate the fissionable materials and produce another burst.

Other Incidents

Other than the above incidents, significant exposures have occurred as a result of an accident in a reactor core experiment at the Argonne National Laboratory in 1952 when a control rod was manually withdrawn too rapidly and four individuals received exposures in the range of 30 to 180 rads; all recovered. Two fatalities occurred at the Los Alamos Scientific Laboratory during criticality experiments in 1945 and 1946. One occurred when an experimenter performing stacking experiments dropped a cube of fissionable material intended for a corner of the assembly into its center, the experimenter thus receiving an estimated total body dose of some 500 rads and subsequently succumbing. In the other, the principal experimenter was using a screwdriver to hold apart the two sections of the reflector for some fissionable material. The screwdriver slipped and the reflector completely surrounded the fissionable material, producing criticality and giving the individual concerned a dose of some 1500 rads, which he did not survive. In the second of these instances, seven other individuals received doses in the range of 20 to 200 rads, and all subsequently apparently fully recovered.

Other incidents in criticality facilities have occurred as a result of accidents and unexpected experimental conditions such as occur in all experimental activities. Since personnel were shielded, no particularly significant results occurred, although in each case cleanup did require time and some expense, which is also frequently the case for normal experiments that go awry. In general all of these incidents have been fully covered in the literature.

NOTORIOUS INCIDENTS

In addition to the actual emergencies described above because they are perhaps typical and are thus of interest for reasons beyond the actual resultant exposures or other damage, it may also be instructive to review two incidents that are rather notorious—not for the radiation injury or damage caused, but for the publicity they received.

One of these involved a chemical explosion and fire in a New York City plant which caused several severe thermal burns. Since the plant had previously processed thorium materials, radiation hazard signs had been appropriately posted even though little alpha-emitting contamination in plant areas was anticipated. No one received radiation injuries or even excessive internal exposures, and little contamination was spread by the incident itself. However, because of the lack of information by the city fire department and other emergency groups concerning the very minor nature of the actual hazard, as well as the fact that plant personnel were not immediately available, the reported hazard was blown up out of all proportion to either the actual problems or perhaps even those which would have occurred had the plant been in operation with considerably more radioactive materials available than was actually the case. The publicity in the very-newspaper-active metropolis, much of it speculative, did little to provide a sane treatment of the incident. Fortunately no untoward incidents resulted from the overly frightened approach to incident control, and the publicity problems eventually subsided.

The other incident was, to the author, a comedy (or, more accurately, a tragedy) of mishandling from beginning to its long end, although some of the nonincident occurrences do have some rather serious overtones.

In this incident, which occurred in Houston, Texas, in 1957, two employees opened a can supposedly containing 10 iridium-192 ($^{192}_{77}$Ir) pellets in a well-shielded and supposedly well-designed cell. However, two of the pellets had crumbled, and the radioactive dust had escaped as was then shown on an air monitor and subsequently by the laboratory's becoming highly contaminated. Plant management apparently became aware of the incident and its results only considerably after its occurrence, and this information was not related to the AEC until still later. In the meantime the two men had thoroughly contaminated their homes and some of their personal effects. A private company was subsequently engaged to evaluate the incident and decontaminate all locations involved.

The two employees received some minor radiation burns from the soft-gamma emission of the iridium contaminating their clothing and hair. Otherwise no effects of radiation were observed at any time, and in 1961 a thorough examination of the employees and their families at the Mayo Clinic showed no radiation effects. The incident was thus actually similar to some of those described previously, as well as some early X-ray cases, and would normally have been treated accordingly.

In this incident, however, sensational newspaper publicity, scare headlines and all, began to appear throughout the country. Even worse, these headlines and the general hysteria and misinformation concerning radiation that had been previously engendered had their impact on the neighborhood of the individuals involved, with the result that both families were essentially ostracized and reportedly began to develop various "symptoms."

Activities of the private company employed did little to put the problem in a less sensational context. Traces of contamination were found in several private homes and automobiles and even on a Scotch terrier belonging to one of the families. The "decontamination" performed was apparently all too often done with a dull knife so that jagged holes in carpets, bedcovers, and automobile cushions were left to view as the "contaminated" parts were hacked away. Even the dog suffered part of his fur, but not all, to be snipped away. Little publicity, certainly, attended any efforts made to remove the material by other and more normal means or even to determine if the levels were such that prompt removal was indicated.

Not only did the newspapers get into the act, but a national-circulation magazine played all of the speculation, including the sensational and "human interest" angles, to the hilt, not only once but subsequently and even after it had become apparent that from a technical standpoint the seriousness of the incident had been very much overrated. Television did its part by presenting the same sensationalism at least once and with the usual summer reruns. Finally, one of the men was taken as a labor union exhibit to a Congressional hearing. Again the newspapers, particularly those of a certain bent, had a field day.

The net result, as has been noted above, was a Mayo Clinic finding after four years that there were no indications of any radiation effects in any of the principals or their families. This received little publicity, especially in those news and opinion media that had been so busily headlining the sensational charges and opinions previously.

What can be learned from the incident is obvious. However, the important technical information which could have been obtained was completely lost by a lack of almost any emergency or evaluation actions until long after the incident and the actions taken during the decontamination procedures. The blazing publicity, which not only caused much more harm to the principals than did their exposures but also affected their families and the community, is a sad story for responsible journalism.

PLANNING FOR RADIATION EMERGENCIES

It is obvious that the simplest way to cope with any emergency, radiation or otherwise, is not to have one. This obviously means the inclusion of emergency prevention considerations in the design and operation of a facility.

It is also just as obviously beyond the scope of this book to attempt any detailed description of emergency planning for a given situation or potential hazard. Such details for a specific problem depend not only on the type of hazard involved and its potential severity but also on such factors as the location of the facility, the ready availability of normal emergency backup forces such as municipal fire services and other equipment, and even the organizational structure of the plant itself. However, a few suggestions may be made of some of the items that experience with emergencies and planning to cope with them have indicated to be appropriate.

The first step in any emergency control planning is the evaluation of the potential emergency itself. This would include such items as the probability of its occurrence; the failures of personnel or equipment that could result in the incident, the amount of material that could be released or the radiation levels that might reasonably be expected to occur; the possibility that the given accident might occur concurrently with, trigger, or be triggered by, an unrelated event; and so on.

The next step in this planning is that for immediate control of the fault itself and thus prompt limitation of the continuation of the incident and its spread or severity. Among items considered may be the provision of localized containment units such as valve covers, automatic stopping of waste discharge or water intake operations, the use of chemical processes to change the state of the radioactive materials involved (which may be as simple as spraying water on leaking gaseous materials), the use of protective and measuring devices for personnel who may be involved, specially designed and manufactured equipment which could quickly be used to blanket or divert released material, and so on. The availability of trained manpower is also of importance.

The final step is restoring normal operation and providing cleanup, continuing medical attention, environmental surveys, and similar needs.

MAXIMUM CREDIBLE ACCIDENT

In evaluating the potentiality of an incident, the maximum possible accident is frequently discussed. Although this need not be all inclusive, it all too

330 Fundamentals of Radiation Protection

often includes an assumption of such a host of concurrently occurring incidents, though each with a very small probability of actually happening, that it appears impracticable to attempt to prevent an accident; thus only the most elementary precautions are taken. Similarly, it is frequently possible to hypothesize the maximum damage from each of the possible effects of an incident to the extent that, again, feasible control measures appear so inadequate that only sketchy control measures are taken. In other cases it is sometimes felt that one hazard is so much more important than others that the latter are neglected in favor of the single one.

In view of the dangers of inadequate planning, either through fatalism or carelessness, a concept that is frequently adopted (particularly with reactors) is that of the *maximum credible accident*. In this case, the occurrence and hazard of the maximum possible incident are reduced to as nearly realistic and probable values as feasible, but the possibility of untoward failures is foreseen.

Criticality control activities recognize the impossibility of absolute assurance that an accident will not occur as a result of the concurrent occurrence of a group of unlikely events, no one of which might individually result in an incident. Hence the so-called "double-contingency" method of evaluation is used in partial determination of the possibility of a maximum credible accident; this method may be applicable to a wide variety of problems. Briefly, a contingency may be defined as the probability that a rather unlikely event will occur at any time as a result of a failure of the control adopted or precaution taken to prevent that event. Under the double-contingency concept, the accident will not occur except for the simultaneous occurrence of two independent contingencies. These contingencies are independent if the occurrence of either one does not cause the other one or affect its probability of occurrence. Simultaneous occurrence implies that corrective action cannot be taken to control the first event prior to the occurrence of the second one, or that both contingencies can occur in a shorter interval of time than that necessary to correct an uncontrolled change in one or more of the variables concerned. It is possible that a contingency can occur under conditions where it does not upset normal operations and cannot be detected by available instrumentation except after an extended period (and then only by a special check). Simultaneity would then be implied by a shorter time interval than the period between such checks. An example of such a condition would be the slow inleakage of a contaminant that would cause a chemical change that would be hazardous if sufficiently extensive.

ACCIDENT POTENTIAL

Evaluation of incident potentialities should include considerations of personnel, equipment, processes, and the environment—in plant and off plant. Of particular importance is an appreciation of the relative severity of various occurrences. The most dangerous incidents are probably those that can result in immediate and severe personnel injury. Incidents which may result only in some exposure above long-term limits or the expense of cleanup are of secondary importance, even though it is realized that incidents of this latter type may actually mean the difference between continuance and elimination of an operation.

The evaluation of an incident potentiality usually involves chemical considerations, the effects of corrosion, the history of equipment in similar service, and particularly the likelihood that an unrelated occurrence would initiate it; for example, heat from a fire might cause a failure which would result in a material release or other radiation incident; the use of water in combatting a fire could also be the triggering effect. It is also necessary to consider the possibility that equipment may be incorrectly restored to service during maintenance, as might be the case when normally separate systems are cross-connected. Failure to replace all shielding properly after special work can be a serious problem. It is usually recognized that sabotage and the fanatic, religious or otherwise, are potential hazards, but it is just as well recognized that little can be done about them except on an administrative level.

Criticality incidents have a major immediate hazard in that a system with little associated radiation may suddenly reach and maintain a dangerously high level. A break in a line carrying high-activity gaseous beta- or alpha-emitting radioisotopes may present an immediate problem in a location in which such problems (as well as those of radiation generally) are negligible. A failure in a stack system may mean widespread contamination of the countryside, and a break in a liquid high-radiation-level holding tank may mean consequent significant contamination of streams, rivers, etc.

IMPORTANCE OF ALARMS

Immediate notification of an incident to all potentially affected employees is of prime importance. Where radiation is the hazard, automatic alarm systems are usually rather easily contrived since the presence of radioactive materials or high radiation levels in locations in which they are normally absent identifies the incident.

Depending on the potential hazard, alarm systems should meet the following requirements:

1. Alarms should be distributed throughout the area where people could be; and the alarm indication, whether of sight, or sound, or both, should be noticeable above normal conditions in each of these locations.

2. The unit should be appropriate to the effect being detected; for example, if the radiation expected is a sharp peak, as would probably be the case for a criticality incident, the unit may require different engineering from that necessary for a slowly increasing field; also, a potential release of material into air would require a different detector than that for a radiation burst.

3. Alarms should be fail-safe in that failure or incipient failure would be indicated. In some cases instrument design is such that a weakening of one component results in complete failure so that the instrument is either available for its job or is completely and recognizably out of service.

4. The system should be so adjusted that false alarms very rarely occur, if at all. This usually means some reduction of maximum sensitivity (and perhaps more alarm units required), as well as an effort to see that the level of detection is never below the field that is planned for routine operation in the facility concerned, even on a short term basis.

5. The alarm should be distinctive and used for no other purpose unless the response to the alarm for the radiation emergency is the same as that for the other usage.

6. The system should not be bypassed or disconnected for any significant period of time without use of temporary alarms, possibly portable ones, which can serve the same purpose for short term usage.

The actions of personnel on hearing (or perhaps seeing) an alarm should be well defined; for example, in the event of a criticality incident the only feasible action for an individual is to run, not walk, away from the reaction site. If its location is known—as might be the case if the only site of fissionable materials is in a single container in a small laboratory—the direction taken should be away from this site. If the location cannot be so defined—as might be the case if it occurred in one of a large number of possible containers spread over a wide area (as was the actual case at the Oak Ridge Y–12 criticality incident)—anyone noting an alarm should get to a place that is safe from the possibility of a reaction as quickly as possible, even if this means running out of the building. Running past the site of the incident might increase exposure but probably not dangerously so; however, if anyone who is adjacent to the reaction does not get away, his exposure could be dangerously high. Similar specific actions can be taken for other types of emergency, such as immediate donning of respiratory equipment if an air release is signalled by the alarm, and so on.

PLANNING FOR IMMEDIATE CONTROL

It is in the immediate control of an incident that difficulties frequently arise, and it is here that planning, especially if backed up by design and operational requirements, may be of the greatest assistance. One characteristic of most of the more serious incidents that have occurred, particularly those involving criticality, is the rather extended period before the full assessment of the problem could be made and the incident brought under control. There may be difficulties in locating the source of the trouble, especially of a criticality incident in a large facility or of a major radioisotope release into a large room. The two methods already described for source location may be used. However, in the case of large sources, it is necessary to consider the possibility of reflection, the effect of shields on site identification, and so on.

Actual location of the source may require comparatively close approach to the general vicinity of the radiation. In this case, if there is any possibility of successive flare-up, it would perhaps be advisable for two employees to approach the source, one attempting to identify the actual location or condition of the source and the other watching a meter to be sure that a second burst does not cause significant injury. Similarly, if the radiation hazard is one of an escaping airborne radioisotope, the approach may require the use of special respiratory equipment.

One consideration in any emergency procedure at radiation incidents is recognition that no individual should receive more than a specific exposure, usually taken as 25 rems, unless efforts to save life are concerned or the result of not taking the risk increases tremendously the resultant possible hazard. Hence key people in emergency control groups, such as certain members of the staff group and perhaps operational management personnel, should initially avoid insofar as possible any exposure except that absolutely necessary so that their services may continue to be available until the emergency is controlled. More permissible exposure is probably used up by such individuals in "sight-seeing" at the time of an accident than may be desirable in a really significant incident.

It is entirely possible for the radioactive material released in a serious incident to be so widely spread as to raise the background radiation levels in a variety of locations, even including medical treatment rooms, to significantly high levels. These may readily be such as to hamper determination of the efficiency of the removal of such contaminating materials from the individuals involved in the incident. In fact, they may actually become so great as to pose a radiation exposure problem to those working in these locations, this including the physicians providing treatment to the injured as well as various emergency control personnel.

AVAILABILITY OF EQUIPMENT

At the time of an incident it is fantastic how much and how rapidly personnel equipment of an emergency type as well as that employed routinely can be used. This results from the large number of people involved as compared to those normally using such equipment and the fact that the emergency itself frequently results in the equipment and supplies quickly becoming too highly contaminated for use. This should be considered in equipment and supply stocking.

Similarly, where there is a possibility that an incident will result in widespread contamination by highly active materials, plans should be made and facilities arranged to provide some locations in which necessary activities will not be hampered by the high radiation fields. This may well include the provision of special facilities, although it may be possible to adapt those used for other purposes so that personnel decontamination or other services and activities may readily be carried on.

The same consideration applies to radiation detection instruments and monitoring devices, and for the same reasons. Generally the radiation levels involved are much higher than those encountered routinely. This may mean that additional high-range instruments and monitoring devices need to be made available as well as additional ones of the types routinely used. One of the problems with equipment set aside for emergency use is that it may deteriorate or otherwise not be available when needed. Thus emergency equipment, particularly if similar to that used routinely, should not be set aside, but the whole group of similar items should be periodically rotated between use and stock with the hope that the item needed for emergency use at any given time is available and probably in working condition.

It is necessary to consider the possibility that emergency equipment and supplies may be in locations where they are not readily available because of high radiation levels or contamination. This may also apply to normal operational equipment that would probably be used in control activities.

Methods for controlling the emergency under various conditions should be thoroughly studied, and action should be taken to obtain the required equipment, especially if it is of special design and not for routine use. At this point economic considerations may play a major role in decision-making. However, in spite of all this preplanning, it is entirely possible that conditions at the incident may themselves necessitate the manufacture and use of specialized items.

EMERGENCY DRILLS

In addition to the procurement of adequate supplies and equipment, probably the best insurance against emergencies causing serious and significant hazards is the use of periodic emergency drills, which should be made as realistic as possible and carried out seriously and completely. Such drills will not only point out flaws in planning but will also prepare people to do the right thing when an actual emergency occurs, even, hopefully, to the extent that they will react automatically. The tougher, the more realistic, and the more complete the drill, the better. Check lists for use in an actual emergency may be developed with the assistance of these drills.

ACTIVITIES AFTER AN ACCIDENT

The actual occurrence of an emergency will promptly set into action initial control procedures and activities to locate and treat obviously injured personnel. Subsequent control procedures, including cleanup and actions to restore operation, will probably be required for an extended period for all types of incidents. Although evaluation of personnel injury potential is not normally required for an extended period, radiation incidents do present special injury aspects. Thus, at the same time that control procedures are being taken, attempts should be made to find who was in the neighborhood of the incident, to determine who may have been most highly exposed, and to provide medical attention for these as well as those others who had more specific and obvious traumatic injuries.

Table 18–1 gives some indication of the times at which various observations may be made and actions taken at the time of a criticality incident. A similar target table may be made for almost any type of incident considered possible. Such information is useful as a guide to future action.

After the immediate emergency actions in caring for personnel and at least identifiying the source of the emergency (perhaps hopefully also bringing it under control), come the longer term activities for which rapid actions may not be so necessary. This will include getting the plant back to normal insofar as possible, determining the extent of damage or con-

TABLE 18–1

POSSIBLE TIME SCHEDULE OF ACTIONS AND ACTIVITIES
AFTER A NUCLEAR ACCIDENT[a]

Time after Accident	Probable Actions, Activities, or Events
10 min	1. All employees have evacuated.
2 hr	1. Source of reaction has been located.
	2. Radiation boundaries have been established.
	3. Checking of personnel monitors has been started to determine those who have been potentially seriously exposed.
	4. Personnel decontamination has been initiated.
	5. Problems of possible fallout are under review.
	6. Public-relations activities have started.
	7. Mutual-assistance agreements have been implemented as necessary.
6 hr	1. Reaction has been stopped or it has been determined that it cannot readily be stopped.
	2. All employees with potentially high exposures have been located, and medical treatment has been started.
	3. Personnel and area dosimetry to determine accurately the exposures of highly exposed personnel is under way.
12 hr	1. Exposures of highly exposed individuals have been determined as well as possible from dosimetry.
	2. Cleanup of reaction has started and plans are being laid for necessary work connected with getting back into operation.
	3. Investigation of accident cause is under way.
	4. Evaluation of technical aspects of the incident has been started.
24 hr	1. All employees have identified their actions at the time of and immediately after the accident as a part of the plant record.
	2. Follow-up activities to determine as well as possible the doses received by those who were not highly exposed have been initiated.

[a] The times listed are not target times but indicate the *maximum* intervals within which the items enumerated should be initiated or completed, it being recognized that the peculiarities of a specific event will affect these times, that strenuous efforts should be made to complete the items as rapidly as possible, and that effective planning and training may always reduce them. The above tabulation also obviously does not include all activities that might be undertaken.

tamination spread inside and outside the plant boundaries, cleaning up, investigating the accident to determine its cause and to consider the preventive measures that may be used subsequently, and perhaps evaluating personal exposures. It is assumed that employees who need medical care or hospitalization are receiving them. Most of these activities are long term extensions of normal plant activities and are amenable to the same types of treatment, recognizing that the radiation levels involved and the operational difficulties encountered may be much greater than is routinely the case.

DETERMINATION OF RADIATION EXPOSURE

Exposure determination may be a problem. Persons whose exposures are above about 50 to 100 rems will eventually show clinical symptoms, which will probably be used in medical treatment, regardless of any other indications. The problem may become more significant, however, for other types of dose measurement, especially those for which long term records are necessary but actual apparent injury is not involved. If the exposure is almost entirely to gamma radiation, routine monitoring devices such as film badges may be entirely adequate, although high exposures may so blacken the film normally used as to make such determinations subject to wide uncertainties. Efforts are being made to solve this problem, and it has been reported that doses up to 1000 rads may be measured with reasonable precision. Similarly high beta doses may be indicated. However, mixed neutron-gamma exposures, such as would result from a criticality accident, present a much more difficult picture, and it would seem that no device which an individual wears can give an exposure indication, even under the most favorable conditions, with a greater precision than about 50 percent (although an accuracy of 25 percent has also been mentioned). The precision is much smaller in an actual incident where conditions of exposure are not ideal.

Thus most workers in the field have agreed that probably the most accurate measure of an individual's exposure to mixed neutron-gamma radiation at a criticality incident is provided by the following procedure:

1. Determine the activity of the individual's blood sodium, which is activated by thermal neutrons; this is considered to be proportional to the thermal neutron flux.

2. Obtain a flux ratio for thermal to fast neutrons from area dosimetry, which requires several neutron detectors spread over a wide area in the operating facilities. The thermal flux may be most conveniently found from bare and cadmium-covered indium or gold foils, the latter being the most widely used. The fast neutron flux is usually determined from foil detectors, as described in Chapter 12. The total neutron dose for a given thermal neutron flux may thus be evaluated. Since blood sodium measurements indicate the thermal neutron flux incident on an individual, his total neutron exposure is also indicated.

3. In addition to neutron exposure, information on gamma exposure is also necessary. Thus a neutron-gamma dose ratio should be obtained by the use of neutron activation detectors and a relatively neutron-insensitive gamma detector in the area monitors. Of the devices that the author has noted, phosphate glass as shielded by lithium or appropriate lithium fluoride thermoluminescent dosimeters are probably less neutron sensitive than most others available. Calibration of any dosimeter used will be necessary.

4. The existence of a significant exposure is shown most quickly by activation of a piece of indium worn or carried by the individual concerned.

MISCELLANEOUS PREPARATION AND SUMMARY

In addition to the items covered above, emergency planning, especially for really serious incidents, will also include attention to such nontechnical considerations as adequate communication over the sometimes wide expanses involved, centralized direction of emergency control efforts, control of entry to eliminate curious spectators and others (this may be a real problem if nonplant personnel become interested and entry to the plant is not specifically limited by fencing or other means), mutual assistance with other facilities or towns, the possibility of offsite problems such as might result from material release into the air or streams, public relations, and so on.

The above has given some indication of the problems that may be encountered as a result of emergencies in plants and some of the actions that may be taken to control them. All of these, however, may be best summed up by the following admonitions:

1. Analyze the potential emergencies.

2. Provide alarm systems and specify employee actions when they are activated.

3. Plan for the limitation and control of emergency conditions.

4. Have available adequate stocks of emergency equipment as well as appropriate items used routinely.

5. Hold emergency drills.

REFERENCES and SUGGESTED READING

Accidental Radiation Excursion at the Y–12 Plant Y–1234, Union Carbide Nuclear Company, Oak Ridge, Tenn., 1958.

Hayes, Daniel F., *A Summary of Accidents and Incidents Involving Radiation in Atomic Energy Activities,* TID–5360, USAEC, Washington, D.C., 1956. (Also the supplements to this document).

Nicholls, C. M., editor, *Progress in Nuclear Energy,* Series IV, *Technology, Engineering, and Safety,* Vol. 5, Pergamon, New York, 1960, pp. 163–228.

Bailey, J. C., *"Criticality-Accident Dosimetry Studies,"* K-1618, Union Carbide Nuclear Corp., U.S.A.E.C., Washington, D.C. Oct. 15, 1964.

Saenger, Eugene L., editor, *Medical Aspects of Radiation Accidents,* U.S.A.E.C., 1963.

QUESTIONS

1. What could cause "holes" in shielding devices?
2. What is the only requirement for a criticality accident to occur?
3. What are the six short term radiation fatalities to date accredited to criticality accidents?
4. What single factor seems to be the cause of most emergencies?
5. Comment on the treatment the news media have given instances of possible accidents.
6. What factors would you consider if you were evaluating the potential of an emergency?
7. What is the double contingency method of evaluation?
8. What ideas should an alarm system incorporate?
9. Make out a sample plan for personnel actions to be followed when an alarm is sounded. Use the first floor of your building as a mock plant.
10. How would you determine mixed neutron-gamma radiation exposure at a criticality site?
11. What is meant by "maximum credible accident"?
12. Draw some conclusions about the press coverage of nuclear-plant emergencies.
13. Describe the Windscale incident and its significance.

CHAPTER NINETEEN

Administration

Since protection of personnel from potential radiation injury is the prime aim of any radiation protection program, administrative methods are necessarily geared to this purpose. However, other than some rather general items, the variety of methods actually used is somewhat bewildering, and any description of administrative methods and procedures must recognize the fact that many different types of programs are apparently successful, even though some of them may appear to be almost the direct opposites of others. As in the rearing of children, almost no method has had a vast majority of failures and no method has been 100 percent successful.

In general a health-physics or radiation protection program produces the best results if its organizational position and structure are closely related to similar programs of the plant and the overall setup is consistent with administration philosophy concerning such matters as individual and line organization responsibilities, the place of a staff organization, the methods of training, and all of the myriad factors that go into making a successful industrial organization or operation. A successful program almost necessarily meets criteria of these types, and, since there are so many successful types of industrial organization, a reason for the success of the various programs mentioned above is apparent.

Limitation of doses to the values expressed as maximum permissible limits, both as directly measured and as indirectly specified by environmental measurements, is accepted as a basic premise in determining the possibility of disadvantageous conditions. Frequently added to this is the additional concept that doses should be maintained at levels that are as low as feasible but always below those stated. Hence, any administrative criteria or methods that are adopted will have assurance that these maximum permissible limits will be met as a primary aim.

Of particular importance in the establishment of any radiation protection program is the need to comply with the directives, regulations, and requirements of various governmental agencies having jurisdiction. Any records, monitoring methods, or other activities of any program must obviously, as a minimum condition, be adequate to satisfy such requirements.

BASIC ASPECTS OF A HEALTH PHYSICS PROGRAM

Because of local conditions and problems there are variations in radiation protection programs, although there are also some rather well-defined aspects of similarity, as is noted subsequently; for example, a hospital may have the specific problem that regular doses much above any permissible limits may be used in diagnosis or therapy, including internally administered radioisotopes. At another extreme, an industrial plant that handles plutonium-239 ($^{239}_{94}$Pu) or uranium-235 ($^{235}_{92}$U) may have criticality problems in addition to routine ones of radiation protection.

Personnel Monitoring

A program for personnel monitoring is almost always necessary. Ordinarily this includes a specific function for film-meter handling, processing, calibrating, and interpreting, although an operation requiring only a few film meters may find it more convenient to contract the processing, calibrating, and interpreting to an outside commercial firm. In a major facility, however, this service will probably be provided by plant personnel either as their principal or only function or as a part of their related activities. In addition, responsibilities may be divided; for example, badge calibration, reading, and interpretation may be a health-physics staff-group function, whereas processing is a laboratory activity and distribution is another group's responsibility. Neutron film is not too widely used; if used, it is usually handled completely by the operating facility.

Personnel monitoring may also require proper collecting, handling, and reading of dosimeters and other devices, such as thermoluminescent devices or neutron-activated foils, if necessary. Where functions are divided, the analyses may be a works laboratory function and the collections may be made by another group, such as that responsible for industrial security. It may also be possible to contract these analyses to an outside commercial firm.

Environmental Monitoring

Environmental and job monitoring are also requisites of a program, these usually being handled by operations personnel as part of their job or by individuals for whom this is their principal or only assignment; these latter can be in a staff health-physics group or in the operating group itself. These provisions also apply to offsite monitoring and shipping checks. In any event, some arrangement should be made for a regular but aperiodic and independent check of monitoring efficiency. If continuous monitors and alarm systems are installed, provision for their routine checking is necessary.

Engineering and Maintenance

As described in Chapter 16, a good program requires a specific health-physics review of new construction or alterations, as performed either by members of the design group itself or by others. Instrument check, calibration, and maintenance are always needed, whether done by a special organization or as a part of an overall instrument maintenance function. Protective equipment must be procured and maintained. Record-keeping and official reports are other musts.

Special Problems

Since each facility that is faced with radiation hazards has peculiarities of its own and since considerable operating economics as well as other advantages are frequently derived from evaluating the special health-physics aspects of these inherent differences, provision should also be made for some research and development activities; for example, with respect to instruments, these activities can vary from short-term minor testing of protective devices or new instruments for specific applicability to a single operation all the way to a full-blown organization for instrument design and development. In addition to work of primary interest to the plant, research of value to both the plant and other organizations and operations is usually advisable. This is especially true if the plant operation, facilities, or available personnel are peculiarly well adapted to the experimental work envisaged.

Administrative Procedures

Characteristic of most plant programs is a set of well-defined procedures and regulations that implement management decisions and interests. Some of these procedures apply to specific health-physics matters, and specific limits may or may not be listed. However, even in this latter case the regulatory criteria are so stated that permissible limits will be met. In other procedures, which may deal with a variety of plant concerns, health-physics matters may be incidental to the principal subject, which may even specify requirements which are unduly restrictive as far as health-physics needs alone are concerned.

RADIATION-PROTECTION GROUPS

A variety of radiation protection organizations in the atomic energy industry both in the United States and abroad have been successful. It is difficult to evaluate degrees of success since each type of organization can show examples of excellent control, despite differences in the way it is applied, the problems encountered, and the effectiveness of its personnel. However, all organizations have the one factor in common that someone, or some group, is assigned the specific responsibility for seeing that radiation protection procedures and criteria in a particular plant are adequate and proper. The number of employees in this type of group today ranges all the way from a part-time job of one individual to an organization comprising as much as 3 to perhaps 5 percent of the total plant population.

A radiologist usually supervises health-physics activities in a hospital. In industry, however, the health-physics group has been found to work successfully as a part of the engineering organization, the medical department, an overall safety or health and safety organization, or even as a staff function reporting directly to the plant manager. The last-named organizational position has ordinarily been temporary only, although there are many examples in which it was the original setup. Health physics has also been successful either as a single self-contained group in some organizational niche or as one function of a combination with industrial hygiene, safety, or other related activities. On the other hand, the importance of this staff group's being in a relatively independent position is shown by the fact that it has not generally been made a part of the production organization or other group with significant health-physics problems even though such organizations may frequently use their own staff group for routine monitoring and similar activities.

Typically, all staff organizations have the responsibilities of knowing the radiation protection problems of a facility, performing needed monitoring, seeing that operating procedures and equipment are adequate for the necessary employee protection, and keeping records both of actual dose data and of environmental conditions as related to exposure possibilities.

As would be expected, however, the actual activities of any such group may vary from plant to plant, depending primarily on the philosophy of the company, the personal characteristics of the individuals concerned, the degree of hazard, and other applicable technical factors. One of these factors may very well be the existence, or lack thereof, of alarms and continuous monitoring equipment that obtain needed information routinely.

GENERAL ACTIVITIES OF A HEALTH PHYSICS ORGANIZATION

Many of the activities that health-physics organizations are called on to carry out are similar to those for either (or both) of industrial safety or industrial hygiene, or even fire protection, since these are also concerned with the safety of personnel and equipment; for example, the same general techniques of location or operations audit, evaluation of available technical data, preoperational emergency planning, and careful investigation of serious incidents are useful in these other disciplines as well as in radiation protection.

However, it should continuously be recognized that adequate criteria for meeting these respective types of problems do not have a one-to-one correspondence with those of radiation control, although the general methods and approaches used may be entirely analogous and usable, either as is or with refinements. Thus levels of confinement or air sampling techniques that are acceptable for most toxic materials may not be adequate for the very small quantities involved in radioisotope work; similarly, it may be simpler to measure small concentrations of radioisotopes than of many toxic materials. The relative degree of hazard of the various materials used in an operation is a significant factor in any hazard evaluation.

The importance of training and education in emergency planning has been discussed. It is probably of equal importance in normal operations since many administrative and perhaps other problems will be markedly reduced if each of those involved has a relatively clear idea of the actual hazards to be encountered in his work, the adequacy of protective measures taken, and his own responsibility for radiation safety. Formal and informal training programs as well as periodic propaganda-type reminders, newspaper publicity, and all of the other routines of promotion can be applied.

Large Staff Organizations

Differences in company operational philosophies as they affect radiation protection practices are frequently expressed in the division of responsibilities and work between the staff group and other plant organizations.

In one form of such responsibility division which has found rather wide acceptance, essentially all of the plant radiation protection activities and responsibilities are entrusted to the staff group or individual on an almost hour-to-hour basis. This is the case in most hospitals and university research activities in which the radiation or radioactive materials used form a small and well-localized part of the overall activities, and these responsibilities can usually be handled by a single individual. On the other hand, in major industrial facilities such organizations are usually large and include a variety of specialists and functions.

It is typical of the approach used in large organizations that an individual is assigned to take care of the health-physics requirements of a given geographical unit or operation with comparatively minor central direction. He usually works through the local operating supervision or he may have certain specific authorities, such as that of making spot decisions on exposure time limitations. In this he may even be empowered to permit gross overexposures or to stop an operation with or without what might be later considered justifiable reasons. In fact ill-defined authority has frequently been a real problem in this general approach to protection.

His records may be only those which he himself decides to keep, other than film meter data and similar information that are routinely obtained for the plant as a whole, and his log may provide the principal data on "normal" facility operations. He estimates potential employee exposure levels for given operations and may subsequently measure dose rates and exposure times. He is expected to do most of the monitoring, including obtaining the results of continuously monitoring equipment. However, he has little responsibility in routine activities beyond advising members of the line organization of measures that he considers desirable or necessary to reduce exposure probabilities and to control personnel exposures. In case of a potential or actual overexposure he is usually responsible for seeing that employees are promptly given proper treatment, and he eventually plays an important role in efforts to determine the cause of the incident. He frequently engages in a small amount of personal "research" in the operation assigned. Typically about half these staff members are college graduates, and the remainder may or may not have attended college.

Central Groups

These staff organizations usually include a central engineering-type group to do certain types of evaluations of new and continuing problems. This is not always the case, however, since for many uses, such as most medical applications and radiography, engineering-type information of specific applicability is normally readily available. Members of this group may thus assist the local specialist or their paths may rarely cross, and they may undertake a fair amount of research, both of local interest and for more widespread use. Special groups are sometimes incorporated into the health-physics organization for a specific purpose and perhaps on a temporary basis; an example may be groups that are concerned with instrument development or maintenance.

Film badge processing and pocket chamber readings are usually made by personnel in the health-physics organization, and the information is compiled in a way considered necessary to meet regulations or other requirements; a sizable central records group is usually a part of the organization. Environmental surveys or offsite monitoring, as well as monitoring of all outgoing and perhaps incoming materials for possible radiation problems, are usually the responsibilities of the staff group. A bioassay group may or may not be part of the health-physics staff or organization, and there may or may not be a central procedure to review engineering designs, drawings, or other factors to see that adequate radiation protection considerations are incorporated therein.

Analysis of Performance

Obviously the actual radiation protection afforded depends rather largely on the judgment, integrity, ability, and efficiency of the individual assigned to an area. Although these duties are usually responsibly performed, this may not always be the case. Even at best, because of the different personalities of the various individuals and their relative autonomy, the actual radiation protection requirements in one part of the plant may differ from those in another part for reasons which do not result entirely from differences in the operations concerned. The boredom attendant on a routine job may lead an individual gradually to settle into a pattern of considering only certain jobs as his total responsibility, and an operating supervisor, or perhaps his employees, may begin a "game" to see how much he "can get away with" before the specialist "catches" him. On the other hand, some staff members may tend to make decisions, for example on exposure limitations, that are at least partly based on their own estimates of the "safety morale" of the individual operating groups or supervisors. It is also possible that the health-physics specialist and the operation supervisor may unofficially decide that only favorable evaluations of each other are made. A possible result is concealment of their possible mutual incompetence.

Where the specialist has relatively complete radiation protection responsibilities, operations employees, including supervisors, may frequently and incorrectly claim to have an almost total lack of knowledge of proper radiation protection requirements, including instrument use, and will wait for the specialist's assistance, stating this to be in accord with their instructions. Sometimes a supervisor, not knowing what the true hazards of an operation are, may unjustifiably assume that the safety factors used in specifying local exposure rules give greater exposure leeway than may be the case and make decisions with predictable results. Another possible effect may be an unfortunate reduction in the operating supervisor's interest in the radiation protection aspects of his work so that what is normally a very fruitful source of ideas and other advantages may be closed.

On the other hand, an active and competent individual may have major opportunities to make significant contributions to radiation protection both in his assigned location and, by extension, elsewhere.

Advantages and Disadvantages

This type of organization presumably permits a close and effective control of radiation problems since a specialist is always available to help operating personnel solve even very minor problems, and shift coverage may be readily available. The comparatively large number of technically trained people in a large installation should permit a more efficient evaluation of special problems of a research or development nature, both those of local interest only and those of widespread application. The organizational importance of the function itself may justify the higher wages for supervision or scientific personnel which theoretically insures their competence. An adequate supply of specialists may be available for meeting emergencies.

This approach naturally results in a comparatively high operating cost, principally because of the large number of people required. In addition a too routine survey job may result in efforts at unionization of the surveyors even though their function is basically that of management.

Small Staff Organization

In the other principal type of organizational approach the prime responsibility for routine radiation protection is given to the line organization as a normal job function, and the comparatively small radiation protection staff group gives major attention to monitoring the adequacy of this routine work. The group may or may not have the responsibility for such various plant-wide activities as film-badge monitoring or bioassay work, laboratory sample counting, or instrument development and repair.

Typically operations personnel do the routine work—including such activities as day-to-day personnel and environmental monitoring; job timing; and the collection of data from operational surveys, hand and foot checks, pocket chamber or dosimeter readings, and continuous air and radiation monitors. Normally members of the staff group make well-documented surveys of locations and facilities on a regular but aperiodic schedule in order to verify and evaluate the effectiveness and competency with which operational personnel discharge their routine radiation protection responsibilities. The wide variety of items to be monitored is usually check listed, with specific attention also directed to potential changes or alterations. It is thus hoped that not only will no significant hazards or changes be missed but also that proper attention will be given to less obvious potential hazards. Members of the staff group are quickly available to assist in solving various problems, to help in more extensive emergencies, or to take part in frequently performed special operations.

Typically these monitoring personnel are backed up by a relatively strong engineering group that also has responsibility for a thorough knowledge of all potential radiation hazards; evaluation of monitoring data and their consequent review with operating supervision and personnel; studies of facilities during engineering, development, and construction; and the provision of close special attention during startup operations.

Staff-Line Cooperation

Necessary to the success of this approach is considerable cooperation between members of the health-physics staff group and operating supervisors concerning problems that exist or are foreseen as well as possible exposure potentials resulting from operational changes or accidents. The necessarily assumed responsibility of the operating group for the radiation safety of its own activities will obviously have the salutary effect of making them better aware of health-physics problems and of their own responsibilities in meeting them.

Because of this close involvement of operational personnel in radiation protection activities and their consequent awareness of the criteria applied, it will be almost a necessity that these criteria and methods be relatively identical throughout the plant with a consequent smaller dependence on personal bias. This may appear as something of an economic detriment since the strict methods of control that are necessary in some locations may not be required in other areas. However, the eventual economic advantages of plantwide criteria may far outweigh the disadvantages, and there is also less probability of untoward incidents resulting from employee job changes.

From a records viewpoint the data obtained by the staff group will probably be rather consistent, primarily because subsequent availability and easy recognition of such data are important parts of the control program. The aperiodicity of contact requires an "as is" evaluation of the operation, and the records obtained will thus tend to reflect its "average" radiation protection aspects rather than special conditions such as may be observed during emergencies. Since the monitor contacts several supervisors and at rather well spaced intervals, less opportunity probably exists to realize the results of too close contact, good or bad. This is especially true if, as is usually the case, the monitor assignments are rotated rather frequently, primarily to gain the advantage of a relatively fresh look at an operation.

Advantages and Disadvantages

This type of organization has considerable appeal, not the least of which is economy of operation; for example, staff groups comprising some 0.2 percent of a plant population have been successful, whereas the more usual fraction has been some 10 times as great. This may lead to the higher salaries, especially for periodic monitoring jobs, that supposedly provide competence.

Under this arrangement such corollary activities as bioassay, other laboratory work, and instrument maintenance are almost necessarily a responsibility of appropriate plant groups. Thus the special technical capabilities of many plantwide operational, maintenance, engineering, and laboratory groups may be brought to bear on radiation protection activities as an important part of their overall jobs, with a consequent increase not only in efficiency but also possibly in economy of operation.

This method has a rather obvious disadvantage in that professional radiation protection personnel probably will not be at a given location when a minor emergency occurs even though someone may be on prompt call. As a result operating supervisors may authorize actions, the consequences of which they may be ill-equipped technically to evaluate. Also, too few technically trained people may be available to handle a significantly serious emergency efficiently; this latter, incidently, was a major observation in the Windscale incident in England. Thus, whereas an observation that the more people there are working on a job the more efficiently and effectively that job is done may not be true for routine operations, it may have sufficient truth in the event of a significant incident or other emergency to justify a larger group than the routine activities will support. Only a limited amount of research or development work will probably be possible, although appropriate engineering and technical plant groups may do at least some work of this type.

It usually requires considerable ingenuity by the supervision of a small health-physics staff group to see that the data obtained and the protection afforded are adequate. In addition considerable dependence must be placed on such activities as film badge or other personnel monitoring, air monitoring, and other methods designed to provide independent information concerning the actual potential hazard of an operation or location, or changes therein.

Decision-Making

As was mentioned previously, a decision on what type of radiation protection organization to use involves many factors, including the overall philosophy of the plant itself, and the degree and type of hazard presented. However, a good general rule of thumb for management is to be sure that enough personnel and close enough control are available at the start of an operation to prevent it from getting out of hand before the activity reaches routine operational status, to see that adequate and appropriate radiation protection requirements are incorporated at the earliest stage necessary or desirable, and to arrange for maintaining a continuing review of the operations concerned with appropriate consequent action taken.

REQUIREMENTS IMPOSED FOR NON-RADIATION-PROTECTION REASONS

Once health protection and regulatory requirements are met the basic purposes of a radiation protection program are fulfilled. However, attention may also be needed for other operating specifications that are allegedly connected with radiation protection, even though they may actually be little, if at all, involved therewith. Some practices may be based on labor relations or union contractual requirements, others may be based on economic considerations, and still others may be based on a parent company philosophy or management directive.

Labor Relations

Particularly important problems of this type are concerned with employee relations, especially those involving labor unions wherein various practices, regardless of their original justification, tend to become fixed "rights" that are unchangeable except by negotiation. Thus actions that were originally established for radiation protection reasons but were later shown by technical data to be no longer required and possibly even undesirable may be maintained as contract requirements and even considered to be radiation protection necessities; for example, it may be more convenient and practical for a company to furnish work clothing for all employees in a certain area than to provide it only for those whose jobs may result in clothing contamination above limits that may have been set arbitrarily, possibly even by labor agreement; although the limits themselves may not have been specifically established for radiation protection reasons, the clothing issue may continue to be considered "radiation protection" clothing.

Another type of problem is illustrated by a possible decision to follow a plan of routine rotation of a large number of employees among several jobs with different exposure potentials so that the total dose received by each will be known to be far below the permissible limits at all times, except possibly in emergency conditions. The alternative and discarded procedure would be the assignment of a few employees to the jobs with the greater exposure potentials, coupling this action with close control to be sure they receive no actual doses above the permissible limits; other employees would thus be assigned to jobs involving essentially no exposure. An obvious administrative problem would be posed in the alternative case by the individuals becoming overexposed or it perhaps being necessary to remove them from a job at midpoint to limit their doses to the appropriate permissible limit. Especially would this be a problem under emergency conditions where the number of employees capable of handling the necessary work may be limited. Although adequate records would normally be required in either case, a rotation system in an operation involving a large number of areas and employees with no more than a very slight possibility of anyone receiving a marked radiation dose, may even justify a decision to provide no specific monitoring devices routinely. Of course emergencies would probably be a different matter.

In any event, such practices as the above-mentioned rotation requirement or clothing specification, nominally based on radiation protection considerations though not indicative of an actual hazard, may readily come to be regarded by an employee as a major radiation protection requirement. He may thus consider his working conditions to be more hazardous than is actually the case, and the practice may become associated with a largely nonexistent need.

It is recognized that labor-relation problems of the above types are not peculiar to radiation protection considerations. They do, however, tend to hinder the establishment of adequate and effective programs based on realistic assessments of the hazard, and they impose economic and other burdens on the operation. This has been particularly true in some of the older atomic energy plants where, in the absence of technical information at plant startup, rather extensive precautions were taken even though some were even then recognized as probably being of marginal importance.

Other Technical Requirements

Technical requirements of operation may also appear to involve protection from radioactivity when such is not the case; for example, radiation levels in counting laboratories must frequently be maintained at levels that are much below those of any health concern for efficient counting; in fact normal background radiation is sometimes considered to be too high. As a matter of housekeeping or production it may be extremely undesirable to transfer materials between areas, and it may be desirable to define clearly the areas in which certain materials are to be used. Thus more stringent requirements against the interarea movement of such materials on clothing, shoes, or hands may be imposed than may be necessary from possible health considerations alone. In fact some facilities require radiation checks of persons entering certain locations to prevent the entry of radioactive material. This is especially true where the presence of radioactive materials may cause damage, as with cameras or photographic film manufacture. Cost, scarcity, or other factors that make the loss of even small amounts of certain radioactive materials a serious concern may be a factor in control decisions. In particular, security accountability usually justifies very strict control of material movement.

Possible Effects

Although they tend to complicate radiation protection programs in their apparent overstating of a potential problem, overstrict control considerations such as those described above are not entirely undesirable from a radiation protection standpoint, since any facility which meets them will also meet the less restrictive requirements of health considerations. However, it is very important to recognize that such strictness is not because of health-physics necessities and to specify the requirements for health safety clearly. Otherwise there can be undesirable consequences such as employee confusion or eventual opposition to necessary health protection requirements, supervisory use of radiation protection as justification for operational desires, possible economic difficulties, and so on.

Misuse of radiation protection criteria have actually tended to militate against adequate protection since they may give an employee a false sense of security to the possible extent of using an unnecessary piece of equipment and ignoring a useful one. Thus some employees consider themselves "protected" from the environmental hazards of alpha-emitters if they wear work clothing, which is comfortable and economically advantageous, whereas they ignore or object to the use of respiratory protection equipment, which is not particularly comfortable but is recognized to be almost the only useful protection against many alpha-emitters of interest. Appropriate recognition of certain radiation problems may also be complicated by rather obvious observations. For example, in certain facilities and operations, air levels at which respiratory protection should be used will normally occur only under conditions in which the alpha-emitting radioactive material is readily visible as surface dust or otherwise, while in fact, clearly visible surface material, especially if it is bound or fixed, may be essentially no hazard at all. On the other hand, an air contamination problem may exist even though the surface is "clean." Yet, to an employee it is not the invisible material in the air that is a problem but the visible surface material, and clothing is the obvious protector. Similarly there is no obvious indication of the hazard of penetrating radiation. Although the educational efforts to correct this situation are obvious, it does indicate that engrained feelings about the efficacy of the five senses in detecting danger cannot be completely ignored, especially when the apparently logical protective measure is ineffective.

RECORDS

The rather extensive radiation protection records maintained today are largely the result of the legal requirements imposed by regulations of the various governmental agencies, although eventual evaluation of data involving low doses also indicate the desirability of long-term maintenance of certain record types. However, it would appear that data record and retention schedules based on legal recommendations are probably at least as great as those necessary for actual health protection interest or medical considerations.

In general, records of personnel exposure shown by such monitors as film badges and pocket chambers should be retained indefinitely along with similar cumulative dose information. Similar retention may be necessary for the results of bioassay data along with any information on how they have been or may be related to actual internal doses. Other data concerning routine potential exposure, such as that represented by hand or shoe counts, should be kept, as should be the results of any routine monitoring of air, water, or work surfaces. Information concerning emergency conditions and the employees involved therein should be maintained and correlated with specific exposure or bioassay data as well as possible; it is recognized that such correlation has been found rather inconclusive quantitatively to date.

Although no specific periods for maintenance of records have been stated as such, a 50-year retention period has been discussed, and a current trend toward developing cumulative exposure data involving *all* radiation exposures (including those received prior to present employment) may necessitate indefinite record retention, with added complications in record-keeping procedures. To date, however, only industrial exposure is considered for such treatment.

Currently, there are rather serious proposals to require that radiation exposure (or dose) records accompany an individual as he goes from job to job. This includes both external dose information and internal dose estimates. Fruition of such plans may obviously increase the record-keeping burden.

HEALTH PHYSICS STAFF

Although a radiologist is a normal and appropriate "health physicist" for a hospital, it is difficult to specify the types of personnel for industrial health-physics staff organizations. No tests have been devised or suggested that would pinpoint special aptitudes, and for health-physics engineering there is certainly nothing available that is analogous to the degree in instrument engineering which is almost a necessity for an instrument development engineer. Probably the best preparation for health-physics engineering or technical work (certainly for advanced work) would be a degree in nuclear engineering or physics with a biology minor; a bachelor's degree with a major in physics or with majors in chemistry, biology, or various engineering equivalents are also appropriate. Several fellowships for graduate study leading to a Master's degree and to the doctorate in health-physics are sponsored by the Atomic Energy Commission, and similar encouragement is also available in various universities and hospitals.

Radiation protection supervision should be trained to the level of a Master's degree in one of the disciplines mentioned above, but personnel with the special training suggested are preferable. For a given facility, technical plant personnel may be an excellent choice, regardless of their original background, although it should probably be in one of the sciences or engineering. These individuals would not only have the distinct advantage of already being familiar with plant operation and the operational philosophy and methods of the company but would have also already shown their competence. In a rather large organization, personnel with varying backgrounds would be desirable to provide the effective condition of combined approaches from several disciplines.

Personnel with the qualifications of foremen or first-line supervisors, provided they are interested in the problem and have an interest in doing something other than mere routine work, are probably first choice for auditors or surveyors because of their proven abilities and adjustment to the plant.

Specific qualifications of staff personnel in a given plant will depend to a certain extent on the type of operation chosen, there probably being a much smaller premium on originality of thought and self-starting capabilities in a large staff operation than may be desirable for a small audit-type staff group. However, technical ability commensurate with the requirements of the job is assumed to be a prime requisite in any case.

Other than these general indications of what has been found useful in many organizations, there will be no attempt to discuss further the type of personnel who have been found specially useful in this type of work except to note that the health physicist should have those personal characteristics that are generally considered desirable. Thus, a minor Stalin is no more successful in this work than in many other jobs, and a "good fellow" who agrees with almost any proposal is probably much more of a problem than is the individual who has good technical reasons for his opinions and firmly sticks to his guns.

CONCLUSION

It may be stated as an overall conclusion that in radiation protection work and administration nothing can take the place of common sense backed with some technical know-how; a free use of instruments; and a willingness to trust the results but only to the extent that the validity of the samplings, the analytical techniques, or other factors that may be more or less well defined can be specified. For example, an individual may work at a location in which he is possibly exposed to several types of radiation fields; however, if his film badge gives no exposure indication, he should probably be considered as having received no dose unless it can be established that he did not use his badge properly or that there was other error in the data. Similarly, probably the best indication that an individual has actually taken radioisotopes into his body is the evidence of his own excretion. Potentiality thus may be recognized but not confused with the actual exposure.

Finally, the health physicist, plant personnel, and others may also remain confident that present permissible limits of exposure are sufficiently low as to provide all of the protection, certainly against somatic effects, that can be expected to be necessary under any condition and that the risk posed by exposures when these limits are met is probably much less than that willingly faced in a wide variety of more prosaic life situations.

SUGGESTED READING

Any good text or analysis of business management or organization, including that for hospitals, will include applicable information.

Descriptions of organizations for safety, fire protection, or industrial hygiene. Some of the publications of the National Safety Council and the National Fire Protection Association include such information.

Organizational and administrative information of specific companies or plants with respect to safety-type activities.

QUESTIONS

1. Of what significance is the statement "the ends justify the means"? the fact that not all radiation-protection plans are identical?
2. In a large operation, what are the usual duties of one assigned to take care of the health-physics problem of a segment of the operation? Can these duties always be clearly defined?
3. What duties are usually under the jurisdiction of the central staff group?
4. What are the differences between a small staff and a large staff organization?
5. What problems may labor relations considerations present to a practically minded and technically oriented health physicist?
6. What special problems in radiation administration might a hospital encounter?
7. Why must radiation protection be an around the clock operation rather than simply a haphazard check?
8. In what ways could the radiation monitoring of a certain area be a reflection of the personal characteristics of the individual in charge?
9. In general, how can individual performances be checked in radiation protection groups?

CHAPTER TWENTY

Nuclear Weapons

It is perhaps unfortunate that the first major acquaintance of the public with nuclear energy was the result of weapon usage at Hiroshima and Nagasaki in 1945, which was the prime factor in ending the very bloody World War II. The some 200,000 casualties of the first two weapon releases have all too frequently been used as a justification for a claim that nuclear weaponry, especially if used by the United States, is more morally reprehensible than other types. Thus it follows that this nation should have an especially great sense of guilt at their introduction. Apparently ignored are the similar moral aspects of any war or any weapon use therein. It is ignored and forgotten that by responsible estimates some one million casualties to Americans and their allies would probably have resulted from what was foreseen as the necessary invasion of Japan to end the conflict. Add to this the at least equal number of probable Japanese casualties, and it might well be concluded that the approximate 100,000 to 120,000 deaths and an equal number of injured was a cheap price to pay for avoiding the probable two million invasion casualties. Similarly forgotten is the fact that the "secret" of nuclear fission and the resulting weapons production could not be indefinitely concealed; scientific discoveries are made and used, regardless of opinion or desire.

In any event, it is from these explosions and evaluations of subsequent tests, principally those of the United States and the United Kingdom, (France, Russia, and Communist China have released no details or data from their testing), that estimates may be made of the results of a nuclear weapon attack; some of these are briefly outlined below, especially as they apply to the radiation consequences. The overall effects of nuclear weapons and general considerations of protection have been rather widely described, both technically and in popular and easily understood documents. However, it may not be amiss to point out that many of the popular presentations are made with "an axe to grind" and they thus reflect a predetermined point of view; unbiased fact can easily come out a poor second to propaganda half-truths.

The effects of nuclear weapons may be considered in two distinct parts. The first concerns results occurring at the time of weapon use and immediately thereafter; normally, only locations near the weapon release site are affected. The second aspect concerns the dispersal of fission products and associated radioactive materials into the upper atmosphere and their subsequent descent as fallout. These provide the principal intermediate and long term effects of nuclear weapons, and locations far distant from the reaction itself may eventually be involved.

ENERGY RELEASE

The "yield" of nuclear weapons, which indicates their power of destruction, is listed in terms of the equivalent number of tons of TNT necessary to produce the same blast effect. The 20 kiloton weapons used at Hiroshima and Nagasaki are considered as having approximately the same explosive effect as 20,000 tons of TNT—if this could indeed be exploded as a unit. The fact that these weapons brought a different order of magnitude of destructiveness to war is indicated by the fact that the "blockbusters" of World War II used explosive considered to be the equivalent of 8 to 20 tons of TNT. Thus, the early Nagasaki-Hiroshima bombs were about a thousand times as effective, explosivewise, as were the largest of the wartime conventional bombs. The energy of a 20-kiloton burst is approximately equal to that received from the sun by a 2 sq mile surface during an average day or that resulting in about 0.25 in. of rain over the city of Washington, D.C. An average earthquake is estimated to be equivalent to about a million (10^6) such bombs.

Since 1 ton of TNT releases about 10^9 cal, 20,000 tons will release about 2×10^{13} cal, or 8.4×10^{13} J, which is the result of fission of about 3×10^{24} atoms, since each fission releases about 200 MeV of energy (3.2×10^{-11} J). This is equivalent to complete fission of some 1120 g of uranium-235 ($^{235}_{92}$U). However, the reaction efficiency is much less than 100 percent, so an actual 20 kiloton weapon must contain more than 1120 g of $^{235}_{92}$U.

Some idea of the reason for the tremendously increased effect of a nuclear weapon as related to other explosives may be obtained by comparing the approximately 200 MeV of energy released in each fission with the maximum values of about 20 MeV per atom in other nuclear reactions and a few electron volts per atom for chemical processes, even explosive ones.

In a typical nuclear explosion approximately 90 percent of the total energy output, or about 180 MeV/atom, is released in a period of less than about 0.1 μsec, the remainder being released subsequently over a much longer period of time and appearing primarily as fallout. Table 20–1 gives an estimate of the energy distribution in these various releases. Additional details on the fallout are given subsequently.

TABLE 20–1

ENERGY RELEASE IN NUCLEAR FISSION REACTIONS

Energy Forms	Energy per Fission (MeV)
Immediate	
Kinetic energy of fission fragments	165 ± 5
Kinetic energy of fission neutrons	5 ± 0.5
Instantaneous gamma-ray energy	7 ± 1
Subtotal	177
Delayed	
Gamma rays from fission products	6 ± 1
Beta particles from fission products	7 ± 1
Neutrons from fission products (beta decay)	10
Subtotal	23
Total	200 ± 6

Note: In weapon usage the overall effect of the kinetic energies of the fission fragments and fission neutrons is the production of blast and shock equivalent to about 100 MeV/fission and of thermal radiation equivalent to about 70 MeV/fission. The delayed energy release becomes a problem as fallout.

NUCLEAR FISSION

A typical nuclear fission reaction has been previously given as follows:

$$\begin{array}{ll}
{}^{1}_{0}n + {}^{235}_{92}U \rightarrow {}^{236}_{92}U \rightarrow {}^{141}_{56}Ba + {}^{92}_{36}Kr + 3{}^{1}_{0}n + Q & \\
{}^{141}_{56}Ba \rightarrow {}^{141}_{57}La + {}^{0}_{-1}e & {}^{92}_{36}Kr \rightarrow {}^{92}_{37}Rb + {}^{0}_{-1}e \\
{}^{141}_{57}La \rightarrow {}^{141}_{58}Ce + {}^{0}_{-1}e & {}^{92}_{37}Rb \rightarrow {}^{92}_{38}Sr + {}^{0}_{-1}e \\
{}^{141}_{58}Ce \rightarrow {}^{141}_{59}Pr + {}^{0}_{-1}e & {}^{92}_{38}Sr \rightarrow {}^{92}_{39}Y + {}^{0}_{-1}e \\
& {}^{92}_{39}Y \rightarrow {}^{92}_{40}Zr + {}^{0}_{-1}e
\end{array}$$

The time required for actual fission of a nucleus once a neutron has entered is extremely short, probably of the order of fractions of a nano-second. Hence, if the neutron emitted by one $^{235}_{92}U$ nucleus as a result of fission causes fission when it enters another nucleus, most of the time between the separate emissions, called generation time, is taken up by the neutron travel between these $^{235}_{92}U$ atoms. This time is on the order of 10^{-8} sec (10 nanosec, or 0.01 μsec), which means that a neutron emitted at an average energy of 2 MeV travels some 20 cm between fissions, both because of the small fission cross section of uranium for high energy neutrons and the large number of elastic impacts required before these neutrons are slowed to thermal velocities at which fission capture is much more probable.

As is apparent from the above typical reaction, one result of the fission caused by entry of a single neutron into a $^{235}_{92}U$ nucleus is the emission of one or more neutrons. As many as six are released in some reactions, although the usual number is two or three (as in the illustration given), and the average for all reactions is about 2.43 neutrons released per fission. However, not all of the neutrons released in one $^{235}_{92}U$ fission cause fission in other $^{235}_{92}U$ nuclei. Some of them escape from the system entirely, and some may be lost through nonfission capture by various nuclei, both of $^{235}_{92}U$ and of other substances, as described in more detail in Chapter 3. By proper adjustment of the dimensions of the fissionable mass, the amount of fissionable material present, and the presence of other materials with appropriate neutron-interaction properties, conditions can be established whereby only one neutron produced by fission in one $^{235}_{92}U$ nucleus causes fission in another $^{239}_{92}U$ nucleus. Under these conditions the system is said to be just critical and the system's multiplication factor—which is defined as the ratio of the number of neutrons causing fission in one generation to the number producing fission in the preceding generation—is unity. If this multiplication factor is less than 1, the system is subcritical, and if it is more than 1, it is supercritical. Thus reactors are operated at the critical point, although they may be slightly supercritical while being brought up to power.

Successful weapons are supercritical systems, with the total amount of energy released and their other effects depending on the total number of fissions produced before the fission process ceases with the explosive scattering of the fissionable material. Accordingly, basic to the design of nuclear fission weapons are high multiplication factors to obtain a fast neutron buildup and the use of methods to hold the fissionable material together long enough to produce many neutron generations. This is illustrated in the following section.

Although $^{235}_{92}U$ is used for illustration, essentially all the conclusions drawn also apply to $^{239}_{94}Pu$ fission.

FISSION BUILDUP

Constant Generation Time

If H is the average number of neutrons released by fission of a single $^{235}_{92}U$ nucleus and J is the average number lost before fission is produced in a second generation, obviously $H - J$ represents the number of fissions in the second generation, and $H - J$ is thus the multiplication factor. Accordingly, if there are N_1 neutrons in the first generation, the number in the second generation, N_2, will be given by $N_2 = N_1(H - J)$.

Similarly
$$N_3 = N_2(H - J).$$

Obviously the number of neutrons in this third generation is $N_1(H - J)^2$, and in the nth generation it is $N_1(H - J)^{n-1}$, which is in accord with normal geometrical progressions. Since the total number of fissions produced, N_T, is obviously the sum of this geometric progression,

$$N_T = \sum_{k=1}^{k=n} N_k = \frac{N_1[(H - J)^n - 1]}{H - J - 1}.$$

A simple approximate illustration may be useful in this discussion of weaponry. Thus, if $H = 2.43$ and $J = 0.43$, then the multiplication factor, $H - J = 2$, and the number of neutrons in the nth generation resulting from an initial N_0 neutrons is

$$N_n = N_0 2^{n-1}.$$

Similarly, the total number of fissions produced is given by the relation

$$N_T = \frac{N_0(2^n - 1)}{2 - 1}.$$

It may readily be calculated that if a single neutron initiates the reaction, about 81 generations will be required to produce the 3×10^{24} fissions of a 20 kiloton blast, and it is readily apparent that over 99 percent of the energy is released in the last 7 generations. If the constant generation time is assumed to be 10^{-8} sec, this 99 percent energy release will occur in about 0.07 μsec.

Variable Generation Time

The above simple analysis assumed that the generation time was well defined and specific. However, in a large number of fission processes the individual generation times will vary to reflect the various probabilities of neutron capture and thus their actual travel times. For a large number of neutrons the generations become "mixed up," and fission production becomes essentially a continuous process even though there may be an average generation time.

In this case the average fractional change in the number of neutrons in a system with time is

$$(H - J - 1)/T,$$

where T is the average generation time. Hence, the time rate of change of the N neutrons in a given system is given by the relation

$$\frac{dN}{dt} = \frac{N(H - J - 1)}{T}.$$

The solution of the above equation is obviously

$$N_T = N_0 e^{(H - J - 1)t/T},$$

where N_T is the total number of fissions produced; if the reaction is initiated by a single fission, N_0 is unity.

Since t is the total time and T is the average generation time, the ratio t/T represents the average number of generations, n, since the start of the reaction. Hence, the above equation may be expressed as

$$N = N_0 e^{(H - J - 1)n}.$$

If the previous values of $H = 2.43$ and $J = 0.43$ are used, this equation becomes

$$N = N_0 e^n$$

Again, starting with a single neutron, it may be shown that approximately 57 generations are required to produce a 20-kiloton blast and that for an average generation time of 10^{-8} sec, over 99 percent of the fissions occur in the last 0.07 μsec. The difference in the number of generations determined by the two methods reflects the fact that, in this latter case, the calculations are based on a continuous buildup at an average rate; in the former one, they are based on a stepwise buildup, with each step being equal to that average. For present purposes this difference is largely of academic interest since the total buildup time is less than 1 μsec and practically all of the energy is shown to be released in less than 0.1 μsec by either method of calculation. In any event, the energy is released in such a small period of time and space that reactions of explosive violence take place.

A major problem in the manufacture and use of a bomb is that of keeping the supercritical mass together long enough to produce the high-explosive characteristics that are desired. Some of the methods used for this purpose have been discussed in the public press and no attempt will be made to mention them further, other than to point out again that a criticality accident in a normal operating facility would not be expected to have anything like the destructiveness of a bomb or even to be explosive in character. This is primarily because the multiplication factor is so small that buildup occurs comparatively slowly, and the heat and other effects produced tend to blow the material apart, thereby stopping the reaction before an explosion can occur.

TYPES OF WEAPON USE

In addition to the air bursts used at Nagasaki and Hiroshima a nuclear weapon may be used to produce underwater, ground-level, or even underground explosions. However, only the air burst will be considered further since from a military standpoint it has the advantage that explosion at a properly chosen height produces the greatest overall destruction of conventional buildings and facilities and the greatest area of injury to personnel. The literature gives the method of computing the height of explosion for most efficient weapon usage for the various effects desired.

Some interest has also been expressed in using high-level explosions against missiles or other weapons since there is good evidence that such an explosion will seriously affect electronic components or other automatic control systems of a missile or plane at great distances; "idiots" will thus be made out of these control devices for a sufficiently long time to cause the weapon to fail in its mission. However, for purposes of this discussion the brief analysis of air bursts is based entirely on one occurring at such a height as to do maximum damage to normal buildings and personnel as described.

Ground-level bursts will produce a greater effect in the immediate vicinity of the bomb itself, thus damaging or destroying underground shelters at a greater distance than would be the case for an air burst. Surface attacks could be of particular value against so-called underground "hardened bases". The fission fragments produced are not the only radioactive materials distributed, since the vaporized ground materials will themselves be subsequently dispersed in the form of radioactive dust, thus increasing the resultant radioactive fallout.

Underwater explosions could have the same effect as all depth-type charges in destroying or damaging ships and submarines over a great distance. They could also make the water of an area somewhat radioactive and could produce a high plume of radioactive vapor which would then be dispersed by the winds and cause fallout.

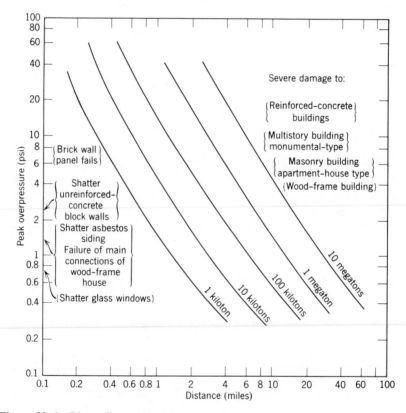

Figure 20–1 Blast effects of nuclear weapons (air burst).

WEAPON CAPABILITIES

Figure 20–1 summarizes the area over which weapons of various yields will produce certain levels of destruction from blast in terms of overpressure. Figure 20–2 gives similar data for thermal effects, and Figure 20–3 provides some data on radiation effects.

For purposes of discussion, a 1-megaton weapon, which has the destruction capability of a million tons of TNT, may be chosen as a basic weapon because, as indicated by the figures, its destructive power is apparently great enough to damage severely, or disable, all but a few conventional targets in the United States and elsewhere. Of course it may not destroy a hardened target or produce severe damage over an extended area, but the same is true for a 5 megaton weapon, which has also been used for reference.

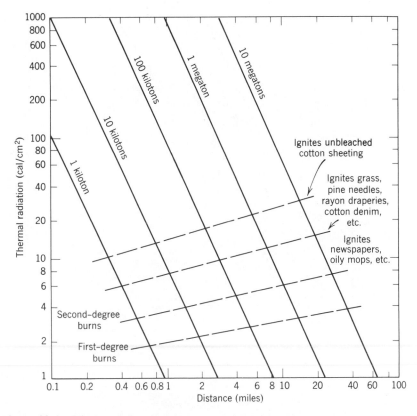

Figure 20–2 Thermal effects of nuclear weapons (air burst).

As first approximations the amount of damage at a given distance from a weapon, expressed in terms of peak overpressure for blast effects and incident energy for thermal and nuclear radiation, are approximately proportional to the weapon yield; this is especially true for blast and for thermal radiation. However, for both neutrons and gammas, especially

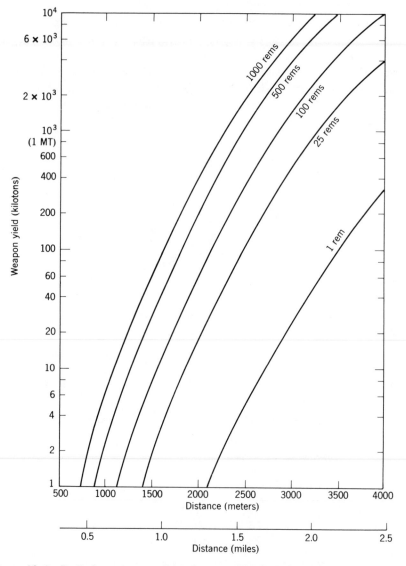

Figure 20–3 Radiation exposure doses from nuclear weapons (air burst).

the former, the actual radiation effects also depend strongly on the type of weapon involved; for gammas particularly the effects for the larger weapons are greater than this simple relation would imply.

Blast effects vary approximately inversely as the cube of the distance from the explosion; and thermal radiation effects, inversely as the square of the distance. For nuclear radiation, however, the decreasing effectiveness with distance is much more rapid, being approximately exponential in form. Thus, whereas the distance at which a given blast effect will be noted is approximately proportional to the cube root of the weapon yield and that for thermal effects is approximately proportional to the square root, the similar relationship for gammas and neutrons is less easily stated but is available from the literature and is summarized in Figure 20–3.

INJURIOUS EFFECTS OF WEAPONS

Table 20–2 lists the approximate distances at which certain injurious effects may be anticipated, on the average, from weapons of various yields. It should be emphasized that these are useful as general guides only, since local conditions, vagaries of the weapon itself, personnel peculiarities, and

TABLE 20–2

EFFECTS OF NUCLEAR AIR BLAST

		Weapon Yield Size	Distance (miles)				
			1 kT	10 kT	100 kT	1 mT	10 mT
Lethal Effects	Blast	Concrete buildings severely damaged	0.19	0.44	1.1	2.5	5.9
	Heat	20% closer than where second degree burns received	0.40	1.20	3.2	8.8	19.2
	Radiation	500 rem exposure	0.60	0.80	1.1	1.5	2.1
Minimum Injury	Blast	Windows shattered; frame house damaged	1.4	2.9	6.5	11.5	30.0
	Heat	First-degree burns	0.70	1.90	5.3	14.0	>33
	Radiation	100 rem exposure	0.70	1.00	1.3	1.8	2.4
No Injury	Blast	20% farther away than window destruction	2.40	5.0	12.0	24.0	55.0
	Heat	20% farther away than distance for first degree burns	0.84	2.28	6.36	16.8	>40
	Radiation	25 rem exposure	0.85	1.2	1.5	2.2	2.9
		1 rem exposure	1.3	1.7	2.4	2.8	3.8

what can only be identified as individual luck will play large parts in the actual results from a weapon. In general these tables attempt to indicate the distances from weapons in the 1 kiloton to 10 megaton range that would probably be lethal, produce minimum injury, and cause no injury to an individual.

Lethal effects are considered to occur at distances from the blast (a) at which unreinforced concrete buildings will be severely damaged, (b) 20 percent closer than the distance at which second degree burns are experienced, or (c) where a 500 rem exposure will be given.

Similarly, minimum injury is considered to occur, or be possible, at distances (a) approximately midway between that where most windows will be destroyed by blast and that where a frame house will be severely damaged, (b) where first degree burns will be readily experienced over most of the body, or (c) where a 100 rem exposure will be received.

Finally, no injury is considered to occur at distances (a) 20 percent further away than that where windows are destroyed, (b) 20 percent further away than that where first degree burns are experienced, or (c) where exposures of no more than approximately 25 rems may be received. Also listed for information are the distances at which exposures of 1 rem may be anticipated. Some of these distances, particularly those for blast effects, are interpolations of tabular values given in the literature, whereas most of the others are based more directly on such tables.

As will be seen from the tables, the greatest possibility of direct injury, as well as damage, results first from fire effects and next from the blast. Both of these are dangerous at much greater distances from the explosion than is the radiation, either gamma or neutron, and the difference becomes greater as the weapon yield is increased. Hence it may be concluded that if an individual is safe from blast and fire at the time of a weapon delivery, he is also safe from immediate injury as far as radiation is concerned. Furthermore, any injury he may get will be much more likely to involve either blast or fire than it will radiation.

Probably the greatest immediate hazard from a weapon is its radiant energy, which is estimated to comprise about 35 percent of the total energy released, with about half of this being transmitted as thermal radiation in approximately the visible, ultraviolet, and infrared ranges. This thermal radiation is sufficiently intense to ignite clothing or other thin materials at much greater distances than necessary to cause burns, although shielding is readily provided by normal materials. Of perhaps greater potential seriousness in a large city is the possibility that the weapon could act as an incendiary, thus igniting fires and possibly initiating a so-called "fire storm" such as swept Hamburg, Germany, in World War II as a result of the use of large quantities of incendiary bombs over a large area. Such a storm not only uses up the oxygen but also completely destroys all combustibles over the area affected. Since such fires could affect locations very much outside the direct flash range or other direct effects, fire is probably the greatest immediate danger of an atomic weapon.

Blast effects are also important but at less distance than fire. The blast effect that will probably cause serious injury at the greatest distance is that of shattering glass windows and blowing the fragments about a room. As with fire, the possibility of injury from blast outside the area of serious destruction depends largely on the location of the individual concerned.

As also shown by Table 20–2, immediate radiation doses at levels necessary to cause any possible clinically observable effects (about 25 rems) or sickness indication (about 100 rems) can occur, in general, only at distances that are much smaller than those probably necessary for thermal radiation to cause actual traumatic injury for most weapons of the yields indicated, and the same is true for blast except for the lowest energy weapon considered. In addition, the distances at which nuclear radiation can cause injurious effects as compared to those from blast or fire become progressively smaller with increased weapon yield.

FISSION VERSUS FUSION WEAPONS

In the preceding discussion of weapon effects no distinction has been made between fission weapons and fusion weapons. For the first type the reactions are those of the fissionable materials—uranium-235, plutonium-239, and possibly uranium-233—individually or in combination, and the results are reportedly indistinguishable, although proper component selection may involve material economies and efficiencies. The fission bomb is the result of the release of energy by the splitting apart of heavy atomic nuclei and the transformation of some of the "excess mass" resulting therefrom to energy of other forms. One result of these reactions is the production of many artificially radioactive isotopes called fission products. One of the fission reactions for $^{235}_{92}U$ has been given.

On the other hand, the fusion weapon, commonly called a hydrogen bomb or thermonuclear weapon, depends on the release of energy as a result of the destruction of mass in the production of helium by combination of various hydrogen isotopes. These involve primarily deuterium ($^{2}_{1}H$), which is a small component of the hydrogen of natural water, and tritium ($^{3}_{1}H$), which must be "manufactured." Some of the hydrogen reactions of particular interest are as follows:

$$
\begin{aligned}
{}_1^2\mathrm{H} + {}_1^2\mathrm{H} &\rightarrow {}_2^3\mathrm{He} + {}_0^1 n, \\
{}_1^2\mathrm{H} + {}_1^2\mathrm{H} &\rightarrow {}_1^1\mathrm{H} + {}_1^3\mathrm{H}, \\
{}_1^3\mathrm{H} + {}_1^2\mathrm{H} &\rightarrow {}_2^4\mathrm{He} + {}_0^1 n, \\
{}_1^3\mathrm{H} + {}_1^3\mathrm{H} &\rightarrow {}_2^4\mathrm{He} + 2{}_0^1 n, \\
{}_1^2\mathrm{H} + {}_2^3\mathrm{He} &\rightarrow {}_2^4\mathrm{He} + {}_1^1\mathrm{H}.
\end{aligned}
$$

The first three of these reactions represent those that can occur in a deuterium system. Thus

$$
5{}_1^2\mathrm{H} \rightarrow {}_2^3\mathrm{He} + {}_2^4\mathrm{He} + {}_1^1\mathrm{H} + 2{}_0^1 n.
$$

In this overall reaction approximately 25 MeV of energy are released. On the average, then, each nucleon (proton or neutron) in this reaction will release about 2.5 MeV, whereas the release of 200 MeV by a ${}_{92}^{235}\mathrm{U}$ fission means that each nucleon releases about 0.8 MeV. Thus on this basis the fusion reaction is about three times as effective as the fission one. Practically it is estimated that the deuterium contained in 1 gallon of water, which may be separated and removed by present techniques at a cost of less than 2 or 3 cents, would release energy in this reaction equivalent to about 300 gal of gasoline used as a fuel.

Tritium may also be produced by neutron bombardment of certain materials, and comparatively large quantities of tritium are thus available. Ordinary hydrogen (${}_1^1\mathrm{H}$) may also take part in a fusion reaction, and this is considered to be the source of the sun's energy. In view of the ready availability of hydrogen, deuterium, and tritium, the technical problem in causing a fusion reaction is that of producing and maintaining conditions under which it may take place.

Such conditions are those that permit the respective atomic nuclei to collide with each other despite the high repulsion of their respective electric charges. In general this means high nuclear velocities (and thus kinetic energies) and sufficient density of nuclei to permit a comparatively high probability of impact and perhaps a large rate of such impacts. These conflicting requirements are the present bugaboo of providing controlled nuclear fusion since the high velocities may be produced by high temperatures, but this has the effect of dispersing the nuclei and consequently producing a low nuclear density. Current efforts at controlled fusion generally employ plasmas, a mixture of charged high energy ions in space, which are placed in strong magnetic fields that have a resultant constricting effect on the paths of the moving ions. The high gravitational field of the sun provides the necessary constricting force to make fusion possible there even with natural hydrogen. In general, however, the conditions necessary to initiate the fusion reaction are much less stringent for the deuterium reaction described above than for a hydrogen reaction producing the same effect, and those for the tritium reaction are less stringent still.

However, since a weapon is not a controlled reaction, it is necessary only to produce a high temperature rapidly, it being recognized that, as with the fission weapon, an eventual result of the reaction will be a wide dispersal of the material. Since one of the characteristics of a fission weapon is high temperature, such weapons are commonly used as starters, or initiators, of the fusion weapon. It is estimated that in the earlier fusion weapons approximately half the energy released was from the fission reaction itself.

No figures have been given indicating weapon sizes for fission and fusion weapons and militarily there is probably little difference. However, for purposes of discussion where it becomes important to distinguish between them, it may be assumed that weapons with yields of 100 kilotons and less are probably fission ones and those of greater yield are fusion weapons.

CLEAN AND DIRTY WEAPONS

Although some neutrons are initially emitted, the fusion reaction does not itself produce radioactive debris or waste similar to the fission products of the fission reaction. Hence the fallout from an air burst of a fusion weapon is primarily that of the fission starter. However, the neutrons in the reaction can activate other materials of the weapon, as is the case with any neutron emission. Since these are comparatively high-energy neutrons, one reaction of particular interest is the fission capture of some of these neutrons by uranium-238 (and uranium-235) in the starter, with the result of both additional fission and release of fission products.

The fusion bomb is essentially what might be called a *clean* weapon as compared to the fission bomb, which is essentially a *dirty* weapon in that it produces large quantities of radioactive fallout materials, some of which have extended half-lives. However, even a clean bomb may be made dirty, and a dirty one even dirtier by two rather simple methods. One is the explosion of the bomb at the earth's surface. In this case the radioactive material would be not only that of the fission products but also that of the metals and other materials in the environment. These not only become radioactive but are also vaporized and pulverized, thus forming a part of the subsequent fallout.

Another simple way of making a dirty bomb from a clean one or to increase the dirtiness of a weapon is to include in its construction some specific nonfissionable, or even fissionable, materials beyond those needed in its manufacture. These materials would be those that become highly radioactive on neutron irradiation at the time of the burst; for example, cobalt is a very strong gamma emitter, and even a "pure" fusion bomb with significant quantities of cobalt could produce significant fallout. Similarly, fission can be produced in uranium-238 itself by energetic neutrons, and the inclusion of additional uranium-238 in a weapon will make a much more radioactive system than if it is not included.

Speculation has evolved around a so-called death ray, or neutron bomb, by which it is speculated that a weapon which becomes a pure neutron source would kill animal life over a large area but would not produce the radioactive debris that has been described. The nearest thing to a neutron bomb as thus described would probably be a fusion reaction without the fission starter. However, this is still in the future as is controlled fusion. Since neutrons themselves have the property of making other materials radioactive, the results of the explosion of even a pure neutron bomb would cause various materials, particularly metals, to become radioactive. A pure gamma emitter would not have this particular effect.

INTRODUCTION TO FALLOUT

The preceding sections have dealt primarily with the immediate effects of a nuclear weapon explosion. It has already been pointed out that the principal aftereffects of concern are those of radioactive fallout, which may involve areas remote from the burst and at considerably later times. For some reason this question of fallout has produced a tremendous amount of concern and anxiety among large portions of the populace—some perhaps desirable, but most of it probably unnecessary.

The fallout itself is composed of particles of various sizes and masses. The smaller particles would be expected to remain aloft for extended periods, perhaps years. These particles are usually brought down to earth in rain, and such occurrences are sometimes called rain-out. In any assessment of fallout results it should be clearly realized that radioactive debris decays and the activity of the airborne material steadily decreases. Thus, if fallout occurs later than predicted, that from a given mass will be less radioactive than if it descended earlier, and the potential radiation for a given location will be less than anticipated.

Of perhaps greatest importance and interest is the fallout which occurs in the first few hours or days after the explosion. Such fallout has generally been described as a rather coarse dust that probably most closely resembles the "fly ash" frequently emitted from the stacks of large steam plants. (The

Japanese fishermen on the Lucky Dragon reportedly stated the fallout affecting them looked like a snow coating.) The most extensive information concerning injury from fallout has come from the experience of the natives in the Marshall Islands who were inadvertently involved in fallout from the 1954 testing of the first fusion bomb in the Pacific; these effects have been already reviewed.

FALLOUT DISTRIBUTION PATTERN

As a general assumption it may be concluded that the radioactive debris from a weapon will be taken by the winds and disseminated over the earth's surface. It is obvious that the course of the wind, or the actual path taken by this debris, or its point of descent cannot be precisely predicted for any given case. However, an idealized condition may be assumed, and the fallout pattern thus proposed has a minimum virtue of at least giving a point of departure for considerations of fallout or other aftereffects, and it may thus provide a reasonably good working estimate of what might be anticipated in an actual incident. As is the case for immediate weapon effects, a 1-megaton weapon exploded in air will be assumed.

Figure 20–4 indicates the approximate overall decay pattern of fission products, this including not only the fission fragments themselves but also presumably that from various other materials activated by the fission neutrons and carried up in the weapon plume. It will be noted that for a considerable time after the reaction itself the activity (and consequently the radiation rate) of this fallout material is reduced by an approximate order of magnitude (factor of 10) after a time-interval factor of 7. Thus, 8 hr after the burst, the radiation level will be only about 10 percent of its value 1 hr after the burst; it will be only 1 percent of that value 2 days after the burst and only 0.1 percent as great after 2 weeks.

Since the areas affected by fallout shortly after a burst depend on distance, direction, and meteorological conditions, neither the specific areas involved in a given incident nor the intensity at a given position can be accurately predicted. However, some indication of fallout conditions may be obtained from idealized fallout patterns that were developed from experimental data. These assume a 1-megaton air burst, a wind of 15 mph,* and average meteorological conditions. Under these conditions the initial fallout pattern will take the approximate form of an ellipse, with very little radioactive material being carried upwind. Obviously no fallout can occur at a given location until the radioactive cloud has reached it,

* A 15/mph wind is considered to be a moderate breeze that will raise dust and loose paper and will also move small tree branches.

374

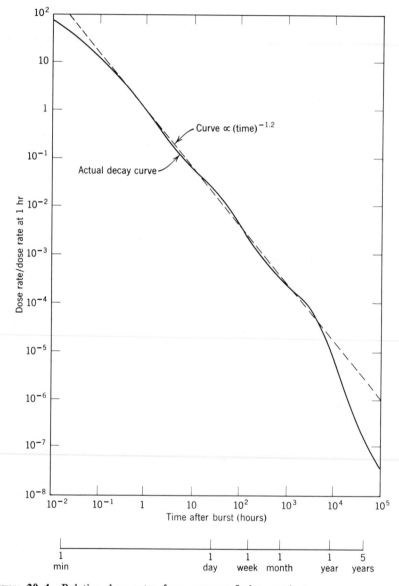

Figure 20-4 Relative dose rates from weapon fission products.

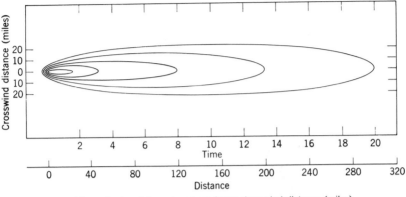

Figure 20–5   Shape of fallout isodose curves

and the actual deposit of material (or fallout) apparently ceases some 6 to 12 hr after its start. Subsequent radiation fields are then caused by material which has already descended. Figure 20–5 indicates the shape of the resulting isodose locations (those receiving the same total dose).

Under the above assumptions Figure 20–6 indicates the total radiation exposure resulting from fallout as a function of downwind distance for various times after a burst, including that of indefinite extension; isodose ellipse widths are also indicated. Since the total fission-product activity is closely proportional to the weapon yield, the total accumulated doses from weapons of larger or smaller yield than the 1 megaton hypothesized may be reasonably well estimated by multiplying the ordinate of Figure 20–6 by the appropriate factor. For fusion weapons the [fission/(fusion + fission)] ratio would indicate the relative fallout problem of the weapon since it is essentially fission alone that produces fallout. Any shelter-type protection of an individual for all or part of any period following the start of exposure will correspondingly reduce the values indicated by this figure.

For purposes of discussion, a location at which a dose of 100 rems could be received by an unprotected person during 2 days may be considered that place at which precautions to limit fallout exposure should certainly be taken. This is the exposure level at which some individuals would be expected to become nauseated if the exposure occurred in a period of a few minutes, the fact of the exposure actually occurring over a period of some 48 hr being itself a significant safety factor as related to minimal sickness. Obviously, a different total dose or other condition may

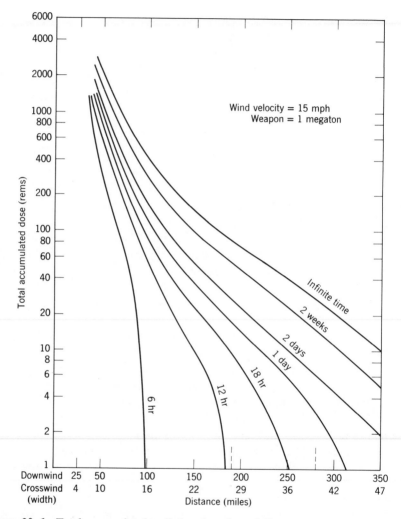

Figure 20–6 Total accumulated radiation dose from fallout.

be chosen as a basis for analysis. It has already been pointed out that there has been no evidence of significant internal exposure from fallout except where the external exposure was very dangerously high.

Under the above assumptions the locations of concern from immediate somatic effects of fallout would be those in an area some 120 miles downwind and with a maximum width of some 18 miles. Inside such an area some fallout protection would be necessary to prevent potential sickness or the individual would need to move outside the area. For smaller exposures than those hypothesized the distances would be correspondingly greater, the respective 25 rem exposure distance being 200 miles downwind and in a 29 mile wide area.

FALLOUT PROTECTION

Table 20–3 gives the shielding protection against fallout as afforded by various types of housing construction in terms of the ratio of the exposure outside the shelter to that inside. The table also gives the approximate downwind distances at which 100 rem doses would be received by individuals remaining in the respective shelters during the first 2 days. However, although the radiation rate outside the shelter after the 2-day period would be only about 1 percent of that 2 days earlier, it may still be substantial if the shelter has a high shielding factor. Therefore, a decision of whether to remain inside a shelter for more than a 2-day period would depend on the actual outside radiation level. Figure 20–7 indicates these initial dose rates at different distances and after various times.

TABLE 20–3

SHIELDING AGAINST FALLOUT

Type of Shelter	Protection Factor	Exposure Distance[a] (miles)
None	0	120
Inside frame residences	1.5–2	90–100
Inside brick veneer or concrete block residences (windows blocked)	2–5	65–90
Partially exposed basements of residences of one-or two-story buildings; Inside residences with masonry walls	2–10	50–90
Residence basements completely below grade (ceiling above shielded or of concrete); Simple basement fallout-shelters.	10–50	30–50
Well-shielded basement fallout shelters	50–250	20–30

[a] Downwind distance from 1 megaton burst at which 100 rems will be received in 2 days by a sheltered individual.

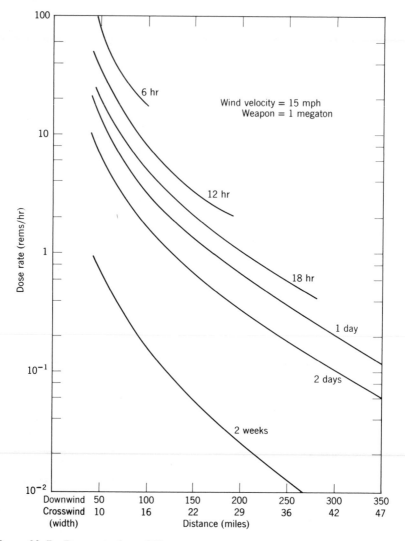

Figure 20–7 Dose rate from fallout.

It will be observed from Table 20–3 that an ordinary frame house will approximately halve the exposure. A brick house, especially the interior rooms and with windows and similar location of low density being shielded, will reduce the radiation exposure by a factor of approximately 10. A below-grade basement will further increase the shielding factor, and if layers of materials of fairly great density, such as books, are placed above the shelter location, a shielding factor of possibly 20 to 50 may be obtained. If reliance is placed on high-shielding shelters near a weapon burst, some consideration may also be given to air filtering, especially during the time of radioactive material descent.

Concrete floors in multistory buildings or in single family units with a basement are much more effective than the usual wood joist floors for shielding people below them from radioactive material on the roof. Similarly, if a method were devised for removing fallout material from the roof, such as washing it down, exposure on the inside of the building and in a basement could be very much reduced. It should be noted also that the detailed methods used by various governmental groups to evaluate buildings recommended for fallout protection generally indicate the need for higher shielding factors than those of Table 20–3. In a specific situation, however, determinations of actual fallout problems are best based on actual measurements made with an appropriate meter and not on general estimates (unless no meter is available).

Consideration may also be given to the possibility that more than one weapon or a larger weapon than that considered will be used. Some scaling factors for analysis of larger or smaller weapons than the 1 megaton hypothesized have been previously indicated. However, as has been mentioned, it seems improbable that a weapon much larger than this would be necessary except to make the destruction area large enough to reduce the possibility of missing a target or to take care of a hardened underground shelter, such as a Strategic Air Command base or other military objective. Some analyses for a 5 megaton weapon have been described in various documents.

The possibility of the fallout patterns of more than one weapon overlapping may also be considered. However, this would be probable only if separate military objectives are very close together. An evaluation of such objectives is obviously beyond the scope of this book, although it is probable that many locations that have considered themselves as at least secondary targets may be far from being gleams in the military eye. An elementary calculation shows that about 1500 1-megaton weapons carefully spaced under the 15 mph wind distribution pattern assumed would be needed to cover the United States surface with fallout so that doses of at least 100 rems would be received everywhere. Fewer larger weapons would be required for the same effect but they would have a total yield of more than 1500 megatons. The improbability of such careful coverage in a practical case, as well as the large areas with negligible targets, obviously means that, even in an all-out attack, large portions of the surface of the United States would be much less heavily irradiated than the minimum hypothesized 100 rems in 2 days. On the other hand, there would be areas in which the fallout radiation would give exposures very much above 100 rems in 2 days.

Table 20–3, which indicates the relative distances at which 100 rems from a 1-megaton weapon would be received in 2 days in different types of shelters, also shows that a well-constructed underground basement with some additional protection on the floor above or provisions for removing contamination from the roof can be at a distance of some 40 miles down-wind from the site of the weapon explosion and still limit exposure to approximately 100 rems in 2 days. Such a location is thus only some 20 to 30 miles from the location of destruction and serious injury or death from the initial effects of the weapon itself. Hence it may be concluded that it is only in this comparatively small elliptical "doughnut" region of some 40 miles by 6 miles maximum dimensions that a shelter with greater shielding properties than is possible with an adapted below-grade basement would be useful for fallout shielding and not at the same time need to be resistant to the blast and fire of an attack site. Obviously, also, the exact point of explosion of a weapon (or epicenter) may not be precisely determinable.

LONG TERM FALLOUT

The above discussion has been concerned primarily with the short term effects of fallout, that occurring in a few days after the burst. However, some of the smaller particles get into the upper atmosphere and remain for long periods of time and travel for great distances. In fact the existence and magnitude of Soviet air testing has been established by observation of fallout at remote distances. Some of these small particles eventually may serve as condensation centers and thus come down in the rain or otherwise descend from the upper atmosphere at various and rather unpredictable times. It was thus freely predicted that the Soviet tests in the fall of 1961 would result in much fallout in the spring of 1963. Actually there was some increase in the amount of radioactive fallout at that time, but it was not generally as much as predicted nor did it cause even a fraction of the extremely low levels established as permissible limits. For some reason, probably meteorological but also reflecting the areas of principal testing, most fallout occurs in the earth's Northern Hemisphere.

The possibility that fallout radioisotopes could be taken up by the vegetation and then be used as a foodstuff either directly by man, as in cereals, or after it has gone through a vegetable-animal cycle, as in milk, has also been considered to be a particular fallout problem. Although there have been cases in which the content of radioactive materials in foodstuffs was much higher than normal, there has been no evidence that any concentrations have been great enough to cause even a possibility of danger.

Strontium-90 ($^{90}_{38}$Sr) and cesium-137 ($^{137}_{55}$Cs) are the principal radio-isotopes of concern entering this food cycle over a long period of time. In addition, iodine-131 ($^{131}_{53}$I) is important in fresh fallout, but its rapid decay makes it unimportant as a long-term fallout problem. However, because of its long half-life, the carbon-14 ($^{14}_{6}$C) produced and being dispersed in the air is usually considered to be the most hazardous product of nuclear testing. This has been mentioned in Chapter 6. Actually carbon-14 is routinely produced in the atmosphere by the impact of neutrons produced in upper air cosmic ray reactions with atmospheric nitrogen. The reaction is considered to be

$$^{14}_{7}\text{N} + ^{1}_{0}n \rightarrow ^{14}_{6}\text{C} + ^{1}_{1}\text{H}.$$

PRACTICAL ACTIONS

There is little that can be done in the way of radiation protection during a nuclear war as far as controlling the location of the burst or the place in which radiation fallout occurs. Thus the construction of shelters is about the only preparation that can be made. However, after a burst a continuing determination of the patterns of fallout, if any, and the resultant radiation levels would be of considerable value in an evaluation of the potential hazard. Direct measurements of the actual radiation levels and contamination studies to evaluate the possibility of significant radioactivity in drinking water and similar items may be undertaken. Of these actvities, probably the most important at the local level is the determination of the local radiation fields and the time at which it will be safe for individuals to work continuously at their normal jobs or other and perhaps more important ones. Conversely, it would be important to evaluate any time limitations that one should meet when he emerges from a shelter.

Of all the jobs of the radiation protection specialist at the time of actual nuclear war probably the most important is not in the field of physical radiation protection but in that of mental assistance; for example, the average person does not realize that water in a storage tank cannot become radioactive regardless of the amount of radioactive material which might settle on the tank; it is only if the contaminating material actually enters the water that it can become contaminated. This is because neutron bombardment is by all odds the principal way in which radioactive material is produced. On the other hand, gamma and beta bombardment have small effects. Recognition of the rapid decay of the radiation and the fact that a person can come up into comparatively high levels for a short time without receiving significant exposure can also be pointed out. Even under conditions in which it may be dangerous to come out for extended periods or the air is highly contaminated with radionuclides, the fact that comparatively simple devices such as cloth filters can remove significant fractions of the airborne radioactive fallout materials or that rain will remove essentially all material from a roof and thus reduce an initial radiation level are not known. These are items that should be very well emphasized in any preplanning along with the current emphasis on invariant tremendous danger, particularly from fallout, and the need for a great deal of shelter protection.

THE FALLOUT SCARE

Probably the most unfortunate aspect of the whole fallout matter is the too often equation in the lay mind, and possibly even in that of some part of the technical community, of permissible limits as being synonymous with injury—which they certainly are not. Thus it may not be amiss to reiterate that levels for recognizable injury are much higher than the figures given for permissible limits of food, air, or almost any other environmental condition. These limits refer to continuous intake and consequent exposure, whereas the actual exposure to fallout would probably be of very short duration. In fact, the source of food may be so selected as to minimize the intake of radioactive materials or even to eliminate it; canned foods would be nonradioactive, and radioactive decay with time would vastly reduce or essentially remove radioactivity from foodstuffs involved in fallout itself.

What may be a much greater danger of the fallout scare than its actual effects, is its tendency to lead individuals, even governments perhaps, to do things which are otherwise foolish; for example, it has been freely suggested that foodstuffs might need to be destroyed if the fallout reached a level somewhat above the permissible limits or that expensive and time-consuming processing equipment to remove radioactive materials, primarily strontium-90, from milk would be required. The removal of milk from the diet of children and mothers would be a much greater calamity than would result from their intake of this food containing strontium-90, cesium-137, or any other radioactive material—except possibly that resulting from a tremendous catastrophe, nuclear or otherwise, which would leave minimal existence precarious for other reasons. Thus it appears that actual danger would not be expected except for fallout that would be many times greater than appears to be remotely possible under test conditions and even probably under conditions of nonannihilation war.

In this scare attitude it appears that the same psychology which is a real problem with X-rays is appearing. This is a failure to recognize that much less danger exists from X-radiation than there is from a failure to discover the conditions which the X-rays may reveal, such as dental caries, broken limbs, tuberculosis, internal disorders, etc. Thus, concern over fallout, perfectly proper in its place, should perhaps be placed in better perspective than has all too frequently been the case heretofore, particularly among many of those whose views regularly appear in the public press.

REFERENCES and SUGGESTED READINGS

Samuel, Glasstone, editor, *The Effects of Nuclear Weapons,* USAEC, Washington, D.C., 1962.

Report of the United National Scientific Committee on the Effects of Atomic Radiation, United Nations, New York, 1958, 1962, 1964, and 1966.

QUESTIONS

1. What is a common way of denoting the power of an atomic, or nuclear, weapon?
2. What is the difference between a fission and a fusion weapon?
3. Explain what the terms "subcritical," "critical," and "supercritical" mean and how they apply to nuclear weaponry.
4. About how long is the buildup time in a fission weapon reaction and over what length of time is about 99 percent of all the energy released?
5. Assuming you live in Aurora, Illinois (approximately 40 miles west of downtown Chicago) and a 10-megaton nuclear bomb was dropped on the Loop, what effect could you expect to you and your frame house?
6. List the following in the order of greatest immediate hazard: blast, radiation, fire.
7. What are "clean" and "dirty" bombs?
8. What protective measures can be taken against fallout?
9. What are the two aspects of the effects of nuclear weapons that must be considered?
10. What is the approximate amount of Uranium-235 ($^{235}_{92}U$) that a 20 kiloton weapon would require if it were 100 percent efficient?
11. Define generation time.
12. At what distance would a person receive a 500 rem exposure from a 100 kiloton weapon?
13. How far downwind (15 mph breeze) would an unshielded individual need to be to receive only a 100 rad exposure from fallout of a 1 megaton device in 48 hrs?

14. Relate the energy yield of nuclear weapons to those of conventional weapons, and to chemical and physical phenomena.
15. How does a criticality accident differ from the detonation of a nuclear weapon?
16. Define multiplication factor.
17. What is the general theory behind the fusion weapon?

CHAPTER TWENTY-ONE

Benefits of Radiation

To this point radiation and radioactive materials have necessarily been considered as hazards that require control and the use of protective methods. However, it is because of the value of this product of the twentieth century and its consequent widespread usefulness that such considerations are needed. In fact, it appears that the limit of the usefulness of radiation and radioactive materials, as well as the reactors and other devices which produce them, is limited more by man's ingenuity than by anything else. In addition, there seems little doubt that the radiation exposures, actual injuries, and even fatalities which have comprised the peacetime price paid for development of these benefits have been very minor, especially in view of the advantages derived. Certainly this is the case when a similar evaluation is made of the cost of benefits derived from other types of advances, even such primitive ones as fire and the wheel, the products of which are even today busily exacting a toll for their use. The effectiveness of nuclear weapons in preserving the peace, albeit an uneasy one, and preventing a major war for over two decades now, is in itself a rather major accomplishment even though it is a question about which many people have very strong emotional opinions.

Hence, it is appropriate in a book designed to indicate the hazards of radiation and to give criteria for protection therefrom to conclude by taking a look, even though a very brief one, at the other side of the coin and indicate some of the benefits that have accrued from this development of our understanding and control of the atom and its nucleus.

MEDICAL USES

Probably the most striking advantages of radiation continue to be in the field of medicine, even as the use of X-rays for such purposes as setting broken bones was its earliest application. Today X-rays are similarly used, but they are also useful in determining the existence of tuberculosis, dental caries, and similarly evaluating other body conditions. Both radioactive materials and X-rays are widely used in various forms of therapy, particularly in treating such diseases as cancer, and some radioisotopes are specifics for certain cases, iodine-131 being such a specific for treatment of thyroid troubles. Beta emitters have also provided useful therapy for such skin difficulties as warts and benign or even malignant skin tumors, and massive radiation has proven to be a powerful palliative in many cases of cancer for which no treatment has thus far proven effective.

Radioisotopes are widely used in medical diagnosis, primarily as tracers. Both radiographic techniques and direct measurement can provide a great variety of information on such items as the size and shape of a bodily organ or malignancy, the circulation of the blood, and the adequacy of the blood supply for proposed treatment or operation.

Specifically in the field of medical research, tracer techniques have also been used in evaluating such diverse items as metabolism, the action of various drugs, the effects of damaging agents including carcinogens and radiation itself, cell permeability, and the building of synthetic molecules. Other biological and physical research, even without medical use as a direct goal, has potential for producing medically important results.

AGRICULTURAL USES

In agriculture, radioisotope usage has provided a method for gaining information concerning such factors as the movement of nutrients in plants, the uptake of fertilizers, soil fertility and structure, the action of growth regulators and herbicides, root growth, the role of water in photosynthesis, and characteristics of plant diseases and fungicides. Plant irradiation has produced genetic effects that are important in accelerating mutation rates, thus increasing the probability of developing new strains of plants with particularly desirable characteristics; they are also useful in evaluating the potential virulence of diseases and perhaps as indications of the types of new diseases that may be anticipated. In fact, some new and more useful strains of plants have been produced from irradiated stock.

Radioisotopes have proven invaluable in animal studies for investigating animal diseases, organic and biochemical processes, the effects of micronutrients, and the physiological availability of diet additives. As with plants, radiation genetics has proven valuable in certain animal studies, as well as in pest control. Radiation and radioisotopes have been used to interfere with the reproductive cycles of certain harmful insects to their eventual detriment. A particularly outstanding example of this type of pest control has been the practical elimination of the screwworm fly from the southern United States at a total cost of some $10,000,000, which was less than half the *annual loss* from this insect alone; such elimination also prevented the spread of the pest into other sections of the country used much more extensively for cattle growing. Radioisotopes have been used in studying insect migration and hibernation habits as well as in determining the effectiveness of various insecticides.

INDUSTRIAL USAGE

The industrial uses of radioisotopes are almost literally limitless. They include such diverse uses as cigarette density gages, oil well acidizing, studies of machine wear and the wear resistance of such everyday products as floor wax, radiographic testing including weld checks, flow markers in oil lines, and thickness or liquid level gaging. Success is also attending efforts to increase the nondecay time for foods by massive radiation. It was estimated that even in 1957 savings resulting from the use of industrial radioisotopes were in the range of three hundred to five hundred million dollars.

POWER FROM REACTORS

In addition to the uses of radioisotopes there are direct peacetime uses for the reactors which produce them. Probably the most important of these at this time is that of providing the energy source for power production, thus replacing plants that use such fossil fuels as oil or coal. Not only are there many advantages in the use of nuclear fuel but increased experience in reactor design and construction has given several United States corporations the ability to provide off-the-shelf reactors with guaranteed performance and at the same time steadily bring down their installation and operating costs. In early 1964 public utilities in Pennsylvania and New York, making cost estimates for power plants to go into operation in 1967 and 1968, respectively, reported overall price advantages for nuclear-fuel power plants even as compared to a more conventional coal-fired plant located at mine mouth. Subsequently the number of proposed nuclear-fueled plants increased to the extent that a veritable "explosion" of such proposals occurred in 1967. Unfortunately, by late 1968, it had become apparent that some of the earlier estimates, particularly of delivery dates on some power-plant components (even such conventional ones as pressure vessels), were perhaps overly optimistic. Accordingly, nuclear plant proposals have become a smaller fraction of the total predicted power addition to the nation's power pool. It is probable that siting and containment requirements, supposedly reflecting the potential hazards of nuclear plants (but probably not entirely divorced from political and other nontechnical matters either), have been important factors in this changing picture.

It was shown earlier that nuclear-fuel power plants have attractive possibilities for providing power at remote locations in which the provision of conventional fuel is a major problem or in which there are other justifications for a power plant that requires infrequent fueling. Thus major credit for early development of nuclear-fuel power plants goes to the military, which could quickly establish a complete power plant in a location remote from "civilization" but of military significance, and the development of a power plant requiring only infrequent fueling for naval submarine use.

There are also obvious advantages in nuclear-fuel power plants both for those nations that have no fossil fuel reserves of their own and must obtain their fuel from other countries and for those in which these fuels are expensive. Thus it was that several European nations made the first major use of nuclear powered plants on other than essentially a developmental basis, and such usage continues to expand.

An interesting switch in estimates of relative plant hazards was also noted in 1967, when, in the midst of the attention given the problems of air pollution by industrial wastes, it became apparent that such problems are practically nonexistent with nuclear plants.

OPERATION PLOWSHARE

Nuclear explosions for military purposes ushered in the nuclear era with its promise, but such explosions have generally been considered the era's threat. However, under the general title of Operation Plowshare, nonmilitary uses of such explosions are being explored and show high promise of aiding man in the improved control of his environment. They have potential for digging canals, such as a replacement for the present Panama Canal or to provide a water link across the Isthmus of Kra; they may prepare harbors where needed or desired and not as located by a whim of nature, and the west coasts of South America and Africa seem to provide many locations in which such construction would be very advantageous;

they may find use in water control, which might make possible the diversion of water from one side of a mountain range with plentiful rainfall to the other where near-desert conditions may exist, as in parts of Australia; they may find application in preparing water storage facilities in deserts or in constructing underground caves; and they may make it possible to obtain oil and gases from geologic formations where this is not otherwise practicable.

Various types of nuclear and other research will also find such high explosives useful. Since the energy-release range of nuclear explosives is some magnitudes above that of conventional explosives, many research projects would be extremely difficult, if not impossible, by conventional means, and even when conventional methods are feasible, the cost of using nuclear methods is estimated to be only a fraction of that required otherwise.

A POSSIBLE FUTURE?

It may be anticipated that continued research with X-rays, fissionable materials in various applications, nuclear reactors, and a whole host of accelerators that produce high-energy nuclear reactions will continually widen man's ability to control and shape his environment. Such work will involve radiation fields and the accompanying production of radioactive materials, which can be exceedingly dangerous if improperly handled; on the other hand, such danger can be kept to a minimum with careful handling.

It is just as easy to become so exercised about the hazards of radiation and thus to feel that its involvement in any usage should be forbidden as it is to recognize the fact that very few radiation injuries have actually happened and thus to become so complacent about the hazard that careless procedures are condoned. Overall, it is almost equally difficult to justify either the extreme of the undue cautiousness which removes many of the benefits of radiation from man's use or that of the carelessness which makes him pay too high a price for its benefits. This all points up again the importance of determining as accurately as possible the true extent of hazards for materials shown to be harmful to man under the conditions of their use. It must be reiterated that it is the facts which should control; opinion, especially any which ignores, overstates, or minimizes applicable data, is not enough or even appropriate.

True, the threat of a nuclear war is an ever-hovering cloud; but it must be remembered that the first weapon brought a final conclusion to a tremendous war that may have cost many thousands of lives, both combattant and noncombattant, before its end otherwise. Similarly, it is probably the free world's competence in nuclear weaponry since 1945 which has enabled it to withstand the continuing threat of war, and it is entirely possible that the possible horror of an all-out nuclear conflict may make such wars unthinkable—or the weapon itself may, by mutual and perhaps nonformalized consent, be shelved, as was the case with the poison gas of a prior era.

Yes, the problems of radiation and its proper control on earth are many. The possibility that it may get out of hand is an ever-present threat, but the rewards which have come from intelligent use of nuclear reactions and nuclear radiation are many, and those that seem to be just over the horizon but realizable with continued careful and effective usage may be even more important.

It is thus in this field of the developing use of the atom that the radiation protection specialist can provide a very important service in seeing that the new art is used safely and carefully but without hysteria and without unnecessary restrictions. He can see that any information obtained is made available as soon as possible with evaluations of the problems inherent in the use of these materials being rapidly assessed and determined. He can also assist the public in a clear understanding of the hazards involved, neither to the extent of "poo-pooing" them on the one hand nor of overestimating them on the other. In all cases, the radiation protection specialist has the basic responsibility to be in all things, true to himself and to the technical truth of his profession, regardless of his emotional opinions. It is to encourage and assist him in this effort that this volume has been prepared.

QUESTIONS

1. Give five uses of radiation that have helped and continue to improve medical care and research.
2. What beneficial (to man) genetic effects of radiation are mentioned in this chapter?
3. Has radiation had an economic effect on industry?
4. What advantages do nuclear reactors have as power sources?
5. What is Operation Plowshare?
6. Could there be any truth to the statement, "Without radiation man could neither have evolved nor now exist"?
7. What isotope is used specifically for thyroid troubles?
8. What are the agricultural uses of radioactive material?
9. What beneficial industrial uses could nuclear explosions have in the future?

General Bibliography

One of the problems of a neophyte in radiation protection is assessing the huge amount of work currently under way and the tremendous number of reports that have thus been produced on all phases of the subject; for example, during 1967 alone *Nuclear Science Abstracts*, a bimonthly publication of the U. S. Atomic Energy Commission, gave abstracts of some 47,055 articles of which 1197 were under the heading of "Health Physics and Safety," 1755 under the heading of "Radiation Effects on Animals," and 296 under the heading of "Radiation Effects on Plants." Original articles of radiation protection interest appear in various of the medical, biological, and nuclear journals as well as the instrument journals and the magazine *Health Physics*. Frequently of interest are also various publications of drug companies and appendages to the medical profession, those of industrial corporations, and others which may be issued for the specific benefit of special groups, such as those in fire-protection work or even for labor union use.

Not only is there available a tremendous volume of literature but much of the work and many of the analyses have been governmentally financed, even to publication, with the result that some of the recommendations made have at least a semiofficial standing. In general, as is treated in Chapter 10, these activities have resulted in the establishment of a rather wide variety of professional and semiprofessional committees and groups, both national and international; these have usually been at least partly composed of representatives from related scientific and professional organizations. Many of the publications of these groups have rather extensive bibliographies, and these, of course, may themselves lead to other references.

Accordingly, at the risk of unhappy repercussions, the author is suggesting, in addition to materials of a regulatory nature, the following as a basic "library" for a potential health physicist, many of those listed being included because of their extensive bibliographies and several are under frequent revision:

1. Handbooks of the National Bureau of Standards (NBS) of the U.S. Department of Commerce. These are listed at the end of this bibliography.

2. Publications of the National Committee on Radiation Protection and Measurements (NCRP), the International Commission on Radiation Protection (ICRP), and the International Commission on Radiological Units and Measurements (ICRU). Many of the publications of these groups also appear as NBS Handbooks (see list below), and others appear under a variety of sponsorships and in a variety of journals.

3. Standards issued by the U.S. Institute of Standards and reports of the National Academy of Sciences-National Research Council, the World Health Organization (WHO), and the International Atomic Energy Authority (IAEA).

4. *Report of the United Nations Scientific Committee on the Effects of Atomic Radiation,* General Assembly Official Records: (a) Thirteenth Session, Supplement No. 17 (A/3838) (1958); (b) Seventeenth Session, Supplement No. 16 (A/5216) (1962); (c) Nineteenth Session, Supplement No. 14 (A/5814) (1964); (d) Twenty-First Session, Supplement No. 14 (A/6314) (1966).

5. *Radiological Health Handbook,* U.S. Public Health Service, U.S. Department of Health, Education, and Welfare, Washington, D.C.,* 1960.

6. *The Effects of Nuclear Weapons,* U.S. Department of Defense, Washington, D.C.,* 1962.

7. *Radiation Hygiene Handbook,* edited by Hanson Blatz, McGraw-Hill, New York, 1959.

8. *Principles of Radiation Protection,* edited by K. Z. Morgan and J. E. Turner, Wiley, New York, 1967.

* Currently under revision.

NATIONAL BUREAU OF STANDARDS HANDBOOKS AND REPORTS OF THE
NCRP AND THE ICRU

| Handbook (Report) | | | | Superseded |
NBS	NCRP	ICRU	Title	by No.
15			X-Ray Protection	20
18			Radium Protection	23
20			X-Ray Protection	41
23			Radium Protection	54
27			Safe Handling of Radioactive Luminous Compounds	42
41			Medical X-Ray Protection up to Two Million Volts	60
42			Safe Handling of Radioactive Isotopes	92
47			Recommendations of the ICRP and ICRU 1950	62
48	8		Control and Removal of Radioactive Contamination in Laboratories	
49	9		Recommendations for Waste Disposal of Phosphorus-32 and Iodine-131 for Medical Users	
50			X-Ray Protection Design	
51	10		Radiological Monitoring Methods and Instruments	
52			Maximum Permissible Amounts of Radioisotopes in the Human Body and Maximum Permissible Concentrations in Air and Water	69
53	12		Recommendations for the Disposal of Carbon-14 Wastes	
54			Protection against Radiations from Radium, Cobalt-60 and Cesium-137	73
55	14		Protection Against Beta-Synchrotron Radiations up to 100 Million Electron Volts	
56			Safe Handling of Cadavers Containing Radioactive Isotopes	65
57			Photographic Dosimetry of X- and Gamma Rays	
58	16		Radioactive Waste Disposal in the Ocean	
59	17		Permissible Dose from External Sources of Ionizing Radiation	
60			X-Ray Protection	76
61			Regulation of Radiation Exposure by Legislative Means (out of print)	
62			Report of ICRU 1956	78
63	20		Protection against Neutron Radiation up to 30 Million Electron Volts	

NATIONAL BUREAU OF STANDARDS HANDBOOKS AND REPORTS OF THE
NCRP AND THE ICRU—Continued

| Handbook (Report) | | | | Superseded |
NBS	NCRP	ICRU	Title	by No.
64			Design of Free-Air Ionization Chambers	
65	21		Safe Handling of Bodies Containing Radioactive Isotopes	
66			Safe Design and Use of Industrial Beta-Ray Sources	
69	22		Maximum Permissible Body Burdens and Maximum Permissible Concentrations of Radionuclides in Air and in Water for Occupational Exposure	
72	23		Measurement of Neutron Flux and Spectra for Physical and Biological Applications	
73	24		Protection against Radiations from Gamma Sources	
75	25		Measurement of Absorbed Dose of Neutrons and of Mixtures of Neutrons and Gamma Rays	
76	26		Medical X-Ray Protection up to Three Million Volts	
78			Report of the ICRU 1959	84–89
79	27		Stopping Powers for Use with Cavity Chambers	
80	28		A Manual of Radioactivity Procedures	
84		10a	Radiation Quantities and Units	ICRU–11
85		10b	Physical Aspects of Irradiation	
86		10c	Radioactivity	ICRU–12
87		10d	Clinical Dosimetry	
88		10e	Radiobiological Dosimetry	
89		10f	Methods of Evaluating Radiological Equipment and Materials	
92	30		Safe Handling of Radioactive Materials	
93			Safety Standard for Non-Medical X-Ray and Sealed Gamma-Ray Sources	
97			Shielding for High-Energy Electron Accelerator Installations	
	29		Exposure to Radiation in an Emergency	
	32		Radiation Protection in Educational Institutions	
	33		Medical X-Ray and Gamma-Ray Protection for Energies up to 10 MeV—Equipment Design and Use	
		11	Radiation Quantities and Units	
		12	Certification of Standardized Radioactive Sources	

Part I. Radiation Protection Standards for Individuals in Controlled Areas[1]

A. *Radiation from sources external to the body*

Type of Exposure	Period of Time	Dose (rem)
Whole body, head and trunk, active blood-forming organs gonads, or lens of eye.	Accumulated dose quarter[3]	5 (N-18)[2] 3[4]
Skin of the whole body and thyroid	Year quarter[3]	30 10[4]
Hands, and forearms, feet and ankles	Year quarter[3]	75 25

B. *Radiation from emitters internal to the body* *

	Dose	
Type of Exposure	rem/year	rem/quarter
Whole body, active blood-forming organs, gonads.	5	3
Thyroid	30 -/5	10
Bone	Body burden of 0.1 microgram of radium-226 or its biological equivalent[5]	—
Other organs	15	5

---

* Part II of this Appendix gives maximum permissible concentrations of radioisotopes in air and water designed to meet these criteria.

[1] An individual under age 18 shall not be employed in or allowed to enter controlled areas in such manner that he will receive doses of radiation in amounts exceeding the

Wome.   3.5 rem per Tri mester,

standards applicable to individuals in uncontrolled areas. Exposures to individuals under age 18 may be averaged over periods not to exceed one calendar quarter.

[2] N equals the age in years at last birthday. An individual employed at age 18 or an individual beyond age 18 who had no accrued unused exposure shall not be exposed during the ensuing year to doses exceeding (a) 1.25 rem for the first calendar quarter, (b) 2.5 rem total for the first two calendar quarters, (c) 3.75 rem total for the first three calendar quarters and (d) 5 rem for the year, but in no case will exposure be more than 3 rem per quarter.

[3] A calendar quarter may be taken as a predetermined period of 13 consecutive weeks or any predetermined quarter year based on the calendar.

[4] Personnel monitoring equipment shall be provided each individual who receives or is likely to receive a dose in any calendar quarter in excess of 10% of these values.

[5] Exposure must be governed such that the individual's body burden does not exceed this value (a) when averaged over any period of 12 consecutive months and (b) after 50 years of occupational exposure.

Part II. Maximum permissible body burdens and maximum permissible concentrations of radionuclides in air and in water for occupational exposure

Radionuclide and type of decay	Organ of reference (critical organ in boldface)	Maximum permissible burden in total body $q(\mu c)$	Maximum permissible concentrations			
			For 40 hour week		For 168 hour week**	
			$(MPC)_w$ $\mu c/cc$	$(MPC)_a$ $\mu c/cc$	$(MPC)_w$ $\mu c/cc$	$(MPC)_a$ $\mu c/cc$
$_1H^3(H_2^3O)$ (β^-) (Sol)	**Body Tissue**	10^3	0.1	2×10^{-5}	0.03	5×10^{-6}
	Total Body	2×10^3	0.2	2×10^{-5}	0.05	7×10^{-6}
(H_2^3) (Immersion)	**Skin**	--------		2×10^{-3}		4×10^{-4}
$_4Be^7$ (ϵ,γ) (Sol)	**GI (LLI)***	--------	0.05	10^{-5}	0.02	4×10^{-6}
	Total Body	600	6	6×10^{-6}	2	2×10^{-6}
	Kidney	800	9	8×10^{-6}	3	3×10^{-6}
	Liver	800	9	8×10^{-6}	3	3×10^{-6}
	Bone	2×10^3	20	2×10^{-5}	7	6×10^{-6}
	Spleen	4×10^3	50	4×10^{-5}	20	2×10^{-6}
(Insol)	**Lung**	--------	0.05	10^{-6}	0.02	4×10^{-7}
	GI (LLI)	--------	0.05	9×10^{-6}	0.02	3×10^{-6}
$_6C^{14}(CO_2)$ (β^-) (Sol)	**Fat**	300	0.02	4×10^{-6}	8×10^{-3}	10^{-6}
	Total Body	400	0.03	5×10^{-6}	0.01	2×10^{-6}
	Bone	400	0.04	6×10^{-6}	0.01	2×10^{-6}

Nuclide	Form	Total Body		(MPC)$_w$	(MPC)$_a$	(MPC)$_w$	(MPC)$_a$
$_9$F^{18} (β^+)	(Immersion)				5×10^{-5}		10^{-5}
	(Sol)	GI (SI)		0.02	5×10^{-6}	8×10^{-3}	2×10^{-6}
		Bone and Teeth	20	0.2	3×10^{-5}	0.06	9×10^{-6}
		Total Body	20	0.3	4×10^{-5}	0.09	10^{-5}
	(Insol)	GI (ULI)		0.01	3×10^{-6}	5×10^{-3}	9×10^{-7}
		Lung			2×10^{-5}		6×10^{-6}
$_{11}$Na^{22} (β^+,γ)	(Sol)	Total Body	10	10^{-3}	2×10^{-7}	4×10^{-4}	6×10^{-8}
		GI (LLI)		0.01	2×10^{-6}	3×10^{-3}	7×10^{-7}
	(Insol)	Lung			9×10^{-9}		3×10^{-9}
		GI (LLI)		9×10^{-4}	2×10^{-7}	3×10^{-4}	5×10^{-8}
$_{11}$Na^{24} (β^-,γ)	(Sol)	GI (SI)		6×10^{-3}	10^{-6}	2×10^{-3}	4×10^{-7}
		Total Body	7	0.01	2×10^{-6}	4×10^{-3}	6×10^{-7}
	(Insol)	GI (LLI)		8×10^{-4}	10^{-7}	3×10^{-4}	5×10^{-8}
		Lung			8×10^{-7}		3×10^{-7}
$_{14}$Si^{31} (β^-,γ)	(Sol)	GI (S)		0.03	6×10^{-6}	9×10^{-3}	2×10^{-6}
		Lung	10	0.1	2×10^{-5}	0.05	7×10^{-6}
		Adrenal	30	0.3	4×10^{-5}	0.1	10^{-5}
		Total Body	30	0.3	4×10^{-5}	0.1	2×10^{-5}
		Testis	40	0.4	5×10^{-5}	0.1	2×10^{-5}
		Ovary	60	0.6	8×10^{-5}	0.2	3×10^{-5}
		Skin	100	1	2×10^{-4}	0.5	7×10^{-5}
	(Insol)	GI (ULI)		6×10^{-3}	10^{-6}	2×10^{-3}	3×10^{-7}
		Lung			10^{-5}		4×10^{-7}

*The abbreviations GI, S, SI, ULI, and LLI refer to gastrointestinal tract, stomach, small intestines, upper large intestine, and lower large intestine, respectively.

**It will be noted that the MPC values for the 168-hour week are not always precisely the same multiples of the MPC for the 40-hour week. Part of this is caused by rounding off the calculated values to one digit, but in some instances it is due to technical differences discussed in the ICRP report. Because of the uncertainties present in much of the biological data and because of individual variations, the differences are not considered significant. The MPC values for the 40-hour week are to be considered as basic for occupational exposure, and the values for the 168-hour week are basic for continuous exposure as in the case of the population at large.

†Recommendations of the International Commission on Radiological Protection (As Amended and Revised 1962) ICRP Publication 6, Pergamon Press, New York, 1964.

Radionuclide and type of decay	Organ of reference (critical organ in boldface)	Maximum permissible burden in total body $q(\mu c)$	Maximum permissible concentrations			
			For 40 hour week		For 168 hour week**	
			$(MPC)_w$ $\mu c/cc$	$(MPC)_a$ $\mu c/cc$	$(MPC)_w$ $\mu c/cc$	$(MPC)_a$ $\mu c/cc$
$_{15}P^{32}$ (β^-) (Sol)	**Bone**	6	5×10^{-4}	7×10^{-8}	2×10^{-4}	2×10^{-8}
	Total Body	30	3×10^{-3}	4×10^{-7}	9×10^{-4}	10^{-7}
	GI (LLI)	---	3×10^{-3}	6×10^{-7}	9×10^{-4}	2×10^{-7}
	Liver	50	5×10^{-3}	6×10^{-7}	2×10^{-3}	2×10^{-7}
	Brain	300	0.02	3×10^{-6}	8×10^{-3}	10^{-6}
(Insol)	**Lung**	---	---	8×10^{-8}	---	3×10^{-8}
	GI (LLI)	---	7×10^{-4}	10^{-7}	2×10^{-4}	4×10^{-8}
$_{16}S^{35}$ (β^-) (Sol)	**Testis**	90	2×10^{-3}	3×10^{-7}	6×10^{-4}	9×10^{-8}
	Total Body	400	7×10^{-3}	10^{-6}	3×10^{-3}	4×10^{-7}
	Bone	800	0.02	2×10^{-6}	5×10^{-3}	8×10^{-7}
	Skin	3×10^{3}	0.07	10^{-5}	0.02	3×10^{-6}
	GI (LLI)	---	0.2	4×10^{-5}	0.05	10^{-5}
(Insol)	**Lung**	---	---	3×10^{-7}	---	9×10^{-8}
	GI (LLI)	---	8×10^{-3}	10^{-6}	3×10^{-3}	5×10^{-7}
$_{17}Cl^{36}(\beta^-)$ (Sol)	**Total Body**	80	2×10^{-3}	4×10^{-7}	8×10^{-4}	10^{-7}
	GI (LLI)	---	0.04	8×10^{-6}	0.01	3×10^{-6}
(Insol)	**Lung**	---	---	2×10^{-8}	---	8×10^{-9}
	GI (LLI)	---	2×10^{-3}	3×10^{-7}	6×10^{-4}	10^{-7}

$_{17}Cl^{38}(\beta^-,\gamma)$ (Sol)	GI (S)		0.01	3×10^{-6}	4×10^{-3}	9×10^{-7}		
	Total Body	9	0.3	4×10^{-6}	0.1	2×10^{-5}		
(Insol)	GI (S)		0.01	2×10^{-6}	4×10^{-3}	7×10^{-7}		
	Lung			10^{-5}		5×10^{-6}		
$_{18}A^{37}(\epsilon)$ (Immersion)	Skin			6×10^{-3}		10^{-3}		
$_{18}A^{41}(\beta^-,\gamma)$ (Immersion)	Total Body			2×10^{-6}		4×10^{-7}		
$_{19}K^{42}(\beta^-,\gamma)$ (Sol)	GI (S)	10	9×10^{-3}	2×10^{-6}	3×10^{-3}	7×10^{-7}		
	Total Body	20	0.02	3×10^{-6}	8×10^{-3}	10^{-6}		
	Brain	20	0.04	6×10^{-6}	0.01	2×10^{-6}		
	Spleen	20	0.04	6×10^{-6}	0.01	2×10^{-6}		
	Muscle	50	0.08	6×10^{-6}	0.02	4×10^{-6}		
	Liver			10^{-5}	0.03	4×10^{-8}		
(Insol)	GI (LLI)		6×10^{-4}	10^{-7}	2×10^{-4}	3×10^{-7}		
	Lung			9×10^{-7}				
$_{20}Ca^{45}(\beta^-)$ (Sol)	Bone	30	3×10^{-4}	3×10^{-8}	9×10^{-5}	10^{-8}		
	Total Body	200	2×10^{-3}	3×10^{-7}	7×10^{-4}	9×10^{-8}		
	GI (LLI)		0.01	3×10^{-6}	4×10^{-3}	10^{-6}		
(Insol)	Lung			10^{-7}		4×10^{-8}		
	GI (LLI)		5×10^{-3}	9×10^{-7}	2×10^{-3}	3×10^{-7}		
$_{20}Ca^{47}(\beta^-,\gamma)$ (Sol)	Bone	5	10^{-3}	2×10^{-7}	5×10^{-4}	6×10^{-8}		
	GI (LLI)		2×10^{-3}	5×10^{-7}	8×10^{-4}	2×10^{-7}		
	Total Body	10	4×10^{-3}	5×10^{-7}	2×10^{-3}	2×10^{-7}		
(Insol)	GI (LLI)		10^{-3}	2×10^{-7}	3×10^{-4}	6×10^{-8}		
	Lung			2×10^{-7}		6×10^{-8}		
$_{21}Sc^{46}(\beta^-,\gamma)$ (Sol)	GI (LLI)	10	10^{-3}	2×10^{-7}	4×10^{-4}	8×10^{-8}		
	Liver	10	6	2×10^{-7}	2	8×10^{-8}		
	Kidney	20	6	3×10^{-7}	2	9×10^{-8}		
	Total Body	60	6	3×10^{-7}	2	10^{-7}		
	Bone		20	10^{-6}	8	4×10^{-7}		

Radionuclide and type of decay	Organ of reference (critical organ in boldface)	Maximum permissible burden in total body $q(\mu c)$	Maximum permissible concentrations			
			For 40 hour week		For 168 hour week**	
			$(MPC)_w$ $\mu c/cc$	$(MPC)_a$ $\mu c/cc$	$(MPC)_w$ $\mu c/cc$	$(MPC)_a$ $\mu c/cc$
$_{21}Sc^{47}$ (β^-, γ) (Insol)	**Lung**	---	---	2×10^{-8}	---	8×10^{-9}
	GI (LLI)	---	10^{-3}	2×10^{-7}	4×10^{-4}	7×10^{-8}
(Sol)	GI (LLI)	---	3×10^{-3}	6×10^{-7}	9×10^{-4}	2×10^{-7}
	Liver	50	100	6×10^{-6}	50	2×10^{-6}
	Kidney	60	200	8×10^{-6}	60	3×10^{-6}
	Bone	60	200	8×10^{-6}	60	3×10^{-6}
	Total Body	80	200	10^{-5}	80	3×10^{-6}
(Insol)	GI (LLI)	---	3×10^{-3}	5×10^{-7}	9×10^{-4}	2×10^{-7}
	Lung	---	---	10^{-6}	---	3×10^{-7}
$_{21}Sc^{48}$ (β^-, γ) (Sol)	GI (LLI)	---	8×10^{-4}	2×10^{-7}	3×10^{-4}	6×10^{-8}
	Total Body	9	50	2×10^{-6}	20	7×10^{-7}
	Liver	9	50	2×10^{-6}	20	7×10^{-7}
	Kidney	10	80	3×10^{-6}	30	10^{-6}
	Bone	30	200	8×10^{-6}	60	3×10^{-6}
(Insol)	GI (LLI)	---	8×10^{-4}	10^{-7}	3×10^{-4}	5×10^{-8}
	Lung	---	---	4×10^{-7}	---	10^{-7}

₂₃V⁴⁸ (β^+, ϵ, γ)	(Sol)	GI (LLI)	—	9×10^{-4}	2×10^{-7}	3×10^{-4}	6×10^{-8}
		Kidney	8	0.03	3×10^{-7}	0.01	9×10^{-8}
		Total Body	10	0.04	4×10^{-7}	0.02	10^{-7}
		Spleen	20	0.06	5×10^{-7}	0.02	2×10^{-7}
		Liver	20	0.09	8×10^{-7}	0.03	3×10^{-7}
		Bone	60	0.2	2×10^{-6}	0.08	7×10^{-7}
	(Insol)	Lung	—	—	6×10^{-8}	—	2×10^{-8}
		GI (LLI)	—	8×10^{-4}	10^{-7}	3×10^{-4}	5×10^{-8}
₂₄Cr⁵¹ (ϵ, γ)	(Sol)	GI (LLI)	—	0.05	10^{-5}	0.02	4×10^{-6}
		Total Body	800	0.6	10^{-5}	0.2	4×10^{-6}
		Lung	10^3	1	2×10^{-5}	0.4	8×10^{-6}
		Prostate	2×10^3	2	3×10^{-5}	0.5	10^{-5}
		Thyroid	4×10^3	3	6×10^{-5}	1	2×10^{-5}
		Kidney	8×10^3	6	10^{-4}	2	4×10^{-5}
	(Insol)	Lung	—	—	2×10^{-6}	—	8×10^{-7}
		GI (LLI)	—	0.05	8×10^{-6}	0.02	3×10^{-6}
₂₅Mn⁵² (β^+, ϵ, γ)	(Sol)	GI (LLI)	—	10^{-3}	2×10^{-7}	3×10^{-4}	7×10^{-8}
		Pancreas	5	0.01	4×10^{-7}	4×10^{-3}	2×10^{-7}
		Liver	6	0.01	5×10^{-7}	4×10^{-3}	2×10^{-7}
		Total Body	9	0.02	8×10^{-7}	7×10^{-3}	3×10^{-7}
	(Insol)	Lung	—	—	10^{-7}	—	5×10^{-8}
		GI (LLI)	—	9×10^{-4}	2×10^{-7}	3×10^{-4}	5×10^{-8}
₂₅Mn⁵⁴ (ϵ, γ)	(Sol)	GI (LLI)	—	4×10^{-3}	8×10^{-7}	10^{-3}	3×10^{-7}
		Liver	20	0.01	4×10^{-7}	4×10^{-3}	10^{-7}
		Total Body	40	0.02	8×10^{-7}	8×10^{-3}	3×10^{-7}
		Pancreas	50	0.02	9×10^{-7}	9×10^{-3}	3×10^{-7}
	(Insol)	Lung	—	—	4×10^{-8}	—	10^{-8}
		GI (LLI)	—	3×10^{-3}	6×10^{-7}	10^{-3}	2×10^{-7}

Radionuclide and type of decay		Organ of reference (critical organ in boldface)	Maximum permissible burden in total body $q(\mu c)$	Maximum permissible concentrations			
				For 40 hour week		For 168 hour week**	
				$(MPC)_w$ $\mu c/cc$	$(MPC)_a$ $\mu c/cc$	$(MPC)_w$ $\mu c/cc$	$(MPC)_a$ $\mu c/cc$
$_{25}Mn^{56}$ (β^-, γ)	(Sol)	**GI (LLI)**	—	4×10^{-3}	8×10^{-7}	10^{-3}	3×10^{-7}
		Pancreas	2	0.2	6×10^{-6}	0.05	2×10^{-6}
		Liver	5	0.4	10^{-5}	0.1	5×10^{-6}
		Total Body	10	0.9	3×10^{-5}	0.3	10^{-5}
	(Insol)	**GI (LLI)**	—	3×10^{-3}	5×10^{-7}	10^{-3}	2×10^{-7}
		Lung	—	—	5×10^{-6}	—	2×10^{-6}
$_{26}Fe^{55}$ (ϵ)	(Sol)	**Spleen**	10^3	0.02	9×10^{-7}	8×10^{-3}	3×10^{-7}
		Liver	2×10^3	0.04	2×10^{-6}	0.01	5×10^{-7}
		Total Body	3×10^3	0.06	2×10^{-6}	0.02	8×10^{-7}
		Lung	4×10^3	0.07	3×10^{-6}	0.03	9×10^{-7}
		GI (LLI)	—	0.08	2×10^{-5}	0.03	6×10^{-6}
		Bone	7×10^3	0.1	5×10^{-6}	0.04	2×10^{-6}
	(Insol)	**Lung**	—	—	10^{-6}	—	3×10^{-7}
		GI (LLI)	—	0.07	10^{-5}	0.02	4×10^{-6}
$_{26}Fe^{59}$ (β^-, γ)	(Sol)	**GI (LLI)**	20	2×10^{-3}	4×10^{-7}	6×10^{-4}	10^{-7}
		Spleen	20	4×10^{-3}	10^{-7}	10^{-3}	5×10^{-8}
		Total Body	30	5×10^{-3}	2×10^{-7}	2×10^{-3}	7×10^{-8}
		Liver	100	6×10^{-3}	2×10^{-7}	2×10^{-3}	7×10^{-8}
		Lung	100	0.02	8×10^{-7}	7×10^{-3}	3×10^{-7}
		Bone		0.03	10^{-6}	0.01	4×10^{-7}

	(Insol)	**Lung**			**5×10^{-8}**		**2×10^{-8}**
		GI (LLI)		**2×10^{-3}**	3×10^{-7}	**5×10^{-4}**	9×10^{-8}
$_{27}Co^{57}$ (ϵ, γ, e^-)	(Sol)	**GI (LLI)**		**0.02**	**3×10^{-6}**	**5×10^{-3}**	**10^{-6}**
		Total Body	200	0.07	6×10^{-6}	0.03	2×10^{-6}
		Pancreas	700	0.2	2×10^{-5}	0.08	7×10^{-6}
		Liver	10^3	0.4	2×10^{-5}	0.1	6×10^{-6}
		Spleen	2×10^3	0.7	6×10^{-5}	0.2	2×10^{-5}
		Kidney	3×10^3	0.9	8×10^{-5}	0.3	3×10^{-5}
	(Insol)	**Lung**			**2×10^{-7}**		**6×10^{-8}**
		GI (LLI)		**0. 01**	2×10^{-6}	**4×10^{-3}**	7×10^{-7}
$_{27}Co^{58m}$ (β^+, ϵ, γ)	(Sol)	**GI (LLI)**		**0. 08**	**2×10^{-5}**	**0. 03**	**6×10^{-6}**
		Total Body	200	2	10^{-4}	0. 6	5×10^{-5}
		Pancreas	800	6	5×10^{-4}	2	2×10^{-4}
		Liver	10^3	9	4×10^{-4}	3	10^{-4}
		Spleen	2×10^3	20	10^{-3}	5	4×10^{-4}
		Kidney	3×10^3	20	2×10^{-3}	8	6×10^{-4}
	(Insol)	**Lung**			**9×10^{-6}**		**3×10^{-6}**
		GI (LLI)		**0. 06**	10^{-5}	**0. 02**	4×10^{-6}
$_{27}Co^{58}$ (β^+, ϵ)	(Sol)	**GI (LLI)**		**4×10^{-3}**	**8×10^{-7}**	**10^{-3}**	**3×10^{-7}**
		Total Body	30	0. 01	10^{-6}	4×10^{-3}	**3×10^{-7}**
		Pancreas	200	0. 06	5×10^{-6}	0. 02	2×10^{-6}
		Liver	200	0. 08	4×10^{-6}	0. 03	10^{-6}
		Spleen	400	0. 1	10^{-5}	0. 05	4×10^{-6}
		Kidney	600	0. 2	2×10^{-5}	0. 07	6×10^{-6}
	(Insol)	**Lung**			**5×10^{-8}**		**2×10^{-8}**
		GI (LLI)		**3×10^{-3}**	5×10^{-7}	**9×10^{-4}**	2×10^{-7}
$_{27}Co^{60}$ (β^-, γ)	(Sol)	**GI (LLI)**		**10^{-3}**	**3×10^{-7}**	**5×10^{-4}**	**10^{-7}**
		Total Body	10	4×10^{-3}	4×10^{-7}	10^{-3}	**10^{-7}**
		Pancreas	70	0. 02	2×10^{-6}	7×10^{-3}	6×10^{-7}
		Liver	90	0. 03	10^{-6}	9×10^{-3}	5×10^{-7}
		Spleen	200	0. 05	4×10^{-6}	0. 02	2×10^{-6}
		Kidney	200	0. 07	6×10^{-6}	0. 03	2×10^{-6}

Radionuclide and type of decay		Organ of reference (critical organ in boldface)	Maximum permissible burden in total body $q(\mu c)$	Maximum permissible concentrations			
				For 40 hour week		For 168 hour week**	
				$(MPC)_w$ $\mu c/cc$	$(MPC)_a$ $\mu c/cc$	$(MPC)_w$ $\mu c/cc$	$(MPC)_a$ $\mu c/cc$
$_{28}Ni^{59}$ (ϵ)	(Insol)	**Lung**	---	---	9×10^{-9}	---	3×10^{-9}
		GI (LLI)	---	10^{-3}	2×10^{-7}	3×10^{-4}	6×10^{-8}
	(Sol)	**Bone**	10^3	6×10^{-3}	5×10^{-7}	2×10^{-3}	2×10^{-7}
		Total Body	3×10^3	0.01	10^{-6}	4×10^{-3}	3×10^{-7}
		Liver	4×10^3	0.02	10^{-6}	6×10^{-3}	5×10^{-7}
		GI (LLI)	---	0.08	2×10^{-5}	0.03	6×10^{-6}
$_{28}Ni^{63}$ (β^-)	(Insol)	**Lung**	---	---	8×10^{-7}	---	3×10^{-7}
		GI (LLI)	---	0.06	10^{-5}	0.02	3×10^{-6}
	(Sol)	**Bone**	200	8×10^{-4}	6×10^{-8}	3×10^{-4}	2×10^{-8}
		Total Body	900	4×10^{-3}	4×10^{-7}	2×10^{-3}	10^{-7}
		Liver	10^3	6×10^{-3}	5×10^{-7}	2×10^{-3}	2×10^{-7}
		GI (LLI)	---	0.03	6×10^{-6}	0.01	2×10^{-6}
$_{28}Ni^{65}$ (β^-, γ)	(Insol)	**Lung**	---	---	3×10^{-7}	---	10^{-7}
		GI (LLI)	---	0.02	4×10^{-6}	7×10^{-3}	10^{-6}
	(Sol)	**GI (ULI)**	---	4×10^{-3}	9×10^{-7}	10^{-3}	3×10^{-7}
		Bone	4	0.1	10^{-5}	0.04	3×10^{-6}
		Total Body	10	0.4	3×10^{-5}	0.1	10^{-5}
		Liver	20	0.5	4×10^{-5}	0.2	10^{-5}

Isotope		Organ	q (μc)	$(MPC)_w$	$(MPC)_a$	$(MPC)_w$	$(MPC)_a$
$_{29}\text{Cu}^{64}\ (\beta^-,\ \beta^+,\ \epsilon)$	(Insol)	**GI (ULI)**	—	3×10^{-3}	5×10^{-7}	10^{-3}	2×10^{-7}
		Lung	—	—	5×10^{-6}	—	2×10^{-6}
	(Sol)	**GI (LLI)**	—	0.01	2×10^{-6}	3×10^{-3}	7×10^{-7}
		Spleen	10	0.08	6×10^{-6}	0.03	2×10^{-6}
		Kidney	30	0.2	2×10^{-6}	0.07	5×10^{-6}
		Total Body	80	0.5	4×10^{-5}	0.2	10^{-5}
		Liver	100	0.6	5×10^{-5}	0.2	2×10^{-5}
		Heart	200	0.9	7×10^{-5}	0.3	3×10^{-5}
		Brain	600	4	3×10^{-4}	1	10^{-4}
$_{30}\text{Zn}^{65}\ (\beta^+,\ \epsilon,\ \gamma)$	(Insol)	**GI (LLI)**	—	6×10^{-3}	10^{-6}	2×10^{-3}	4×10^{-7}
		Lung	—	—	7×10^{-6}	—	3×10^{-6}
	(Sol)	**Total Body**	60	3×10^{-3}	10^{-7}	10^{-3}	4×10^{-8}
		Prostate	70	4×10^{-3}	10^{-7}	10^{-3}	4×10^{-8}
		Liver	80	4×10^{-3}	10^{-7}	10^{-3}	5×10^{-8}
		Kidney	100	6×10^{-3}	2×10^{-6}	2×10^{-3}	7×10^{-8}
		GI (LLI)	—	6×10^{-3}	10^{-6}	2×10^{-3}	4×10^{-7}
		Pancreas	200	7×10^{-3}	3×10^{-7}	3×10^{-3}	9×10^{-8}
		Muscle	200	0.01	4×10^{-7}	4×10^{-3}	10^{-7}
		Ovary	300	0.01	5×10^{-7}	4×10^{-3}	2×10^{-7}
		Testis	400	0.02	6×10^{-7}	6×10^{-3}	2×10^{-7}
		Bone	700	0.04	6×10^{-6}	0.01	4×10^{-7}
$_{30}\text{Zn}^{69m}\ (\gamma,\ e^-,\ \beta^-)$	(Insol)	**Lung**	—	—	6×10^{-8}	—	2×10^{-7}
		GI (LLI)	—	5×10^{-3}	9×10^{-7}	2×10^{-3}	10^{-7}
	(Sol)	**GI (LLI)**	—	2×10^{-3}	4×10^{-7}	7×10^{-4}	2×10^{-7}
		Prostate	0.7	0.01	4×10^{-7}	4×10^{-3}	10^{-7}
		Pancreas	5	0.07	3×10^{-6}	0.02	9×10^{-7}
		Liver	8	0.1	5×10^{-6}	0.05	2×10^{-6}
		Kidney	10	0.2	8×10^{-6}	0.07	3×10^{-6}
		Ovary	10	0.2	8×10^{-6}	0.07	3×10^{-6}
		Total Body	30	0.5	2×10^{-5}	0.2	6×10^{-6}
		Testis	30	0.5	2×10^{-5}	0.2	6×10^{-6}
		Bone	40	0.6	2×10^{-5}	0.2	7×10^{-6}
		Muscle	100	2	8×10^{-5}	0.7	3×10^{-5}

Radionuclide and type of decay		Organ of reference (critical organ in boldface)	Maximum permissible burden in total body $q(\mu c)$	Maximum permissible concentrations			
				For 40 hour week		For 168 hour week**	
				$(MPC)_w$ $\mu c/cc$	$(MPC)_a$ $\mu c/cc$	$(MPC)_w$ $\mu c/cc$	$(MPC)_a$ $\mu c/cc$
$_{30}Zn^{69}(\beta^-)$	(Insol)	**GI (LLI)**	---	2×10^{-3}	3×10^{-7}	6×10^{-4}	10^{-7}
		Lung	---	---	2×10^{-6}	---	8×10^{-7}
	(Sol)	GI (S)	---	0.05	10^{-5}	0.02	4×10^{-6}
		Prostate	0.8	0.2	7×10^{-6}	0.07	2×10^{-6}
		Pancreas	5	1	5×10^{-5}	0.5	2×10^{-5}
		Liver	10	3	10^{-4}	1	4×10^{-5}
		Ovary	20	4	10^{-4}	1	5×10^{-5}
		Kidney	20	4	2×10^{-4}	2	5×10^{-5}
		Testis	30	9	3×10^{-4}	3	10^{-4}
		Bone	40	10	4×10^{-4}	3	10^{-4}
		Total Body	50	10	5×10^{-4}	5	2×10^{-4}
		Muscle	200	60	2×10^{-3}	20	7×10^{-4}
$_{31}Ga^{72}(\beta^-,\gamma)$	(Insol)	**GI (S)**	---	0.05	9×10^{-6}	0.02	3×10^{-6}
		Lung	---	---	5×10^{-5}	---	2×10^{-5}
	(Sol)	**GI (LLI)**	---	10^{-3}	2×10^{-7}	4×10^{-4}	8×10^{-8}
		Liver	5	9	4×10^{-6}	3	10^{-6}
		Total Body	10	20	8×10^{-6}	6	3×10^{-6}
		Bone	10	20	10^{-5}	8	3×10^{-6}
		Spleen	10	20	10^{-5}	8	4×10^{-6}
		Kidney	10	20	10^{-5}	8	4×10^{-6}

$_{32}Ge^{71}$ (ϵ)	(Insol)	**GI (LLI)**		**10^{-3}**	**2×10^{-7}**	**4×10^{-4}**	**6×10^{-8}**
		Lung			10^{-6}		4×10^{-7}
	(Sol)	**GI (LLI)**		**0.05**	**10^{-5}**	**0.02**	**4×10^{-6}**
		Kidney	100	10	5×10^{-5}	4	2×10^{-5}
		Liver	10^3	100	6×10^{-4}	50	2×10^{-4}
		Total Body	2×10^3	200	9×10^{-4}	70	3×10^{-4}
	(Insol)	**Lung**			**6×10^{-6}**		**2×10^{-6}**
		GI (LLI)		**0.05**	8×10^{-6}	**0.02**	3×10^{-6}
$_{33}As^{73}$ (ϵ, γ)	(Sol)	**GI (LLI)**		**0.01**	3×10^{-6}	**5×10^{-3}**	10^{-6}
		Total Body	300	0.2	**2×10^{-6}**	0.06	**7×10^{-7}**
		Kidney	600	0.3	4×10^{-6}	0.1	10^{-6}
		Liver	10^3	0.5	6×10^{-6}	0.2	2×10^{-6}
	(Insol)	**Lung**			**4×10^{-7}**		**10^{-7}**
		GI (LLI)		**0.01**	2×10^{-6}	**5×10^{-3}**	8×10^{-7}
$_{33}As^{74}$(β^-, β^+, ϵ, γ)	(Sol)	**GI (LLI)**		**2×10^{-3}**	**3×10^{-7}**	**5×10^{-4}**	**10^{-7}**
		Total Body	40	0.07	8×10^{-7}	0.02	3×10^{-7}
		Kidney	80	0.1	2×10^{-6}	0.05	6×10^{-7}
		Liver	100	0.2	3×10^{-6}	0.08	10^{-6}
	(Insol)	Lung			**10^{-7}**		**4×10^{-8}**
		GI (LLI)		**2×10^{-3}**	3×10^{-7}	**5×10^{-4}**	9×10^{-8}
$_{33}As^{76}$ (β^-, γ)	(Sol)	**GI (LLI)**		**6×10^{-4}**	**10^{-7}**	**2×10^{-4}**	**4×10^{-8}**
		Total Body	20	0.4	5×10^{-6}	0.1	2×10^{-6}
		Kidney	20	0.6	8×10^{-6}	0.2	3×10^{-6}
		Liver	40	1	10^{-5}	0.4	5×10^{-6}
	(Insol)	**GI (LLI)**		**6×10^{-4}**	**10^{-7}**	**2×10^{-4}**	**3×10^{-8}**
		Lung			6×10^{-7}		2×10^{-7}
$_{33}As^{77}$(β^-, γ)	(Sol)	**GI (LLI)**		**2×10^{-3}**	**5×10^{-7}**	**8×10^{-4}**	**2×10^{-7}**
		Total Body	80	2	2×10^{-5}	0.5	7×10^{-6}
		Kidney	100	2	2×10^{-5}	0.7	8×10^{-6}
		Liver	200	4	5×10^{-5}	1	2×10^{-5}

Radionuclide and type of decay		Organ of reference (critical organ in boldface)	Maximum permissible burden in total body $q(\mu c)$	Maximum permissible concentrations			
				For 40 hour week		For 168 hour week**	
				$(MPC)_w$ $\mu c/cc$	$(MPC)_a$ $\mu c/cc$	$(MPC)_w$ $\mu c/cc$	$(MPC)_a$ $\mu c/cc$
$_{34}Se^{75}(\epsilon, \gamma)$	(Insol)	**GI (LLI)**	---	2×10^{-3}	4×10^{-7}	8×10^{-4}	10^{-7}
		Lung	---	---	2×10^{-6}	---	6×10^{-7}
	(Sol)	**Kidney**	90	9×10^{-3}	10^{-6}	3×10^{-3}	4×10^{-7}
		Total Body	100	0.01	10^{-6}	3×10^{-3}	5×10^{-7}
		Liver	100	0.01	2×10^{-6}	4×10^{-3}	5×10^{-7}
		Spleen	200	0.02	3×10^{-6}	8×10^{-3}	10^{-6}
		GI (LLI)	---	0.07	2×10^{-5}	0.03	6×10^{-6}
	(Insol)	**Lung**	---	---	10^{-7}	---	4×10^{-8}
		GI (LLI)	---	8×10^{-3}	10^{-6}	3×10^{-3}	5×10^{-7}
$_{34}Br^{82}(\beta^-, \gamma)$	(Sol)	**Total Body**	10	8×10^{-3}	10^{-6}	3×10^{-3}	4×10^{-7}
		GI (SI)	---	8×10^{-3}	2×10^{-6}	3×10^{-3}	6×10^{-7}
		GI (LLI)	---	10^{-3}	2×10^{-7}	4×10^{-4}	6×10^{-8}
	(Insol)	**Lung**	---	---	6×10^{-7}	---	2×10^{-7}
$_{36}Kr^{85}m(\beta^-, \gamma)$	(Immersion)	**Total Body**	---	---	6×10^{-6}	---	10^{-6}
$_{36}Kr^{85}(\beta^-)$	(Immersion)	**Total Body**	---	---	10^{-5}	---	3×10^{-6}
$_{36}Kr^{87}(\beta^-, \gamma)$	(Immersion)	**Total Body**	---	---	10^{-6}	---	2×10^{-7}

Radionuclide	Sol.	Organ of reference	q (μc)	$(MPC)_w$ (40 hr)	$(MPC)_a$ (40 hr)	$(MPC)_w$ (168 hr)	$(MPC)_a$ (168 hr)
$_{37}\text{Rb}^{86}$ (β^-, γ)	(Sol)	Total Body	30	2×10^{-3}	3×10^{-7}	7×10^{-4}	10^{-7}
		Pancreas	30	2×10^{-3}	3×10^{-7}	7×10^{-4}	10^{-7}
		Liver	40	3×10^{-3}	4×10^{-7}	10^{-3}	10^{-7}
		Spleen	50	3×10^{-3}	5×10^{-7}	2×10^{-3}	2×10^{-7}
		Muscle	70	5×10^{-3}	7×10^{-7}	5×10^{-3}	2×10^{-7}
		GI (LLI)		0.01	3×10^{-6}		10^{-6}
	(Insol)	Lung			7×10^{-8}	2×10^{-4}	2×10^{-8}
		GI (LLI)		7×10^{-4}	10^{-7}		4×10^{-8}
$_{37}\text{Rb}^{87}$ (β^-)	(Sol)	Pancreas	200	3×10^{-3}	5×10^{-7}	10^{-3}	2×10^{-7}
		Total Body	200	4×10^{-3}	6×10^{-7}	2×10^{-3}	2×10^{-7}
		Liver	200	5×10^{-3}	7×10^{-7}	2×10^{-3}	2×10^{-7}
		Muscle	400	7×10^{-3}	10^{-6}	2×10^{-3}	4×10^{-7}
		Spleen	400	0.1	2×10^{-5}	0.03	8×10^{-6}
	(Insol)	Lung			7×10^{-8}	2×10^{-3}	2×10^{-8}
		GI (LLI)		5×10^{-3}	9×10^{-7}		3×10^{-7}
$_{38}\text{Sr}^{85m}$ (ϵ, γ)	(Sol)	GI (SI)	50	0.2	4×10^{-5}	0.07	10^{-5}
		Total Body	70	3	2×10^{-4}	1	8×10^{-5}
		Bone		5	4×10^{-4}	2	10^{-4}
	(Insol)	GI (SI)		0.2	3×10^{-5}	0.07	10^{-5}
		Lung			9×10^{-5}		3×10^{-5}
$_{38}\text{Sr}^{85}$ (ϵ, γ)	(Sol)	Total Body	60	3×10^{-3}	2×10^{-7}	10^{-3}	8×10^{-8}
		Bone	70	4×10^{-3}	4×10^{-7}	2×10^{-3}	10^{-7}
		GI (LLI)		7×10^{-3}	2×10^{-6}	2×10^{-3}	5×10^{-7}
	(Insol)	Lung			10^{-7}		4×10^{-8}
		GI (LLI)		5×10^{-3}	9×10^{-7}		3×10^{-7}
$_{38}\text{Sr}^{89}$ (β^-)	(Sol)	Bone	4	3×10^{-4}	3×10^{-8}	10^{-4}	10^{-8}
		Total Body	40	10^{-3}	3×10^{-7}	4×10^{-4}	9×10^{-8}
		GI (LLI)		2×10^{-3}	2×10^{-7}	7×10^{-4}	6×10^{-8}
	(Insol)	Lung			4×10^{-8}		10^{-8}
		GI (LLI)		8×10^{-4}	10^{-7}	3×10^{-4}	5×10^{-8}

Radionuclide and type of decay		Organ of reference (critical organ in boldface)	Maximum permissible burden in total body $q(\mu c)$	Maximum permissible concentrations			
				For 40 hour week		For 168 hour week**	
				$(MPC)_w$ $\mu c/cc$	$(MPC)_a$ $\mu c/cc$	$(MPC)_w$ $\mu c/cc$	$(MPC)_a$ $\mu c/cc$
$†_{38}Sr^{90}$ (β^-)	(Sol)	**Bone**	2.0	10^{-5}	10^{-9}	4×10^{-6}	4×10^{-10}
		Total Body	3.0	2×10^{-5}	2×10^{-9}	7×10^{-6}	7×10^{-10}
		GI (LLI)	--	10^{-3}	3×10^{-7}	5×10^{-4}	10^{-7}
	(Insol)	**Lung**	--	--	5×10^{-9}	--	2×10^{-9}
		GI (LLI)	--	10^{-3}	2×10^{-7}	4×10^{-4}	6×10^{-8}
$_{38}Sr^{91}$ (β^-, γ)	(Sol)	**GI (LLI)**	--	2×10^{-3}	4×10^{-7}	7×10^{-4}	2×10^{-7}
		Bone	3	0.02	2×10^{-6}	7×10^{-3}	5×10^{-7}
		Total Body	9	0.07	6×10^{-6}	0.02	2×10^{-6}
	(Insol)	**GI (LLI)**	--	10^{-3}	3×10^{-7}	5×10^{-4}	9×10^{-8}
		Lung	--	--	10^{-6}	--	4×10^{-7}
$_{38}Sr^{92}$ (β^-, γ)	(Sol)	**GI (ULI)**	--	2×10^{-3}	4×10^{-7}	7×10^{-4}	2×10^{-7}
		Bone	2	0.05	4×10^{-6}	0.02	2×10^{-6}
		Total Body	8	0.2	2×10^{-5}	0.07	6×10^{-6}
	(Insol)	**GI (ULI)**	--	2×10^{-3}	3×10^{-7}	6×10^{-4}	10^{-7}
		Lung	--	--	3×10^{-6}	--	10^{-6}
$_{39}Y^{90}$ (β^-)	(Sol)	**GI (LLI)**	--	6×10^{-4}	10^{-7}	2×10^{-4}	4×10^{-8}
		Bone	3	10	5×10^{-7}	4	2×10^{-7}
		Total Body	20	80	3×10^{-6}	30	10^{-6}

Radionuclide		Organ of reference	q				
(continued)	(Insol)	GI (LLI)	- - -	6×10^{-4}	10^{-7}	2×10^{-4}	3×10^{-8}
		Lung	- - -	- - -	3×10^{-7}	- - -	10^{-7}
$_{39}$Y^{91m} (β^-, γ)	(Sol)	GI (SI)	- - -	0.1	2×10^{-5}	0.03	8×10^{-6}
		Bone	5	10^{3}	6×10^{-5}	400	2×10^{-5}
		Total Body	20	6×10^{3}	2×10^{-4}	2×10^{3}	8×10^{-5}
	(Insol)	GI (SI)	- - -	0.1	2×10^{-5}	0.03	6×10^{-6}
		Lung	- - -	- - -	4×10^{-5}	- - -	10^{-5}
$_{39}$Y^{91} (β^-, γ)	(Sol)	GI (LLI)	- - -	8×10^{-4}	2×10^{-7}	3×10^{-4}	6×10^{-8}
		Bone	5	0.8	4×10^{-8}	0.3	10^{-8}
		Total Body	30	5	2×10^{-5}	2	8×10^{-8}
	(Insol)	GI (LLI)	- - -	8×10^{-4}	3×10^{-8}	3×10^{-4}	10^{-8}
		Lung	- - -	- - -	10^{-7}	- - -	5×10^{-8}
$_{39}$Y^{92} (β^-, γ)	(Sol)	GI (ULI)	- - -	2×10^{-3}	4×10^{-7}	6×10^{-4}	10^{-7}
		Bone	2	100	6×10^{-6}	40	2×10^{-6}
		Total Body	10	800	3×10^{-5}	300	10^{-5}
	(Insol)	GI (ULI)	- - -	2×10^{-3}	3×10^{-5}	6×10^{-4}	10^{-7}
		Lung	- - -	- - -	3×10^{-6}	- - -	10^{-6}
$_{39}$Y^{93} (β^-, γ, e^-)	(Sol)	GI (LLI)	- - -	8×10^{-4}	2×10^{-7}	3×10^{-4}	6×10^{-8}
		Bone	2	50	2×10^{-6}	20	7×10^{-7}
		Total Body	10	250	10^{-5}	90	4×10^{-6}
	(Insol)	GI (LLI)	- - -	8×10^{-4}	10^{-7}	3×10^{-4}	5×10^{-8}
		Lung	- - -	- - -	10^{-6}	- - -	4×10^{-7}
$_{40}$Zr^{93} (β^-, γ, e^-)	(Sol)	GI (LLI)	- - -	0.02	5×10^{-6}	8×10^{-3}	2×10^{-6}
		Bone	100	3	10^{-7}	0.9	4×10^{-8}
		Kidney	300	6	3×10^{-7}	2	9×10^{-8}
		Spleen	500	10	4×10^{-7}	3	10^{-7}
		Total Body	900	20	8×10^{-7}	6	3×10^{-7}
		Liver	10^{3}	30	10^{-6}	9	4×10^{-7}
	(Insol)	Lung	- - -	- - -	3×10^{-7}	- - -	10^{-7}
		GI (LLI)	- - -	0.02	4×10^{-6}	8×10^{-3}	10^{-6}

Radionuclide and type of decay	Organ of reference (critical organ in boldface)	Maximum permissible burden in total body $q(\mu c)$	Maximum permissible concentrations			
			For 40 hour week		For 168 hour week**	
			$(MPC)_w$ $\mu c/cc$	$(MPC)_a$ $\mu c/cc$	$(MPC)_w$ $\mu c/cc$	$(MPC)_a$ $\mu c/cc$
$_{40}Zr^{95}$ (β^-, γ, e^-)						
(Sol)	**GI (LLI)**	---	2×10^{-3}	4×10^{-7}	6×10^{-4}	10^{-7}
	Total Body	20	3	10^{-7}	1	4×10^{-8}
	Bone	30	4	2×10^{-7}	2	6×10^{-8}
	Kidney	30	4	2×10^{-7}	2	6×10^{-8}
	Liver	40	6	3×10^{-7}	2	9×10^{-8}
	Spleen	40	7	3×10^{-7}	2	10^{-7}
(Insol)	**Lung**	---		3×10^{-8}		10^{-8}
	GI (LLI)	---	2×10^{-3}	3×10^{-7}	6×10^{-4}	10^{-7}
$_{40}Zr^{97}$ (β^-, γ)						
(Sol)	**GI (LLI)**	---	5×10^{-4}	10^{-7}	2×10^{-4}	4×10^{-8}
	Bone	5	60	3×10^{-6}	20	10^{-6}
	Kidney	8	100	5×10^{-6}	40	2×10^{-6}
	Total Body	9	100	5×10^{-6}	40	2×10^{-6}
	Liver	10	200	7×10^{-6}	60	3×10^{-6}
	Spleen	10	200	8×10^{-6}	60	3×10^{-6}
(Insol)	**GI (LLI)**	---	5×10^{-4}	9×10^{-8}	2×10^{-4}	3×10^{-8}
	Lung	---		6×10^{-7}		2×10^{-7}

		Organ	q (μc)	(MPC)_w	(MPC)_a	(MPC)_w	(MPC)_a
$_{41}$Nb^{93m} (γ, e$^-$)	(Sol)	GI (LLI)		0.01	3×10^{-6}	4×10^{-3}	9×10^{-7}
		Bone	200	3	10^{-7}	1	4×10^{-8}
		Kidney	300	5	2×10^{-7}	2	7×10^{-8}
		Spleen	400	5	2×10^{-7}	2	8×10^{-8}
		Liver	400	6	3×10^{-7}	2	9×10^{-8}
		Total Body	500	8	3×10^{-7}	3	10^{-7}
	(Insol)	Lung		0.01	2×10^{-6}		5×10^{-8}
		GI (LLI)			2×10^{-6}	4×10^{-3}	7×10^{-7}
$_{41}$Nb^{95} (β^-, γ)	(Sol)	GI (LLI)		3×10^{-3}	6×10^{-7}	10^{-3}	2×10^{-7}
		Total Body	40	10	5×10^{-7}	4	2×10^{-7}
		Liver	60	20	7×10^{-7}	6	3×10^{-7}
		Kidney	60	20	8×10^{-7}	6	3×10^{-7}
		Bone	80	20	9×10^{-7}	7	3×10^{-7}
		Spleen	80	20	10^{-6}	7	3×10^{-8}
	(Insol)	Lung			5×10^{-7}		2×10^{-7}
		GI (LLI)		3×10^{-3}		4×10^{-3}	
$_{41}$Nb^{97} (β^-, γ)	(Sol)	GI (ULI)		0.03	6×10^{-6}	9×10^{-3}	2×10^{-6}
		Bone	10	2×10^{3}	9×10^{-5}	700	3×10^{-5}
		Kidney	20	4×10^{3}	2×10^{-4}	10^{3}	6×10^{-5}
		Total Body	20	4×10^{3}	2×10^{-4}	10^{3}	6×10^{-5}
		Liver	30	4×10^{3}	2×10^{-4}	2×10^{3}	7×10^{-5}
		Spleen	30	5×10^{3}	2×10^{-4}	2×10^{3}	7×10^{-6}
	(Insol)	GI (ULI)		0.03	5×10^{-6}	9×10^{-3}	2×10^{-6}
		Lung			2×10^{-5}		7×10^{-6}
$_{42}$Mo^{99} (β^-, γ)	(Sol)	Kidney	8	5×10^{-3}	7×10^{-7}	2×10^{-3}	3×10^{-7}
		GI (LLI)		7×10^{-3}	2×10^{-6}	2×10^{-3}	5×10^{-7}
		Liver	20	0.01	2×10^{-6}	5×10^{-3}	6×10^{-7}
		Total Body	40	0.02	3×10^{-6}	8×10^{-4}	10^{-6}
	(Insol)	GI (LLI)		10^{-3}	2×10^{-7}	4×10^{-4}	7×10^{-8}
		Lung			5×10^{-7}		2×10^{-7}

Radionuclide and type of decay		Organ of reference (critical organ in boldface)	Maximum permissible burden in total body $q(\mu c)$	Maximum permissible concentrations			
				For 40 hour week		For 168 hour week**	
				$(MPC)_w$ $\mu c/cc$	$(MPC)_a$ $\mu c/cc$	$(MPC)_w$ $\mu c/cc$	$(MPC)_a$ $\mu c/cc$
$_{43}Tc^{96m}$ (ϵ, γ, e^-)	(Sol)	**GI (LLI)**	---	0.4	8×10^{-5}	0.1	3×10^{-5}
		Kidney	60	3	4×10^{-4}	1	10^{-4}
		Total Body	70	4	4×10^{-4}	1	10^{-4}
		Liver	800	40	4×10^{-3}	14	10^{-3}
		Lung	2×10^3	130	0.01	40	5×10^{-3}
		Bone	10^4	700	0.08	200	0.03
		Skin	2×10^4	800	0.09	300	0.03
	(Insol)	**Lung**	---	0.3	3×10^{-5}	0.1	10^{-5}
		GI (LLI)	---		5×10^{-5}		2×10^{-5}
$_{43}Tc^{96}$ (ϵ, γ)	(Sol)	**GI (LLI)**	---	3×10^{-3}	6×10^{-7}	10^{-3}	2×10^{-7}
		Kidney	10	0.03	3×10^{-6}	0.01	10^{-6}
		Total Body	10	0.03	4×10^{-6}	0.01	10^{-6}
		Liver	200	0.4	4×10^{-5}	0.1	10^{-5}
		Lung	500	1	10^{-4}	0.4	4×10^{-5}
		Bone	2×10^3	4	5×10^{-4}	1	2×10^{-4}
		Skin	10^4	20	3×10^{-3}	9	9×10^{-4}
	(Insol)	**GI (LLI)**	---	10^{-3}	3×10^{-7}	5×10^{-4}	8×10^{-8}
		Lung	---		3×10^{-7}		9×10^{-8}

Radionuclide	Solubility	Organ of reference					
$_{43}\text{Tc}^{97m}$ ($\epsilon,\ \gamma,\ e^-$)	(Sol)	**GI (LLI)**	—	**0.01**	**2×10^{-6}**	**4×10^{-3}**	**8×10^{-7}**
		Kidney	20	0.03	4×10^{-6}	0.01	10^{-6}
		Total Body	200	0.4	4×10^{-5}	0.1	2×10^{-5}
		Liver	200	0.4	5×10^{-5}	0.1	2×10^{-5}
		Skin	500	1	10^{-4}	0.3	4×10^{-5}
		Bone	700	1	10^{-4}	0.5	5×10^{-5}
		Lung	2×10^{3}	4	4×10^{-4}	1	2×10^{-4}
	(Insol)	**GI (LLI)**	—	**5×10^{-3}**	**2×10^{-7}**	**2×10^{-3}**	**5×10^{-8}**
		Lung	—	—	9×10^{-7}	—	3×10^{-7}
$_{43}\text{Tc}^{97}$ (ϵ)	(Sol)	GI (LLI)	—	0.05	10^{-5}	0.02	4×10^{-6}
		Kidney	**60**	**0.1**	**10^{-5}**	**0.04**	**4×10^{-6}**
		Liver	800	2	2×10^{-4}	0.5	6×10^{-5}
		Total Body	10^{3}	2	2×10^{-4}	0.6	7×10^{-5}
		Bone	6×10^{3}	10	10^{-3}	4	4×10^{-4}
		Lung	9×10^{3}	20	2×10^{-3}	6	7×10^{-4}
		Skin	3×10^{4}	60	6×10^{-3}	20	2×10^{-3}
	(Insol)	**GI (LLI)**	—	**0.02**	**3×10^{-7}**	**8×10^{-3}**	**10^{-7}**
		Lung	—	—	4×10^{-6}	—	10^{-6}
$_{43}\text{Tc}^{99m}$ ($\beta^-,\ \gamma$)	(Sol)	**GI (ULI)**	—	**0.2**	**4×10^{-5}**	**0.06**	**10^{-5}**
		Total Body	200	2	2×10^{-4}	0.8	9×10^{-5}
		Kidney	800	7	8×10^{-4}	3	3×10^{-4}
		Liver	10^{4}	100	0.01	30	4×10^{-3}
		Lung	2×10^{4}	200	0.02	70	8×10^{-3}
		Bone	10^{5}	10^{3}	0.1	400	0.04
		Skin	10^{5}	10^{3}	0.1	400	0.04
	(Insol)	**GI (ULI)**	—	**0.08**	**10^{-5}**	**0.03**	**5×10^{-6}**
		Lung	—	—	8×10^{-5}	—	3×10^{-5}

Radionuclide and type of decay	Organ of reference (critical organ in boldface)	Maximum permissible burden in total body $q(\mu c)$	Maximum permissible concentrations			
			For 40 hour week		For 168 hour week **	
			$(MPC)_w$ $\mu c/cc$	$(MPC)_a$ $\mu c/cc$	$(MPC)_w$ $\mu c/cc$	$(MPC)_a$ $\mu c/cc$
$_{43}Tc^{99}$ (β^-) (Sol)	**GI (LLI)**	--	0.01	2×10^{-6}	3×10^{-3}	7×10^{-7}
	Kidney	10	0.02	3×10^{-6}	8×10^{-3}	9×10^{-7}
	Liver	200	0.3	4×10^{-5}	0.1	10^{-5}
	Total Body	200	0.4	4×10^{-5}	0.1	10^{-5}
	Skin	400	0.7	7×10^{-5}	0.2	3×10^{-5}
	Bone	500	0.9	9×10^{-5}	0.3	3×10^{-5}
	Lung	2×10^3	4	4×10^{-4}	1	10^{-4}
(Insol)	**GI (LLI)**	---	5×10^{-3}	6×10^{-8}	2×10^{-3}	2×10^{-8}
	Lung			8×10^{-7}		3×10^{-7}
$_{44}Ru^{97}$ (ϵ, γ, e^-) (Sol)	**GI (LLI)**	---	0.01	2×10^{-6}	4×10^{-3}	8×10^{-7}
	Kidney	30	0.4	5×10^{-6}	0.1	10^{-6}
	Total Body	100	2	3×10^{-5}	0.7	9×10^{-6}
	Bone	900	10	2×10^{-4}	5	6×10^{-5}
(Insol)	**GI (LLI)**	---	0.01	2×10^{-6}	3×10^{-3}	6×10^{-7}
	Lung			2×10^{-6}		7×10^{-7}
$_{44}Ru^{103}$ (β^-, γ, e^-) (Sol)	**GI (LLI)**	---	2×10^{-3}	5×10^{-7}	8×10^{-4}	2×10^{-7}
	Kidney	20	0.08	10^{-6}	0.03	3×10^{-7}
	Total Body	50	0.2	3×10^{-6}	0.08	9×10^{-7}
	Bone	100	0.6	7×10^{-6}	0.2	2×10^{-6}

Radionuclide	Solubility	Organ of reference	q (μc)	$(MPC)_w$ occ.	$(MPC)_a$ occ.	$(MPC)_w$ pop.	$(MPC)_a$ pop.
$_{44}Ru^{105}$ (β^-, γ, e^-)	(Insol)	Lung	—	—	8×10^{-8}	—	3×10^{-8}
	(Insol)	GI (LLI)	—	2×10^{-3}	4×10^{-7}	8×10^{-4}	10^{-7}
	(Sol)	GI (ULI)	—	3×10^{-3}	7×10^{-7}	10^{-3}	2×10^{-7}
	(Sol)	Kidney	2	0.3	3×10^{-6}	0.09	10^{-6}
	(Sol)	Total Body	20	3	4×10^{-5}	0.9	10^{-5}
	(Sol)	Bone	40	6	8×10^{-5}	2	3×10^{-5}
	(Insol)	GI (ULI)	—	3×10^{-3}	5×10^{-7}	10^{-3}	2×10^{-7}
	(Insol)	Lung	—	—	4×10^{-6}	—	10^{-6}
$_{44}Ru^{106}$ (β^-, γ)	(Sol)	GI (LLI)	—	4×10^{-4}	8×10^{-8}	10^{-4}	3×10^{-8}
	(Sol)	Kidney	3	0.01	10^{-7}	4×10^{-3}	5×10^{-8}
	(Sol)	Bone	10	0.04	5×10^{-7}	0.01	2×10^{-7}
	(Sol)	Total Body	10	0.06	7×10^{-7}	0.02	3×10^{-7}
	(Insol)	Lung	—	—	6×10^{-9}	—	2×10^{-9}
	(Insol)	GI (LLI)	—	3×10^{-4}	6×10^{-8}	10^{-4}	2×10^{-8}
$_{45}Rh^{103m}$ (γ, e^-)	(Sol)	GI (S)	—	0.4	8×10^{-5}	0.1	3×10^{-5}
	(Sol)	Kidney	200	20	10^{-3}	7	4×10^{-4}
	(Sol)	Spleen	200	30	2×10^{-3}	10	6×10^{-4}
	(Sol)	Total Body	400	40	3×10^{-3}	20	10^{-3}
	(Sol)	Liver	700	80	5×10^{-3}	30	2×10^{-3}
	(Sol)	Bone	10^3	100	9×10^{-3}	50	3×10^{-3}
	(Insol)	GI (S)	—	0.3	6×10^{-5}	0.1	2×10^{-5}
	(Insol)	Lung	—	—	3×10^{-4}	—	10^{-4}
$_{45}Rh^{105}$ (β^-, γ)	(Sol)	GI (LLI)	—	4×10^{-3}	8×10^{-7}	10^{-3}	3×10^{-7}
	(Sol)	Bone	200	8×10^{-3}	5×10^{-5}	3×10^{-3}	2×10^{-5}
	(Sol)	Kidney	40	0.1	9×10^{-6}	0.05	3×10^{-6}
	(Sol)	Spleen	60	0.2	10^{-5}	0.07	5×10^{-6}
	(Sol)	Total Body	100	0.4	2×10^{-5}	0.1	7×10^{-6}
	(Sol)	Liver	200	0.6	4×10^{-5}	0.2	10^{-5}
	(Insol)	GI (LLI)	—	3×10^{-3}	5×10^{-7}	10^{-3}	2×10^{-7}
	(Insol)	Lung	—	—	2×10^{-6}	—	8×10^{-7}

Radionuclide and type of decay		Organ of reference (critical organ in boldface)	Maximum permissible burden in total body $q(\mu c)$	Maximum permissible concentrations			
				For 40 hour week		For 168 hour week**	
				$(MPC)_w$ $\mu c/cc$	$(MPC)_a$ $\mu c/cc$	$(MPC)_w$ $\mu c/cc$	$(MPC)_a$ $\mu c/cc$
$_{46}Pd^{103}$ (ϵ, γ, e⁻)	(Sol)	**GI (LLI)**	--	0.01	2×10^{-6}	3×10^{-3}	8×10^{-7}
		Kidney	20	0.02	10^{-6}	7×10^{-3}	5×10^{-7}
		Spleen	100	0.1	8×10^{-6}	0.04	3×10^{-6}
		Liver	100	0.1	8×10^{-6}	0.04	3×10^{-6}
		Total Body	300	0.4	2×10^{-5}	0.1	8×10^{-6}
	(Insol)	**Lung**	--	--	7×10^{-7}	--	3×10^{-7}
		GI (LLI)	--	8×10^{-3}	10^{-6}	3×10^{-3}	5×10^{-7}
$_{46}Pd^{109}$ (β^-, γ, e⁻)	(Sol)	**GI (LLI)**	--	3×10^{-3}	6×10^{-7}	9×10^{-4}	2×10^{-7}
		Kidney	7	0.06	4×10^{-6}	0.02	10^{-6}
		Spleen	30	0.3	2×10^{-5}	0.09	5×10^{-6}
		Liver	40	0.3	2×10^{-5}	0.1	7×10^{-6}
		Total Body	50	0.4	3×10^{-5}	0.1	9×10^{-6}
	(Insol)	**GI (LLI)**	--	2×10^{-3}	4×10^{-7}	7×10^{-4}	10^{-7}
		Lung	--		3×10^{-6}		10^{-6}
$_{47}Ag^{105}$ (ϵ, γ)	(Sol)	**GI (LLI)**	--	3×10^{-3}	6×10^{-7}	10^{-3}	2×10^{-7}
		Total Body	30	0.6	3×10^{-6}	0.2	10^{-6}
		Kidney	30	0.7	3×10^{-6}	0.2	10^{-6}
		Liver	70	1	6×10^{-6}	0.5	2×10^{-6}
		Bone	200	4	2×10^{-5}	1	6×10^{-6}

Element		Organ	q (µc)	$(MPC)_w$	$(MPC)_a$	$(MPC)_w$	$(MPC)_a$
(continued)	(Insol)	Lung		—	8×10^{-8}	—	3×10^{-8}
		GI (LLI)		3×10^{-3}	5×10^{-7}	10^{-3}	2×10^{-7}
$_{47}$Ag^{110m} (β^-, γ)	(Sol)	GI (LLI)		9×10^{-4}	2×10^{-7}	3×10^{-4}	7×10^{-8}
		Kidney	10	0.2	8×10^{-7}	0.06	3×10^{-7}
		Total Body	10	0.2	9×10^{-7}	0.07	3×10^{-7}
		Liver	20	0.4	2×10^{-6}	0.1	5×10^{-7}
		Bone	40	0.7	3×10^{-6}	0.2	10^{-6}
	(Insol)	Lung		—	10^{-8}	—	3×10^{-9}
		GI (LLI)		9×10^{-4}	2×10^{-7}	3×10^{-4}	5×10^{-8}
$_{47}$Ag^{111} (β^-, γ)	(Sol)	GI (LLI)		10^{-3}	3×10^{-7}	4×10^{-4}	10^{-7}
		Kidney	20	0.7	3×10^{-6}	0.2	10^{-6}
		Total Body	50	1	6×10^{-6}	0.5	2×10^{-6}
		Bone	60	2	8×10^{-6}	0.6	3×10^{-6}
		Liver	80	2	10^{-5}	0.8	3×10^{-6}
	(Insol)	GI (LLI)		10^{-3}	2×10^{-7}	4×10^{-4}	8×10^{-8}
		Lung		—	3×10^{-7}	—	9×10^{-8}
$_{48}$Cd^{109} (ϵ, γ, e$^-$)	(Sol)	GI (CLLI)		5×10^{-3}	10^{-6}	2×10^{-3}	4×10^{-7}
		Liver	20	0.05	5×10^{-8}	0.02	2×10^{-8}
		Kidney	20	0.05	6×10^{-8}	0.02	2×10^{-8}
		Total Body	200	0.5	5×10^{-7}	0.2	2×10^{-7}
	(Insol)	Lung		—	7×10^{-8}	—	3×10^{-8}
		GI (LLI)		5×10^{-3}	9×10^{-7}	2×10^{-3}	3×10^{-7}
$_{48}$Cd^{115m} (β^-, γ, e$^-$)	(Sol)	GI (LLI)		7×10^{-4}	2×10^{-7}	3×10^{-4}	6×10^{-8}
		Liver	3	0.03	4×10^{-8}	0.01	10^{-8}
		Kidney	4	0.04	4×10^{-8}	0.01	2×10^{-8}
		Total Body	30	0.4	4×10^{-7}	0.1	10^{-7}
	(Insol)	Lung		—	4×10^{-8}	—	10^{-8}
		GI (LLI)		7×10^{-4}	10^{-7}	3×10^{-4}	4×10^{-8}

Radionuclide and type of decay	Organ of reference (critical organ in boldface)	Maximum permissible burden in total body $q(\mu c)$	Maximum permissible concentrations			
			For 40 hour week		For 168 hour week**	
			$(MPC)_w$ $\mu c/cc$	$(MPC)_a$ $\mu c/cc$	$(MPC)_w$ $\mu c/cc$	$(MPC)_a$ $\mu c/cc$
$_{48}Cd^{115}$ (β^-, γ, e^-) (Sol)	**GI (LLI)**	---	10^{-3}	2×10^{-7}	3×10^{-4}	8×10^{-8}
	Liver	3	0.6	6×10^{-7}	0.2	2×10^{-7}
	Kidney	5	0.8	8×10^{-7}	0.3	3×10^{-7}
	Total Body	30	5	5×10^{-6}	2	2×10^{-6}
(Insol)	**GI (LLI)**	---	10^{-3}	2×10^{-7}	4×10^{-4}	6×10^{-8}
	Lung	---	---	6×10^{-7}	---	2×10^{-7}
$_{49}In^{113m}$ (γ, e^-) (Sol)	**GI (ULI)**	---	0.04	8×10^{-6}	0.01	3×10^{-6}
	Kidney	30	200	2×10^{-4}	70	6×10^{-5}
	Spleen	30	200	2×10^{-4}	70	6×10^{-5}
	Liver	50	300	3×10^{-4}	100	9×10^{-5}
	Total Body	70	400	4×10^{-4}	200	10^{-4}
	Bone	90	600	5×10^{-4}	200	2×10^{-4}
	Skin	100	900	8×10^{-4}	300	3×10^{-4}
	Thyroid	500	3×10^3	3×10^{-3}	10^3	10^{-3}
(Insol)	**GI (ULI)**	---	0.04	7×10^{-6}	0.01	2×10^{-6}
	Lung	---	---	5×10^{-5}	---	2×10^{-5}

$_{49}In^{114m}$ (β^-, ϵ, γ, e^-)	(Sol)	GI (LLI)	---	5×10^{-4}	10^{-7}	2×10^{-4}	4×10^{-8}
		Kidney	6	0.1	10^{-7}	0.04	4×10^{-8}
		Spleen	7	0.1	10^{-7}	0.04	4×10^{-8}
		Liver	10	0.2	2×10^{-7}	0.07	6×10^{-8}
		Bone	10	0.3	2×10^{-7}	0.09	8×10^{-8}
		Skin	20	0.4	3×10^{-7}	0.1	10^{-7}
		Total Body	20	0.4	4×10^{-7}	0.1	10^{-7}
		Thyroid	50	0.9	8×10^{-7}	0.3	3×10^{-7}
	(Insol)	Lung	---	---	2×10^{-8}	---	7×10^{-9}
		GI (LLI)	---	5×10^{-4}	8×10^{-8}	2×10^{-4}	3×10^{-8}
$_{49}In^{115m}$ (β^-, γ, e^-)	(Sol)	GI (ULI)	---	0.01	2×10^{-6}	4×10^{-3}	8×10^{-7}
		Kidney	30	80	7×10^{-5}	30	2×10^{-5}
		Spleen	30	80	7×10^{-5}	30	2×10^{-5}
		Liver	50	100	10^{-4}	40	4×10^{-5}
		Total Body	80	200	2×10^{-4}	60	6×10^{-5}
		Thyroid	80	200	2×10^{-4}	70	6×10^{-5}
		Bone	90	200	2×10^{-4}	70	6×10^{-5}
		Skin	100	300	3×10^{-4}	100	10^{-4}
	(Insol)	GI (ULI)	---	0.01	2×10^{-6}	4×10^{-3}	6×10^{-7}
		Lung	---	---	2×10^{-5}	---	6×10^{-6}
$_{49}In^{115}$ (β^-)	(Sol)	GI (LLI)	---	3×10^{-3}	6×10^{-7}	9×10^{-4}	2×10^{-7}
		Kidney	30	0.3	2×10^{-7}	0.1	9×10^{-8}
		Spleen	40	0.4	3×10^{-7}	0.1	10^{-7}
		Liver	50	0.5	4×10^{-7}	0.1	10^{-7}
		Bone	60	0.6	5×10^{-7}	0.2	2×10^{-7}
		Skin	80	0.8	7×10^{-7}	0.3	2×10^{-7}
		Total Body	100	1	10^{-6}	0.4	3×10^{-7}
		Thyroid	3×10^3	30	2×10^{-5}	9	8×10^{-6}
	(Insol)	Lung	---	---	3×10^{-8}	---	10^{-8}
		GI (LLI)	---	3×10^{-3}	5×10^{-7}	9×10^{-4}	2×10^{-7}

Radionuclide and type of decay		Organ of reference (critical organ in boldface)	Maximum permissible burden in total body $q(\mu c)$	Maximum permissible concentrations			
				For 40 hour week		For 168 hour week**	
				$(MPC)_w$ $\mu c/cc$	$(MPC)_a$ $\mu c/cc$	$(MPC)_w$ $\mu c/cc$	$(MPC)_a$ $\mu c/cc$
$_{50}Sn^{113}$ (ϵ, γ, e^-)	(Sol)	GI (LLI)	—	2×10^{-3}	5×10^{-7}	9×10^{-4}	2×10^{-7}
		Bone	30	0.02	4×10^{-7}	6×10^{-3}	10^{-7}
		Total Body	60	0.04	8×10^{-7}	0.01	3×10^{-7}
		Prostate	70	0.04	9×10^{-7}	0.02	3×10^{-7}
		Liver	400	0.3	5×10^{-6}	0.09	2×10^{-6}
		Thyroid	10^3	0.9	2×10^{-5}	0.3	6×10^{-6}
	(Insol)	**Lung**	—	—	5×10^{-8}	—	2×10^{-8}
		GI (LLI)	—	2×10^{-3}	4×10^{-7}	8×10^{-4}	10^{-7}
$_{50}Sn^{125}$ (β^-, γ, e^-)	(Sol)	GI (LLI)	—	5×10^{-4}	10^{-7}	2×10^{-4}	4×10^{-8}
		Bone	7	0.02	3×10^{-7}	6×10^{-3}	10^{-7}
		Prostate	10	0.03	6×10^{-7}	9×10^{-3}	2×10^{-7}
		Total Body	20	0.05	10^{-6}	0.02	4×10^{-7}
		Liver	100	0.3	7×10^{-6}	0.1	2×10^{-6}
		Thyroid	300	0.8	2×10^{-5}	0.3	5×10^{-6}
	(Insol)	**Lung**	—	—	8×10^{-8}	—	3×10^{-8}
		GI (LLI)	—	5×10^{-4}	9×10^{-8}	2×10^{-4}	3×10^{-8}
$_{51}Sb^{122}$ (β^-, γ)	(Sol)	GI (LLI)	—	8×10^{-4}	2×10^{-7}	3×10^{-4}	6×10^{-8}
		Total Body	20	0.3	4×10^{-6}	0.1	10^{-6}
		Lung	40	0.5	6×10^{-6}	0.2	2×10^{-6}
		Bone	40	0.5	6×10^{-6}	0.2	2×10^{-6}
		Liver	10^3	10	2×10^{-4}	4	5×10^{-5}
		Thyroid	3×10^3	40	4×10^{-4}	10	2×10^{-4}

This page continues a table of maximum permissible concentrations. Column headers (from a preceding page) are, in order: organ of reference, q (μc), $(MPC)_w$, $(MPC)_a$, $(MPC)_w$, $(MPC)_a$.

Isotope	Sol.	Organ of reference	q (μc)	$(MPC)_w$	$(MPC)_a$	$(MPC)_w$	$(MPC)_a$
(continued)	(Insol)	GI (LLI)	—	8×10^{-4}	10^{-7}	3×10^{-4}	5×10^{-8}
		Lung	—		4×10^{-7}		10^{-7}
$_{51}$Sb^{124} (β^-, γ)	(Sol)	GI (LLI)	—	7×10^{-4}	2×10^{-7}	2×10^{-4}	5×10^{-8}
		Total Body	10	0.02	2×10^{-7}	6×10^{-3}	7×10^{-8}
		Lung	20		3×10^{-7}	8×10^{-3}	10^{-7}
		Bone	30	0.04	5×10^{-7}	0.01	2×10^{-7}
		Liver	800	1	10^{-5}	0.4	4×10^{-6}
		Thyroid	10^{4}	20	2×10^{-4}	6	7×10^{-5}
	(Insol)	Lung	—		2×10^{-8}		7×10^{-9}
		GI (LLI)	—	7×10^{-4}	10^{-7}	2×10^{-4}	4×10^{-8}
$_{51}$Sb^{125} (β^-, γ, e^-)	(Sol)	GI (LLI)	—	3×10^{-3}	6×10^{-7}	10^{-3}	2×10^{-7}
		Lung	40	0.04	5×10^{-7}	0.01	2×10^{-7}
		Total Body	60	0.05	6×10^{-7}	0.02	2×10^{-7}
		Bone	70	0.06	8×10^{-7}	0.02	2×10^{-7}
		Liver	3×10^{3}	3	3×10^{-5}	0.9	10^{-5}
		Thyroid	7×10^{4}	60	7×10^{-4}	20	2×10^{-4}
	(Insol)	Lung	—		3×10^{-8}		9×10^{-9}
		GI (LLI)	—	3×10^{-3}	5×10^{-7}	10^{-3}	2×10^{-7}
$_{52}$Te^{125m} (γ, e^-)	(Sol)	Kidney	29	5×10^{-3}	4×10^{-7}	2×10^{-3}	10^{-7}
		GI (LLI)	—	5×10^{-3}	10^{-6}	2×10^{-3}	4×10^{-7}
		Testis	20	6×10^{-3}	5×10^{-7}	2×10^{-3}	2×10^{-7}
		Spleen	50	0.02	10^{-6}	6×10^{-3}	4×10^{-7}
		Liver	100	0.04	3×10^{-6}	0.01	9×10^{-7}
		Total Body	100	0.04	3×10^{-6}	0.01	10^{-6}
		Bone	100		3×10^{-6}	0.02	10^{-6}
		Thyroid	500	0.2	10^{-5}	0.05	4×10^{-6}
	(Insol)	Lung	—		10^{-7}		4×10^{-8}
		GI (LLI)	—	3×10^{-3}	6×10^{-7}	10^{-3}	2×10^{-7}

Radionuclide and type of decay	Organ of reference (critical organ in boldface)	Maximum permissible burden in total body $q(\mu c)$	Maximum permissible concentrations			
			For 40 hour week		For 168 hour week**	
			$(MPC)_w$ $\mu c/cc$	$(MPC)_a$ $\mu c/cc$	$(MPC)_w$ $\mu c/cc$	$(MPC)_a$ $\mu c/cc$
$_{52}Te^{127m}$ (β^-, γ, e^-) (Sol)	**Kidney**	7	2×10^{-3}	10^{-7}	6×10^{-4}	5×10^{-8}
	Testis	7	2×10^{-3}	10^{-7}	7×10^{-4}	5×10^{-8}
	GI (LLI)		2×10^{-3}	5×10^{-7}	8×10^{-4}	2×10^{-7}
	Spleen	20	6×10^{-3}	5×10^{-7}	2×10^{-3}	2×10^{-7}
	Bone	50	0.01	9×10^{-7}	4×10^{-3}	3×10^{-7}
	Liver	50	0.02	10^{-6}	5×10^{-3}	4×10^{-7}
	Total Body	60	0.02	10^{-6}	6×10^{-3}	4×10^{-7}
	Thyroid	200	0.05	4×10^{-6}	0.02	10^{-6}
(Insol)	**Lung**			4×10^{-8}		10^{-8}
	GI (LLI)		2×10^{-3}	3×10^{-7}	5×10^{-4}	9×10^{-8}
$_{52}Te^{127}$ (β^-) (Sol)	**GI (LLI)**		8×10^{-3}	2×10^{-6}	3×10^{-3}	6×10^{-7}
	Kidney	20	0.1	10^{-5}	0.05	4×10^{-6}
	Testis	20	0.2	10^{-5}	0.05	4×10^{-6}
	Spleen	50	0.5	4×10^{-5}	0.2	10^{-5}
	Total Body	80	0.8	6×10^{-5}	0.3	2×10^{-5}
	Bone	100	1	7×10^{-5}	0.3	2×10^{-5}
	Liver	100	1	9×10^{-5}	0.4	3×10^{-5}
	Thyroid	100	1	10^{-4}	0.5	4×10^{-5}
(Insol)	**GI (LLI)**		5×10^{-3}	9×10^{-7}	2×10^{-3}	3×10^{-7}
	Lung			8×10^{-6}		3×10^{-6}

Radionuclide		Organ of reference					
$_{52}Te^{129m}$ (β^-, γ, e^-)	(Sol)	GI (LLI)	—	10^{-3}	2×10^{-7}	3×10^{-4}	7×10^{-8}
		Kidney	3	10^{-3}	8×10^{-8}	4×10^{-4}	3×10^{-8}
		Testis	3	10^{-3}	9×10^{-8}	4×10^{-4}	3×10^{-8}
		Spleen	10	4×10^{-3}	3×10^{-7}	10^{-3}	9×10^{-8}
		Total Body	20	6×10^{-3}	5×10^{-7}	2×10^{-3}	2×10^{-7}
		Liver	20	8×10^{-3}	6×10^{-7}	3×10^{-3}	2×10^{-7}
		Bone	20	9×10^{-3}	6×10^{-7}	3×10^{-3}	2×10^{-7}
		Thyroid	70	0.03	2×10^{-6}	8×10^{-3}	6×10^{-7}
	(Insol)	Lung	—	—	3×10^{-8}	2×10^{-4}	10^{-8}
		GI (LLI)	—	6×10^{-4}	10^{-7}	2×10^{-4}	4×10^{-8}
$_{52}Te^{129}$ (β^-, γ, e^-)	(Sol)	GI (S)	—	0.02	5×10^{-6}	8×10^{-3}	2×10^{-6}
		Kidney	5	0.4	3×10^{-5}	0.1	10^{-5}
		Testis	6	0.4	3×10^{-5}	0.2	10^{-5}
		Spleen	20	1	10^{-4}	0.5	3×10^{-5}
		Total Body	20	2	10^{-4}	0.5	4×10^{-5}
		Liver	40	3	2×10^{-4}	1	7×10^{-5}
		Bone	40	3	2×10^{-4}	1	8×10^{-5}
		Thyroid	60	4	3×10^{-4}	1	10^{-4}
	(Insol)	GI (ULI)	—	0.02	4×10^{-6}	8×10^{-3}	10^{-6}
		Lung	—	—	2×10^{-5}	—	7×10^{-6}
$_{52}Te^{131m}$ (β^-, γ, e^-)	(Sol)	GI (LLI)	—	2×10^{-3}	4×10^{-7}	6×10^{-4}	10^{-7}
		Kidney	4	0.01	10^{-6}	5×10^{-3}	3×10^{-7}
		Total Body	10	0.04	3×10^{-6}	0.01	10^{-6}
		Spleen	20	0.05	4×10^{-6}	0.02	10^{-6}
		Liver	30	0.09	7×10^{-6}	0.03	2×10^{-6}
		Bone	50	0.1	10^{-5}	0.05	4×10^{-6}
		Thyroid	50	0.2	10^{-5}	0.06	4×10^{-6}
	(Insol)	GI (LLI)	—	10^{-3}	2×10^{-7}	4×10^{-4}	6×10^{-8}
		Lung	—	—	6×10^{-7}	—	2×10^{-7}

Radionuclide and type of decay	Organ of reference (critical organ in boldface)	Maximum permissible burden in total body $q(\mu c)$	Maximum permissible concentrations			
			For 40 hour week		For 168 hour week**	
			$(MPC)_w$ $\mu c/cc$	$(MPC)_a$ $\mu c/cc$	$(MPC)_w$ $\mu c/cc$	$(MPC)_a$ $\mu c/cc$
$_{52}Te^{132}$ (β^-, γ, e^-) (Sol)	**GI (LLI)**	--	9×10^{-4}	2×10^{-7}	3×10^{-4}	7×10^{-8}
	Kidney	3	5×10^{-3}	4×10^{-7}	2×10^{-3}	10^{-7}
	Testis	5	7×10^{-3}	5×10^{-7}	2×10^{-3}	2×10^{-7}
	Total Body	10	0.02	10^{-6}	5×10^{-3}	4×10^{-7}
	Spleen	10	0.03	2×10^{-6}	6×10^{-3}	4×10^{-7}
	Liver	20	0.05	4×10^{-6}	0.01	8×10^{-7}
	Bone	30	0.05	4×10^{-6}	0.02	10^{-6}
	Thyroid	50	0.07	5×10^{-6}	0.02	10^{-6}
(Insol)	**GI (LLI)**	--	6×10^{-4}	10^{-7}	2×10^{-4}	4×10^{-8}
	Lung	--	--	2×10^{-7}	--	7×10^{-8}
$_{53}I^{126}$ (β^-, ϵ, γ) (Sol)	**Thyroid**	1	5×10^{-5}	8×10^{-9}	2×10^{-5}	3×10^{-9}
	Total Body	90	6×10^{-3}	9×10^{-7}	2×10^{-3}	3×10^{-7}
	GI (LLI)	--	0.05	10^{-5}	0.02	4×10^{-6}
(Insol)	**Lung**	--	--	3×10^{-7}	--	10^{-7}
	GI (LLI)	--	3×10^{-3}	5×10^{-7}	9×10^{-4}	2×10^{-7}
$_{53}I^{129}$ (β^-, γ, e^-) (Sol)	**Thyroid**	3	10^{-5}	2×10^{-9}	4×10^{-6}	6×10^{-10}
	Total Body	200	2×10^{-3}	2×10^{-7}	5×10^{-4}	7×10^{-8}
	GI (LLI)	--	0.1	3×10^{-5}	0.04	9×10^{-6}
(Insol)	**Lung**	--	--	7×10^{-8}	--	2×10^{-8}
	GI (LLI)	--	6×10^{-3}	10^{-6}	2×10^{-3}	4×10^{-7}

Radionuclide		Organ					
$_{53}I^{131}$ (β^-, γ, e^-)	(Sol)	Thyroid	0.7	6×10^{-5}	9×10^{-9}	2×10^{-5}	3×10^{-9}
		Total Body	50	5×10^{-3}	8×10^{-7}	2×10^{-3}	3×10^{-7}
		GI (LLI)	—	0.03	7×10^{-6}	0.01	2×10^{-6}
	(Insol)	GI (LLI)	—	2×10^{-3}	3×10^{-7}	6×10^{-4}	10^{-7}
		Lung	—	—	3×10^{-7}	—	10^{-7}
$_{53}I^{132}$ (β^-, γ, e^-)	(Sol)	Thyroid	0.3	2×10^{-3}	2×10^{-7}	6×10^{-4}	8×10^{-8}
		GI (SI)	—	0.01	3×10^{-6}	4×10^{-3}	9×10^{-7}
		Total Body	10	0.1	2×10^{-5}	0.04	6×10^{-6}
	(Insol)	GI (ULI)	—	5×10^{-3}	9×10^{-7}	2×10^{-3}	3×10^{-7}
		Lung	—	—	7×10^{-6}	—	2×10^{-6}
$_{53}I^{133}$ (β^-, γ, e^-)	(Sol)	Thyroid	0.3	2×10^{-4}	3×10^{-8}	7×10^{-5}	10^{-8}
		GI (SI)	—	0.02	4×10^{-6}	6×10^{-3}	10^{-6}
		Total Body	20	0.02	4×10^{-6}	9×10^{-3}	10^{-6}
	(Insol)	GI (LLI)	—	10^{-3}	2×10^{-7}	4×10^{-4}	7×10^{-8}
		Lung	—	—	10^{-6}	—	4×10^{-7}
$_{53}I^{134}$ (β^-, γ)	(Sol)	Thyroid	0.2	4×10^{-3}	5×10^{-7}	10^{-3}	2×10^{-7}
		GI (S)	—	0.02	4×10^{-6}	6×10^{-3}	10^{-6}
		Total Body	10	0.3	5×10^{-6}	0.1	2×10^{-5}
	(Insol)	GI (S)	—	0.02	3×10^{-6}	6×10^{-3}	10^{-6}
		Lung	—	—	2×10^{-5}	—	7×10^{-6}
$_{53}I^{135}$ (β^-, γ, e^-)	(Sol)	Thyroid	0.3	7×10^{-4}	10^{-7}	2×10^{-4}	4×10^{-8}
		GI (SI)	—	0.01	3×10^{-6}	5×10^{-3}	10^{-6}
		Total Body	20	0.05	7×10^{-6}	0.02	3×10^{-6}
	(Insol)	GI (LLI)	—	2×10^{-3}	4×10^{-7}	7×10^{-4}	10^{-7}
		Lung	—	—	3×10^{-6}	—	10^{-6}
$_{54}Xe^{131m}$ (γ, e^-)	(Immersion)	Total Body	—	—	2×10^{-5}	—	4×10^{-6}
$_{54}Xe^{133}$ (γ, e^-)	(Immersion)	Total Body	—	—	10^{-5}	—	3×10^{-6}
$_{54}Xe^{135}$ (β^-, γ)	(Immersion)	Total Body	—	—	4×10^{-6}	—	10^{-6}

Radionuclide and type of decay	Organ of reference (critical organ in boldface)	Maximum permissible burden in total body $q(\mu c)$	For 40 hour week (MPC)$_w$ $\mu c/cc$	For 40 hour week (MPC)$_a$ $\mu c/cc$	For 168 hour week** (MPC)$_w$ $\mu c/cc$	For 168 hour week** (MPC)$_a$ $\mu c/cc$
$_{55}$Cs^{131} (ϵ) (Sol)	**Total Body**	700	0.07	10^{-5}	0.02	4×10^{-6}
	Liver	800	0.09	10^{-5}	0.03	4×10^{-6}
	Spleen	10^3	0.1	2×10^{-5}	0.04	6×10^{-6}
	Kidney	10^3	0.1	2×10^{-5}	0.05	7×10^{-6}
	Muscle	2×10^3	0.2	3×10^{-5}	0.07	10^{-5}
	GI (SI)		0.5	10^{-4}	0.2	4×10^{-5}
	Bone	8×10^3	0.9	10^{-4}	0.3	4×10^{-5}
	Lung	10^4	1	10^{-4}	0.4	5×10^{-5}
(Insol)	**Lung**			3×10^{-6}		10^{-6}
	GI (LLI)		0.03	5×10^{-6}	9×10^{-3}	2×10^{-6}
$_{55}$Cs^{134m} (β^-, γ, e$^-$) (Sol)	**GI (S)**	100	0.2	4×10^{-5}	0.06	10^{-5}
	Total Body	100	0.7	10^{-4}	0.3	4×10^{-5}
	Liver	200	1	10^{-4}	0.3	5×10^{-5}
	Spleen	200	2	2×10^{-4}	0.5	7×10^{-5}
	Kidney	200	2	2×10^{-4}	0.6	8×10^{-5}
	Muscle	600	4	6×10^{-4}	0.6	9×10^{-5}
	Bone	2×10^3	10	2×10^{-3}	1	2×10^{-4}
	Lung				4	6×10^{-4}
(Insol)	**GI (ULI)**		0.03	6×10^{-6}	0.01	2×10^{-6}
	Lung			3×10^{-5}		10^{-5}

Nuclide	Sol.	Organ	q (μc)	$(MPC)_w$ (168 hr)	$(MPC)_a$ (168 hr)	$(MPC)_w$ (40 hr)	$(MPC)_a$ (40 hr)
$_{55}\text{Cs}^{134}$ (β^-, γ)	(Sol)	Total Body	20	3×10^{-4}	4×10^{-8}	9×10^{-5}	10^{-8}
		Liver	30	4×10^{-4}	6×10^{-8}	10^{-4}	2×10^{-8}
		Muscle	30	4×10^{-4}	6×10^{-8}	2×10^{-4}	2×10^{-8}
		Spleen	40	6×10^{-4}	9×10^{-8}	2×10^{-4}	3×10^{-8}
		Kidney	90	10^{-3}	2×10^{-7}	4×10^{-4}	6×10^{-8}
		Bone	200	2×10^{-3}	3×10^{-7}	7×10^{-4}	10^{-7}
		Lung	300	4×10^{-3}	5×10^{-7}	10^{-4}	2×10^{-7}
		GI (SI)	---	0.01	3×10^{-6}	5×10^{-3}	10^{-6}
	(Insol)	Lung	---	---	10^{-8}	---	4×10^{-9}
		GI (LLI)	---	10^{-3}	2×10^{-7}	4×10^{-4}	7×10^{-8}
$_{55}\text{Cs}^{135}$ (β^-)	(Sol)	Liver	200	3×10^{-3}	5×10^{-7}	10^{-3}	2×10^{-7}
		Spleen	300	4×10^{-3}	5×10^{-7}	10^{-3}	2×10^{-7}
		Total Body	300	4×10^{-3}	6×10^{-7}	10^{-3}	2×10^{-7}
		Bone	400	6×10^{-3}	8×10^{-7}	2×10^{-3}	3×10^{-7}
		Muscle	500	6×10^{-3}	9×10^{-7}	2×10^{-3}	3×10^{-7}
		Kidney	600	9×10^{-3}	10^{-6}	3×10^{-3}	4×10^{-7}
		Lung	2×10^3	0.03	4×10^{-6}	0.01	10^{-6}
		GI (LLI)	---	0.1	3×10^{-5}	0.05	10^{-5}
	(Insol)	GI (LLI)	---	---	10^{-6}	2×10^{-3}	4×10^{-7}
		Lung	---	7×10^{-3}	9×10^{-8}	---	3×10^{-8}
$_{55}\text{Cs}^{136}$ (β^-, γ)	(Sol)	Total Body	30	2×10^{-3}	4×10^{-7}	9×10^{-4}	10^{-7}
		Liver	60	5×10^{-3}	7×10^{-7}	2×10^{-3}	2×10^{-7}
		Spleen	80	7×10^{-3}	10^{-6}	2×10^{-3}	4×10^{-7}
		Muscle	90	8×10^{-3}	10^{-6}	3×10^{-3}	4×10^{-7}
		Kidney	100	0.02	5×10^{-6}	3×10^{-3}	4×10^{-7}
		GI (SI)	---	0.03	4×10^{-6}	8×10^{-3}	2×10^{-6}
		Bone	400	0.06	9×10^{-7}	0.01	2×10^{-6}
		Lung	800	---	2×10^{-7}	0.02	3×10^{-6}
	(Insol)	Lung	---	---	2×10^{-7}	---	6×10^{-8}
		GI (LLI)	---	2×10^{-3}	3×10^{-7}	6×10^{-4}	10^{-7}

Radionuclide and type of decay	Organ of reference (critical organ in boldface)	Maximum permissible burden in total body $q(\mu c)$	Maximum permissible concentrations			
			For 40 hour week		For 168 hour week **	
			$(MPC)_w$ $\mu c/cc$	$(MPC)_a$ $\mu c/cc$	$(MPC)_w$ $\mu c/cc$	$(MPC)_a$ $\mu c/cc$
$_{55}Cs^{137}$ (β^-, γ, e^-) (Sol)	**Total Body**	30	4×10^{-4}	6×10^{-8}	2×10^{-4}	2×10^{-8}
	Liver	40	5×10^{-4}	8×10^{-8}	2×10^{-4}	3×10^{-8}
	Spleen	50	6×10^{-4}	9×10^{-8}	2×10^{-4}	3×10^{-8}
	Muscle	50	7×10^{-4}	10^{-7}	2×10^{-4}	4×10^{-8}
	Bone	100	10^{-3}	2×10^{-7}	5×10^{-4}	7×10^{-8}
	Kidney	100	10^{-3}	2×10^{-7}	5×10^{-4}	8×10^{-8}
	Lung	300	5×10^{-3}	6×10^{-7}	2×10^{-3}	2×10^{-7}
	GI (SI)	--	0.02	5×10^{-6}	8×10^{-3}	2×10^{-6}
(Insol)	**Lung**	--	--	10^{-8}	--	5×10^{-9}
	GI (LLI)	--	10^{-3}	2×10^{-7}	4×10^{-4}	8×10^{-8}
$_{56}Ba^{131}$ (ϵ, γ) (Sol)	**GI (LLI)**	--	5×10^{-3}	10^{-6}	2×10^{-3}	4×10^{-7}
	Total Body	50	0.1	2×10^{-6}	0.03	7×10^{-7}
	Bone	80	0.1	3×10^{-6}	0.05	10^{-6}
	Liver	10^4	20	4×10^{-4}	7	10^{-4}
	Muscle	2×10^4	40	7×10^{-4}	10	2×10^{-4}
	Lung	2×10^4	40	7×10^{-4}	10	2×10^{-4}
	Spleen	3×10^4	60	10^{-3}	20	4×10^{-4}
	Kidney	4×10^4	70	10^{-3}	20	5×10^{-4}
(Insol)	**Lung**	--	--	4×10^{-7}	--	10^{-7}
	GI (LLI)	--	5×10^{-3}	9×10^{-7}	2×10^{-3}	3×10^{-7}

$_{56}Ba^{140}$ (β^-, γ)	(Sol)	GI (LLI)	---	8×10^{-4}	2×10^{-7}	3×10^{-4}	6×10^{-8}
		Bone	4	6×10^{-3}	10^{-7}	2×10^{-3}	4×10^{-8}
		Total Body	9	0.01	3×10^{-7}	5×10^{-3}	10^{-7}
		Liver	10^3	2	5×10^{-5}	0.9	2×10^{-5}
		Lung	3×10^3	4	9×10^{-5}	2	3×10^{-5}
		Muscle	3×10^3	5	10^{-4}	2	4×10^{-5}
		Spleen	4×10^3	6	10^{-4}	2	4×10^{-5}
		Kidney	4×10^3	8	2×10^{-4}	3	5×10^{-5}
	(Insol)	**Lung**	---		4×10^{-8}		10^{-8}
		GI (LLI)	---	7×10^{-4}	10^{-7}	2×10^{-4}	4×10^{-8}
$_{57}La^{140}$ (β^-, γ)	(Sol)	GI (LLI)	---	7×10^{-4}	2×10^{-7}	2×10^{-4}	5×10^{-8}
		Liver	9	50	2×10^{-6}	20	7×10^{-7}
		Bone	10	60	2×10^{-6}	20	8×10^{-7}
		Total Body	10	60	2×10^{-6}	20	9×10^{-7}
	(Insol)	GI (LLI)	---	7×10^{-4}	10^{-7}	2×10^{-4}	4×10^{-8}
		Lung	---		4×10^{-7}		10^{-7}
$_{58}Ce^{141}$ (β^-, γ)	(Sol)	GI (LLI)	---	3×10^{-3}	6×10^{-7}	9×10^{-4}	2×10^{-7}
		Liver	30	10	4×10^{-7}	3	2×10^{-7}
		Bone	40	10	6×10^{-7}	5	2×10^{-7}
		Kidney	70	20	9×10^{-7}	7	3×10^{-7}
		Total Body	90	30	10^{-6}	10	4×10^{-7}
	(Insol)	**Lung**	---		2×10^{-7}		5×10^{-8}
		GI (LLI)	---	3×10^{-3}	5×10^{-7}	9×10^{-4}	2×10^{-7}
$_{58}Ce^{143}$ (β^-, γ)	(Sol)	GI (LLI)	---	10^{-3}	3×10^{-7}	4×10^{-4}	9×10^{-8}
		Liver	7	50	2×10^{-6}	20	7×10^{-7}
		Bone	10	70	3×10^{-6}	20	10^{-6}
		Kidney	20	100	5×10^{-6}	40	2×10^{-6}
		Total Body	20	100	6×10^{-6}	50	2×10^{-6}
	(Insol)	GI (LLI)	---	10^{-3}	2×10^{-7}	4×10^{-4}	7×10^{-8}
		Lung	---		6×10^{-7}		2×10^{-7}

Radionuclide and type of decay		Organ of reference (critical organ in boldface)	Maximum permissible burden in total body $q(\mu c)$	Maximum permissible concentrations			
				For 40 hour week		For 168 hour week**	
				$(MPC)_w$ $\mu c/cc$	$(MPC)_a$ $\mu c/cc$	$(MPC)_w$ $\mu c/cc$	$(MPC)_a$ $\mu c/cc$
$_{58}Ce^{144}$ $(\alpha, \beta^-, \gamma)$	(Sol)	**GI (LLI)**	--	3×10^{-4}	8×10^{-8}	10^{-4}	3×10^{-8}
		Bone	5	0.2	10^{-8}	0.08	3×10^{-9}
		Liver	6	0.3	10^{-8}	0.1	4×10^{-9}
		Kidney	10	0.5	2×10^{-8}	0.2	7×10^{-9}
		Total Body	20	0.7	3×10^{-8}	0.3	10^{-8}
	(Insol)	**Lung**	--	--	6×10^{-9}	--	2×10^{-9}
		GI (LLI)	--	3×10^{-4}	6×10^{-8}	10^{-4}	2×10^{-8}
$_{59}Pr^{142}$ (β^-, γ)	(Sol)	**GI (LLI)**	--	9×10^{-4}	2×10^{-7}	3×10^{-4}	7×10^{-8}
		Bone	7	80	4×10^{-6}	30	10^{-6}
		Liver	9	100	4×10^{-6}	40	2×10^{-6}
		Kidney	20	200	8×10^{-6}	60	3×10^{-6}
		Total Body	20	300	10^{-5}	90	4×10^{-6}
	(Insol)	**GI (LLI)**	--	9×10^{-4}	2×10^{-7}	3×10^{-4}	5×10^{-8}
		Lung	--	--	10^{-6}	--	4×10^{-7}
$_{59}Pr^{143}$ (β^-)	(Sol)	**GI (LLI)**	--	10^{-3}	3×10^{-7}	5×10^{-4}	10^{-7}
		Bone	20	10	5×10^{-7}	4	2×10^{-7}
		Liver	20	20	7×10^{-7}	5	2×10^{-7}
		Kidney	40	30	10^{-6}	9	4×10^{-7}
		Total Body	60	40	2×10^{-6}	10	6×10^{-7}

Note: the column headings for this table are printed on the preceding page and are not repeated here. Based on the data, the numeric columns are the maximum permissible burden q (μc) and two pairs of maximum permissible concentrations (MPC)$_w$ / (MPC)$_a$.

Radionuclide		Organ of reference	q (μc)	(MPC)$_w$	(MPC)$_a$	(MPC)$_w$	(MPC)$_a$
(continued)	(Insol)	Lung	—	—	2×10^{-7}	—	6×10^{-8}
		GI (LLI)	—	10^{-3}	3×10^{-7}	5×10^{-4}	9×10^{-8}
$_{60}$Nd^{144} (α)	(Sol)	Bone	0.1	2×10^{-3}	8×10^{-11}	7×10^{-4}	3×10^{-11}
		GI (LLI)	—	2×10^{-3}	5×10^{-7}	8×10^{-4}	2×10^{-7}
		Kidney	0.3	4×10^{-3}	2×10^{-10}	10^{-3}	5×10^{-11}
		Liver	0.7	0.01	4×10^{-10}	3×10^{-3}	2×10^{-10}
		Total Body	1	0.01	6×10^{-10}	5×10^{-3}	2×10^{-10}
	(Insol)	Lung	—	—	3×10^{-10}	—	10^{-10}
		GI (LLI)	—	2×10^{-3}	4×10^{-7}	8×10^{-4}	10^{-7}
$_{60}$Nd^{147} ($\alpha,\ \beta^-,\ \gamma$)	(Sol)	GI (LLI)	—	2×10^{-3}	4×10^{-7}	6×10^{-4}	10^{-7}
		Liver	10	8	6×10^{-7}	3	10^{-7}
		Kidney	20	10	8×10^{-6}	5	2×10^{-7}
		Bone	20	20	2×10^{-6}	6	3×10^{-7}
		Total Body	50	40	2×10^{-7}	10	6×10^{-7}
	(Insol)	Lung	—	—	3×10^{-7}	—	8×10^{-8}
		GI (LLI)	—	2×10^{-3}	2×10^{-6}	6×10^{-4}	10^{-7}
$_{60}$Nd^{149} ($\beta^-,\ \gamma$)	Sol	GI (LLI)	—	8×10^{-3}	10^{-5}	3×10^{-3}	6×10^{-7}
		Liver	3	300	3×10^{-5}	100	5×10^{-6}
		Kidney	5	600	3×10^{-5}	200	9×10^{-6}
		Bone	7	700	9×10^{-5}	300	10^{-5}
		Total Body	20	2×10^{3}	10^{-6}	700	3×10^{-5}
	(Insol)	GI (ULI)	—	8×10^{-3}	9×10^{-6}	3×10^{-3}	5×10^{-7}
		Lung	—	—	10^{-6}	—	3×10^{-6}
$_{61}$Pm^{147} ($\alpha,\ \beta^-$)	(Sol)	GI (LLI)	—	6×10^{-3}	2×10^{-6}	2×10^{-3}	5×10^{-7}
		Bone	60	1	6×10^{-8}	0.5	2×10^{-8}
		Kidney	200	4	2×10^{-7}	2	7×10^{-8}
		Total Body	300	7	3×10^{-7}	2	10^{-7}
		Liver	300	8	4×10^{-7}	3	10^{-7}
	(Insol)	Lung	—	—	10^{-7}	—	3×10^{-8}
		GI (LLI)	—	6×10^{-3}	10^{-6}	2×10^{-3}	4×10^{-7}

Radionuclide and type of decay	Organ of reference (critical organ in boldface)	Maximum permissible burden in total body $q(\mu c)$	Maximum permissible concentrations			
			For 40 hour week		For 168 hour week**	
			$(MPC)_w$ $\mu c/cc$	$(MPC)_a$ $\mu c/cc$	$(MPC)_w$ $\mu c/cc$	$(MPC)_a$ $\mu c/cc$
$_{61}Pm^{149}(\beta^-,\gamma)$ (Sol)	**GI (LLI)**	---	**10^{-3}**	**3×10^{-7}**	**4×10^{-4}**	**10^{-7}**
	Bone	20	70	3×10^{-6}	20	10^{-6}
	Kidney	30	100	6×10^{-6}	40	2×10^{-6}
	Total Body	40	200	7×10^{-6}	50	2×10^{-6}
	Liver	50	200	10^{-5}	80	3×10^{-6}
(Insol)	**GI (LLI)**	---	**10^{-3}**	**2×10^{-7}**	**4×10^{-4}**	**8×10^{-8}**
	Lung	---		7×10^{-7}		3×10^{-7}
$_{62}Sm^{147}(\alpha)$ (Sol)	**Bone**	0.1	**2×10^{-3}**	**7×10^{-11}**	**6×10^{-4}**	**2×10^{-11}**
	GI (LLI)	---	2×10^{-3}	4×10^{-7}	7×10^{-4}	2×10^{-7}
	Kidney	0.6	8×10^{-3}	4×10^{-10}	3×10^{-3}	10^{-10}
	Liver	0.6	9×10^{-3}	4×10^{-10}	3×10^{-3}	10^{-10}
	Total Body	0.9	0.01	5×10^{-10}	4×10^{-3}	2×10^{-10}
(Insol)	**Lung**	---		**3×10^{-10}**		**9×10^{-11}**
	GI (LLI)	---	2×10^{-3}	4×10^{-7}	7×10^{-4}	10^{-7}
$_{62}Sm^{151}(\beta^-,\gamma)$ (Sol)	**GI (LLI)**	---	**0.01**	**2×10^{-6}**	**4×10^{-3}**	**8×10^{-7}**
	Bone	100	2	6×10^{-8}	0.5	2×10^{-8}
	Kidney	300	4	2×10^{-7}	2	6×10^{-8}
	Liver	300	5	2×10^{-7}	2	7×10^{-8}
	Total Body	500	7	3×10^{-7}	2	10^{-7}

Nuclide (type of decay)		Organ	q / col 1	col 2	col 3	col 4	col 5
(continued from preceding page)	(Insol)	Lung	---	---	10^{-7}	---	5×10^{-8}
		GI (LLI)	---	0.01	2×10^{-6}	4×10^{-3}	7×10^{-7}
$_{62}\mathrm{Sm}^{153}\ (\beta^-,\gamma)$	(Sol)	GI (LLI)	---	2×10^{-3}	5×10^{-7}	8×10^{-4}	2×10^{-7}
		Liver	20	70	3×10^{-6}	30	10^{-6}
		Bone	30	100	6×10^{-6}	50	2×10^{-6}
		Kidney	50	200	10^{-5}	80	4×10^{-6}
		Total Body	70	300	4×10^{-7}	100	5×10^{-6}
	(Insol)	GI (LLI)	---	2×10^{-3}	10^{-6}	8×10^{-4}	10^{-7}
		Lung	---	---	---	---	5×10^{-7}
$_{63}\mathrm{Eu}^{152}\ (9.2\ \mathrm{hr})\ (\beta^-,\epsilon,\gamma)$	(Sol)	GI (LLI)	---	2×10^{-3}	4×10^{-7}	6×10^{-4}	5×10^{-7}
		Liver	8	200	9×10^{-6}	70	3×10^{-6}
		Bone	10	300	10^{-5}	90	6×10^{-6}
		Kidney		300	2×10^{-5}	100	10^{-5}
		Total Body	20	500	3×10^{-7}	200	10^{-5}
	(Insol)	GI (LLI)	---	2×10^{-3}	3×10^{-6}	6×10^{-4}	4×10^{-7}
		Lung	---	---	---	---	10^{-6}
$_{63}\mathrm{Eu}^{152}\ (13\ \mathrm{yr})\ (\beta^-,\epsilon,\gamma)$	(Sol)	GI (LLI)	---	2×10^{-3}	5×10^{-7}	8×10^{-4}	2×10^{-7}
		Kidney	20	0.3	10^{-8}	0.09	4×10^{-9}
		Total Body	30	0.5	2×10^{-8}	0.2	7×10^{-9}
		Bone	30	0.6	3×10^{-8}	0.2	8×10^{-9}
		Liver	80	1.0	6×10^{-8}	0.4	2×10^{-8}
	(Insol)	GI (LLI)	---	2×10^{-3}	2×10^{-8}	8×10^{-4}	6×10^{-9}
		Lung	---	---	4×10^{-7}	---	10^{-7}
$_{63}\mathrm{Eu}^{154}\ (\beta^-,\epsilon,\gamma)$	(Sol)	GI (LLI)	---	6×10^{-4}	10^{-7}	2×10^{-4}	5×10^{-8}
		Kidney	5	0.09	4×10^{-9}	0.03	10^{-9}
		Bone	5	0.09	4×10^{-9}	0.03	10^{-9}
		Total Body	20	0.2	10^{-8}	0.08	4×10^{-9}
		Liver	30	0.5	2×10^{-8}	0.2	7×10^{-9}
	(Insol)	Lung	---	---	10^{-7}	---	4×10^{-8}
		GI (LLI)	---	6×10^{-4}	7×10^{-9}	2×10^{-4}	2×10^{-9}

Radionuclide and type of decay	Organ of reference (critical organ in boldface)	Maximum permissible burden in total body $q(\mu c)$	Maximum permissible concentrations			
			For 40 hour week		For 168 hour week**	
			(MPC)$_w$ $\mu c/cc$	(MPC)$_a$ $\mu c/cc$	(MPC)$_w$ $\mu c/cc$	(MPC)$_a$ $\mu c/cc$
$_{63}$Eu^{155}(β^-, γ) (Sol)	**GI (LLI)**	---	6×10^{-3}	10^{-6}	2×10^{-3}	4×10^{-7}
	Kidney	70	2	9×10^{-8}	0.7	3×10^{-8}
	Bone	80	2	10^{-7}	0.8	3×10^{-8}
	Total Body	100	4	2×10^{-7}	1	5×10^{-8}
	Liver	200	5	2×10^{-7}	2	8×10^{-8}
(Insol)	**Lung**	---		7×10^{-8}		3×10^{-8}
	GI (LLI)	---	6×10^{-3}	10^{-6}	2×10^{-3}	4×10^{-7}
$_{64}$Gd^{153}(ϵ, γ, e$^-$) (Sol)	**GI (LLI)**	---	6×10^{-3}	10^{-6}	2×10^{-3}	5×10^{-7}
	Bone	90	5	2×10^{-7}	2	8×10^{-8}
	Total Body	100	7	3×10^{-7}	2	2×10^{-7}
	Liver	100	7	3×10^{-7}	2	10^{-7}
(Insol)	**Lung**	---		9×10^{-8}		3×10^{-8}
	GI (LLI)	---	6×10^{-3}	10^{-6}	2×10^{-3}	4×10^{-7}
$_{64}$Gd^{159}(β^-, γ) (Sol)	**GI (LLI)**	---	2×10^{-3}	5×10^{-7}	8×10^{-4}	2×10^{-7}
	Bone	20	200	9×10^{-6}	70	3×10^{-6}
	Liver	40	400	2×10^{-5}	200	7×10^{-6}
	Total Body	50	700	3×10^{-5}	200	10^{-5}
(Insol)	**GI (LLI)**	---	2×10^{-3}	4×10^{-7}	8×10^{-4}	10^{-7}
	Lung	---		3×10^{-6}		10^{-6}

$_{65}$Tb^{160} (β^-, γ)	(Sol)	GI (LLI)	—	10^{-3}	3×10^{-7}	4×10^{-4}	10^{-7}
		Bone	20	2	10^{-7}	0.8	3×10^{-8}
		Kidney	20	3	10^{-7}	1	4×10^{-8}
		Total Body	20	3	10^{-7}	1	5×10^{-8}
	(Insol)	Lung	—	—	3×10^{-8}	—	10^{-8}
		GI (LLI)	—	10^{-3}	2×10^{-7}	4×10^{-4}	8×10^{-8}
$_{66}$Dy^{165} (β^-, γ)	(Sol)	GI (LLI)	—	0.01	3×10^{-6}	4×10^{-3}	9×10^{-7}
		Bone	10	10^3	5×10^{-5}	400	2×10^{-5}
		Total Body	40	4×10^3	2×10^{-4}	10^3	6×10^{-5}
		Liver	60	6×10^3	3×10^{-4}	2×10^3	9×10^{-5}
	(Insol)	GI (ULI)	—	0.01	2×10^{-6}	4×10^{-3}	7×10^{-7}
		Lung	—	—	2×10^{-5}	—	6×10^{-6}
$_{66}$Dy^{166} (β^-, γ, e^-)	(Sol)	GI (LLI)	—	10^{-3}	2×10^{-7}	4×10^{-4}	8×10^{-8}
		Bone	5	10	6×10^{-7}	4	2×10^{-7}
		Total Body	30	70	3×10^{-6}	20	10^{-6}
		Liver	30	80	4×10^{-6}	30	10^{-6}
	(Insol)	GI (LLI)	—	10^{-3}	2×10^{-7}	4×10^{-4}	7×10^{-8}
		Lung	—	—	3×10^{-7}	—	10^{-7}
$_{67}$Ho^{166} (β^-, γ, e^-)	(Sol)	GI (LLI)	—	9×10^{-4}	2×10^{-7}	3×10^{-4}	7×10^{-8}
		Bone	5	40	2×10^{-6}	10	6×10^{-7}
		Kidney	20	200	7×10^{-6}	50	2×10^{-6}
		Total Body	30	200	10^{-5}	80	4×10^{-7}
		Liver	40	300	10^{-5}	100	4×10^{-6}
	(Insol)	GI (LLI)	—	9×10^{-4}	2×10^{-7}	3×10^{-4}	6×10^{-8}
		Lung	—	—	10^{-6}	—	3×10^{-7}
$_{68}$Er^{169} (β^-, γ)	(Sol)	GI (LLI)	—	3×10^{-3}	6×10^{-7}	9×10^{-4}	2×10^{-7}
		Bone	30	30	10^{-6}	10	5×10^{-7}
		Total Body	50	50	2×10^{-6}	20	8×10^{-7}
		Kidney	70	60	3×10^{-6}	20	10^{-6}
		Liver	200	200	9×10^{-6}	70	3×10^{-6}

Radionuclide and type of decay	Organ of reference (critical organ in boldface)	Maximum permissible burden in total body $q(\mu c)$	Maximum permissible concentrations			
			For 40 hour week		For 168 hour week**	
			$(MPC)_w$ $\mu c/cc$	$(MPC)_a$ $\mu c/cc$	$(MPC)_w$ $\mu c/cc$	$(MPC)_a$ $\mu c/cc$
$_{68}Er^{171}$ (β^-, γ, e^-) (Insol)	**Lung**	—	—	4×10^{-7}	—	10^{-7}
	GI (LLI)	—	3×10^{-3}	6×10^{-7}	9×10^{-4}	2×10^{-7}
(Sol)	GI (ULI)	—	3×10^{-3}	7×10^{-7}	10^{-3}	2×10^{-7}
	Bone	9	300	10^{-5}	90	4×10^{-6}
	Kidney	30	800	4×10^{-5}	300	10^{-5}
	Total Body	30	900	4×10^{-5}	300	10^{-5}
$_{69}Tm^{170}$ (β^-, ϵ, γ, e^-) (Insol)	GI (ULI)	—	3×10^{-3}	6×10^{-7}	10^{-3}	2×10^{-7}
	Lung	—	—	5×10^{-6}	—	2×10^{-6}
(Sol)	GI (LLI)	—	10^{-3}	3×10^{-7}	5×10^{-4}	10^{-7}
	Bone	9	0.8	4×10^{-8}	0.3	10^{-8}
	Kidney	30	4	2×10^{-7}	1	6×10^{-8}
	Total Body	60	5	2×10^{-7}	2	7×10^{-8}
$_{69}Tm^{171}$ (β^-) (Insol)	**Lung**	—	—	3×10^{-8}	—	10^{-8}
	GI (LLI)	—	10^{-3}	2×10^{-7}	5×10^{-4}	8×10^{-8}
(Sol)	**GI (LLI)**	—	0.01	3×10^{-6}	5×10^{-3}	10^{-6}
	Bone	90	3	10^{-7}	0.9	4×10^{-8}
	Kidney	700	20	8×10^{-7}	6	3×10^{-7}
	Total Body	700	20	8×10^{-7}	6	3×10^{-7}

Isotope		Organ of reference	q (μc)	(MPC)$_w$	(MPC)$_a$	(MPC)$_w$	(MPC)$_a$
$_{70}$Yb^{175} (β^-, γ)	(Insol)	Lung		0.01	2×10^{-7}	5×10^{-3}	8×10^{-8}
		GI (LLI)		3×10^{-3}	3×10^{-6}	10^{-3}	9×10^{-7}
	(Sol)	GI (LLI)		3×10^{-3}	7×10^{-7}	10^{-3}	2×10^{-7}
		Bone	30	60	3×10^{-6}	20	9×10^{-7}
		Kidney	30	80	3×10^{-6}	30	10^{-6}
		Total Body	100	300	10^{-5}	100	4×10^{-6}
$_{71}$Lu^{177} (β^-, γ)	(Insol)	GI (LLI)		3×10^{-3}	6×10^{-7}	10^{-3}	2×10^{-7}
		Lung			10^{-6}		4×10^{-7}
	(Sol)	GI (LLI)		3×10^{-3}	6×10^{-7}	10^{-3}	2×10^{-7}
		Bone	20	30	10^{-6}	10	4×10^{-7}
		Total Body	100	200	7×10^{-6}	60	2×10^{-6}
		Kidney	200	200	10^{-5}	80	3×10^{-6}
$_{72}$Hf^{181} (β^-, γ)	(Insol)	GI (LLI)		2×10^{-3}	5×10^{-7}	7×10^{-4}	2×10^{-7}
		Lung			7×10^{-7}		2×10^{-7}
	(Sol)	GI (LLI)		2×10^{-3}	4×10^{-7}	7×10^{-4}	2×10^{-7}
		Spleen	4	0.9	4×10^{-8}	0.3	10^{-8}
		Liver	10	2	10^{-7}	0.8	4×10^{-8}
		Total Body	40	9	4×10^{-7}	3	10^{-7}
		Kidney	50	10	5×10^{-7}	4	2×10^{-7}
		Bone	100	20	9×10^{-7}	7	3×10^{-7}
$_{73}$Ta^{182} (β^-, γ)	(Insol)	Lung			7×10^{-7}		3×10^{-7}
		GI (LLI)		10^{-3}	4×10^{-7}	4×10^{-4}	10^{-7}
	(Sol)	GI (LLI)		10^{-3}	3×10^{-7}	4×10^{-4}	9×10^{-8}
		Liver	7	0.9	4×10^{-8}	0.3	10^{-8}
		Kidney	20	2	8×10^{-8}	0.7	3×10^{-8}
		Total Body	20	2	9×10^{-8}	0.7	3×10^{-8}
		Spleen	30	4	10^{-7}	1	5×10^{-8}
		Bone	50	6	3×10^{-7}	2	9×10^{-8}
	(Insol)	Lung			2×10^{-8}		7×10^{-9}
		GI (LLI)		10^{-3}	2×10^{-7}	4×10^{-4}	7×10^{-8}

Radionuclide and type of decay		Organ of reference (critical organ in boldface)	Maximum permissible burden in total body $q(\mu c)$	For 40 hour week $(MPC)_w$ $\mu c/cc$	$(MPC)_a$ $\mu c/cc$	For 168 hour week** $(MPC)_w$ $\mu c/cc$	$(MPC)_a$ $\mu c/cc$
$_{74}W^{181}$ (ϵ, γ)	(Sol)	**GI (LLI)**	—	0.01	2×10^{-6}	4×10^{-3}	8×10^{-7}
		Liver	70	0.6	2×10^{-5}	0.2	8×10^{-6}
		Total Body	100	0.9	3×10^{-5}	0.3	10^{-5}
		Bone	200	2	7×10^{-5}	0.7	2×10^{-5}
	(Insol)	**Lung**	—		10^{-7}		4×10^{-8}
		GI (LLI)	—	0.01	2×10^{-6}	3×10^{-3}	6×10^{-7}
$_{74}W^{185}$ (β^-)	(Sol)	**GI (LLI)**	—	4×10^{-3}	8×10^{-7}	10^{-3}	3×10^{-7}
		Bone	30		10^{-5}	0.09	3×10^{-6}
		Liver	40	0.3	2×10^{-5}	0.09	5×10^{-6}
		Total Body	100	0.4	5×10^{-5}	0.1	2×10^{-5}
	(Insol)	**Lung**	—	1	10^{-7}	0.5	4×10^{-8}
		GI (LLI)	—	3×10^{-3}	6×10^{-7}	10^{-3}	2×10^{-7}
$_{74}W^{187}$ (β^-, γ)	(Sol)	**GI (LLI)**	—	2×10^{-3}	4×10^{-7}	7×10^{-4}	2×10^{-7}
		Total Body	30	0.5	2×10^{-5}	0.2	7×10^{-6}
		Liver	30	0.6	2×10^{-5}	0.2	8×10^{-6}
		Bone	60	1	4×10^{-5}	0.4	10^{-5}
	(Insol)	**GI (LLI)**	—	2×10^{-3}	3×10^{-7}	6×10^{-4}	10^{-7}
		Lung	—		2×10^{-6}		6×10^{-7}

	q	(MPC)_w	(MPC)_a	(MPC)_w	(MPC)_a
$_{75}Re^{183}$ (ϵ, γ)					
(Sol) GI (LLI)	—	0.02	4×10^{-6}	6×10^{-3}	10^{-6}
Total Body	80	0.02	3×10^{-6}	8×10^{-3}	9×10^{-7}
Thyroid	300	0.09	10^{-5}	0.03	3×10^{-6}
Liver	800	0.2	3×10^{-5}	0.08	8×10^{-6}
Skin	4×10^3	1	10^{-4}	0.4	4×10^{-5}
Bone	2×10^4	6	6×10^{-4}	2	2×10^{-4}
(Insol) Lung	—	—	2×10^{-7}	—	5×10^{-8}
GI (LLI)	—	8×10^{-3}	10^{-6}	3×10^{-3}	5×10^{-7}
$_{75}Re^{186}$ (β^-, γ)					
(Sol) GI (LLI)	—	3×10^{-3}	6×10^{-7}	9×10^{-4}	2×10^{-7}
Thyroid	20	0.01	2×10^{-6}	5×10^{-3}	5×10^{-7}
Skin	30	0.02	2×10^{-6}	7×10^{-3}	8×10^{-7}
Total Body	50	0.04	4×10^{-6}	0.01	10^{-6}
Liver	300	0.2	3×10^{-5}	0.08	9×10^{-6}
Bone	800	0.6	7×10^{-5}	0.2	2×10^{-5}
(Insol) Lung	—	—	2×10^{-7}	—	8×10^{-8}
GI (LLI)	—	10^{-3}	5×10^{-7}	5×10^{-4}	2×10^{-7}
$_{75}Re^{187}$ (β^-)					
(Sol) GI (LLI)	—	0.07	2×10^{-5}	0.03	6×10^{-6}
Skin	300	0.08	9×10^{-6}	0.03	3×10^{-6}
Thyroid	900	0.2	3×10^{-5}	0.08	9×10^{-6}
Total Body	2×10^3	0.4	2×10^{-5}	0.2	2×10^{-5}
Liver	6×10^3	2	10^{-4}	0.5	6×10^{-5}
Bone	4×10^4	9	10^{-3}	3	4×10^{-4}
(Insol) Lung	—	—	5×10^{-7}	—	2×10^{-7}
GI (LLI)	—	0.04	7×10^{-6}	0.02	2×10^{-6}
$_{75}Re^{188}$ (β^-, γ)					
(Sol) GI (LLI)	—	2×10^{-3}	4×10^{-7}	6×10^{-4}	10^{-7}
Thyroid	7	0.02	2×10^{-6}	7×10^{-3}	7×10^{-7}
Skin	20	0.05	5×10^{-6}	0.02	2×10^{-6}
Total Body	20	0.06	7×10^{-6}	0.02	2×10^{-6}
Liver	200	0.5	5×10^{-5}	0.2	2×10^{-5}
Bone	300	0.9	10^{-4}	0.3	3×10^{-5}
(Insol) Lung	—	—	2×10^{-7}	—	6×10^{-8}
GI (LLI)	—	9×10^{-4}	10^{-6}	3×10^{-4}	4×10^{-7}

Radionuclide and type of decay	Organ of reference (critical organ in boldface)	Maximum permissible burden in total body $q(\mu c)$	Maximum permissible concentrations			
			For 40 hour week		For 168 hour week**	
			$(MPC)_w$ $\mu c/cc$	$(MPC)_a$ $\mu c/cc$	$(MPC)_w$ $\mu c/cc$	$(MPC)_a$ $\mu c/cc$
$_{76}Os^{185}$ (ϵ, γ, e^-) (Sol)	**GI (LLI)**	—	2×10^{-3}	5×10^{-7}	7×10^{-4}	2×10^{-7}
	Kidney	8	0.04	10^{-6}	0.01	5×10^{-7}
	Total Body	40	0.2	6×10^{-6}	0.06	2×10^{-6}
	Liver	50	0.2	8×10^{-6}	0.08	3×10^{-6}
(Insol)	Lung	—		5×10^{-8}		2×10^{-8}
	GI (LLI)	—	2×10^{-3}	3×10^{-7}	7×10^{-4}	10^{-7}
$_{76}Os^{191m}$ (β^-, γ, e^-) (Sol)	**GI (LLI)**	100	0.07	2×10^{-5}	0.03	6×10^{-6}
	Kidney		2	8×10^{-5}	0.8	3×10^{-5}
	Total Body	300	7	2×10^{-4}	2	8×10^{-5}
	Liver	600	10	5×10^{-4}	4	2×10^{-4}
(Insol)	Lung	—		9×10^{-6}		3×10^{-6}
	GI (LLI)	—	0.07	10^{-5}	0.02	4×10^{-6}
$_{76}Os^{191}$ (β^-, γ, e^-) (Sol)	**GI (LLI)**	20	5×10^{-3}	10^{-6}	2×10^{-3}	4×10^{-7}
	Kidney		0.1	4×10^{-6}	0.04	10^{-6}
	Total Body	100	0.6	4×10^{-5}	0.2	8×10^{-6}
	Liver	100	0.7	3×10^{-5}	0.2	9×10^{-6}
(Insol)	Lung	—		4×10^{-7}		10^{-7}
	GI (LLI)	—	5×10^{-3}	8×10^{-7}	2×10^{-3}	3×10^{-7}

Nuclide	Sol/Insol	Organ of reference	q (µc)	(MPC)w 40 hr	(MPC)a 40 hr	(MPC)w 168 hr	(MPC)a 168 hr
$_{76}Os^{193}$ (β^-)	(Sol)	GI (LLI)		2×10^{-3}	4×10^{-7}	6×10^{-4}	10^{-7}
		Kidney	10	0.1	4×10^{-6}	0.04	2×10^{-6}
		Total Body	50	0.6	2×10^{-5}	0.2	7×10^{-6}
		Liver	70	0.9	3×10^{-5}	0.3	10^{-5}
	(Insol)	GI (LLI)		2×10^{-3}	3×10^{-7}	5×10^{-4}	9×10^{-8}
		Lung			10^{-6}		5×10^{-7}
$_{77}Ir^{190}$ (ϵ,γ)	(Sol)	GI (LLI)		6×10^{-3}	10^{-6}	2×10^{-3}	4×10^{-7}
		Liver	40	0.04	2×10^{-6}	0.02	5×10^{-7}
		Kidney	40	0.04	2×10^{-6}	0.02	6×10^{-7}
		Spleen	40	0.05	2×10^{-6}	0.02	6×10^{-7}
		Total Body	50	0.06	2×10^{-6}	0.02	8×10^{-7}
	(Insol)	Lung			4×10^{-7}		10^{-7}
		GI (LLI)		5×10^{-3}	9×10^{-7}	2×10^{-3}	3×10^{-7}
$_{77}Ir^{192}$ (β^-,γ)	(Sol)	GI (LLI)		10^{-3}	3×10^{-7}	4×10^{-4}	9×10^{-8}
		Kidney	6	4×10^{-3}	10^{-7}	10^{-3}	4×10^{-8}
		Spleen	7	4×10^{-3}	10^{-7}	10^{-3}	5×10^{-8}
		Liver	8	5×10^{-3}	2×10^{-7}	2×10^{-3}	6×10^{-8}
		Total Body	20	0.01	4×10^{-7}	4×10^{-3}	10^{-7}
	(Insol)	Lung			3×10^{-8}		9×10^{-9}
		GI (LLI)		10^{-3}	2×10^{-7}	4×10^{-4}	6×10^{-8}
$_{77}Ir^{194}$ (β^-)	(Sol)	GI (LLI)		10^{-3}	2×10^{-7}	3×10^{-4}	8×10^{-8}
		Kidney	7	0.08	3×10^{-6}	0.03	10^{-6}
		Liver	8	0.09	3×10^{-6}	0.03	10^{-6}
		Spleen	8	0.09	4×10^{-6}	0.03	4×10^{-6}
		Total Body	20	0.3		0.1	
	(Insol)	Lung			2×10^{-7}		5×10^{-8}
		GI (LLI)		9×10^{-4}	10^{-6}	3×10^{-4}	4×10^{-7}
$_{78}Pt^{191}$ (ϵ,γ)	(Sol)	GI (LLI)		4×10^{-3}	8×10^{-7}	10^{-3}	3×10^{-7}
		Kidney	10	0.04	10^{-6}	0.01	5×10^{-7}
		Total Body	30	0.1	4×10^{-6}	0.03	10^{-6}
		Liver	30	0.1	4×10^{-6}	0.04	2×10^{-6}
		Spleen	70	0.2	8×10^{-6}	0.08	3×10^{-6}

Radionuclide and type of decay		Organ of reference (critical organ in boldface)	Maximum permissible burden in total body $q(\mu c)$	Maximum permissible concentrations			
				For 40 hour week		For 168 hour week**	
				$(MPC)_w$ $\mu c/cc$	$(MPC)_a$ $\mu c/cc$	$(MPC)_w$ $\mu c/cc$	$(MPC)_a$ $\mu c/cc$
$_{78}Pt^{193m}$ (ϵ, γ)	(Insol)	**GI (LLI)**	---	3×10^{-3}	6×10^{-7}	10^{-3}	2×10^{-7}
		Lung	---	---	8×10^{-7}	---	3×10^{-7}
	(Sol)	**GI (LLI)**	---	0.03	7×10^{-6}	0.01	2×10^{-6}
		Kidney	100	0.3	10^{-5}	0.1	4×10^{-6}
		Total Body	300	0.8	3×10^{-5}	0.3	10^{-5}
		Liver	300	1	4×10^{-5}	0.3	10^{-5}
		Spleen	600	2	7×10^{-5}	0.7	2×10^{-5}
	(Insol)	**GI (LLI)**	---	0.03	5×10^{-6}	0.01	2×10^{-6}
		Lung	---	---	7×10^{-6}	---	2×10^{-6}
$_{78}Pt^{193}$ (ϵ)	(Sol)	**Kidney**	70	0.03	10^{-6}	9×10^{-3}	4×10^{-7}
		GI (LLI)	---	0.05	10^{-5}	0.02	4×10^{-6}
		Spleen	500	0.2	6×10^{-6}	0.06	2×10^{-6}
		Total Body	500	0.2	6×10^{-6}	0.06	2×10^{-6}
		Liver	600	0.2	9×10^{-6}	0.09	3×10^{-6}
	(Insol)	**Lung**	---	---	3×10^{-7}	---	10^{-7}
		GI (LLI)	---	0.05	8×10^{-6}	0.02	3×10^{-6}

Nuclide	Sol.	Organ	Mass (g)				
$_{78}Pt^{197m}$ (β^-, γ, e^-)	(Sol)	GI (ULI)		**0.03**	6×10^{-6}	**0.01**	2×10^{-6}
		Kidney	5	0.8	3×10^{-5}	0.3	10^{-5}
		Liver	20	3	10^{-4}	1	4×10^{-5}
		Spleen	30	5	2×10^{-4}	2	7×10^{-5}
		Total Body	40	6	2×10^{-4}	2	7×10^{-5}
	(Insol)	GI (ULI)		**0.03**	5×10^{-5}	**9×10^{-3}**	2×10^{-6}
		Lung			2×10^{-5}		8×10^{-6}
$_{78}Pt^{197}$ (β^-, γ)	(Sol)	GI (LLI)		**4×10^{-3}**	8×10^{-7}	**10^{-3}**	3×10^{-7}
		Kidney	10	0.1	5×10^{-6}	0.05	2×10^{-6}
		Liver	40	0.6	2×10^{-5}	0.2	7×10^{-6}
		Spleen	70	0.8	3×10^{-5}	0.3	10^{-5}
		Total Body	80	1	4×10^{-5}	0.3	10^{-5}
	(Insol)	GI (LLI)		**3×10^{-3}**	6×10^{-7}	**10^{-3}**	2×10^{-7}
		Lung			4×10^{-6}		10^{-6}
$_{79}Au^{196}$ (β^-, γ, e^-)	(Sol)	GI (LLI)		**5×10^{-3}**	10^{-6}	**2×10^{-3}**	4×10^{-7}
		Total Body	40	0.07	3×10^{-6}	0.03	9×10^{-7}
		Kidney	50	0.09	4×10^{-6}	0.03	10^{-6}
		Spleen	200	0.3	10^{-5}	0.1	4×10^{-6}
		Liver	200	0.3	10^{-5}	0.1	4×10^{-6}
	(Insol)	Lung			6×10^{-7}		2×10^{-7}
		GI (LLI)		**4×10^{-3}**	8×10^{-7}	**10^{-3}**	3×10^{-7}
$_{79}Au^{198}$ (β^-, γ)	(Sol)	GI (LLI)		**2×10^{-3}**	3×10^{-7}	**5×10^{-4}**	10^{-7}
		Kidney	20	0.07	3×10^{-6}	0.02	9×10^{-7}
		Total Body	30	0.1	4×10^{-6}	0.04	2×10^{-6}
		Spleen	60	0.2	8×10^{-6}	0.07	3×10^{-6}
		Liver	80	0.3	10^{-5}	0.1	4×10^{-6}
	(Insol)	GI (LLI)		**10^{-3}**	2×10^{-7}	**5×10^{-4}**	8×10^{-8}
		Lung			6×10^{-7}		2×10^{-7}

Radionuclide and type of decay		Organ of reference (critical organ in boldface)	Maximum permissible burden in total body $q(\mu c)$	Maximum permissible concentrations			
				For 40 hour week		For 168 hour week**	
				$(MPC)_w$ $\mu c/cc$	$(MPC)_a$ $\mu c/cc$	$(MPC)_w$ $\mu c/cc$	$(MPC)_a$ $\mu c/cc$
$_{79}Au^{199}$ (β^-, γ)	(Sol)	**GI (LLI)**	--	5×10^{-3}	10^{-6}	2×10^{-3}	4×10^{-7}
		Kidney	70	0.2	8×10^{-6}	0.07	3×10^{-6}
		Total Body	100	0.3	10^{-5}	0.1	4×10^{-6}
		Spleen	200	0.6	2×10^{-5}	0.2	8×10^{-6}
		Liver	300	0.8	3×10^{-5}	0.3	10^{-5}
	(Insol)	**GI (LLI)**	--	4×10^{-3}	8×10^{-7}	2×10^{-3}	3×10^{-7}
		Lung	--	--	2×10^{-6}	--	6×10^{-7}
$_{80}Hg^{197m}$ (ϵ, γ, e^-)	(Sol)	**Kidney**	4	6×10^{-3}	7×10^{-7}	2×10^{-3}	3×10^{-7}
		GI (LLI)	--	0.02	4×10^{-6}	7×10^{-3}	10^{-6}
		Spleen	40	0.05	7×10^{-6}	0.02	2×10^{-6}
		Liver	50	0.07	9×10^{-6}	0.02	3×10^{-6}
		Total Body	70	0.09	10^{-5}	0.03	4×10^{-6}
	(Insol)	**GI (LLI)**	--	5×10^{-3}	8×10^{-7}	2×10^{-3}	3×10^{-7}
		Lung	--	--	4×10^{-6}	--	10^{-6}
$_{80}Hg^{197}$ (ϵ, γ, e^-)	(Sol)	**Kidney**	20	9×10^{-3}	10^{-6}	3×10^{-3}	4×10^{-7}
		GI (LLI)	--	0.06	10^{-5}	0.02	4×10^{-6}
		Spleen	200	0.08	10^{-5}	0.03	4×10^{-6}
		Liver	200	0.1	10^{-5}	0.03	4×10^{-6}
		Total Body	200	0.1	2×10^{-5}	0.04	5×10^{-6}

Isotope	Sol.	Organ	q				
$_{80}Hg^{203}$ (β^-, γ, e^-)	(Insol)	GI (LLI)		0.01	3×10^{-6}	5×10^{-3}	9×10^{-7}
		Lung			5×10^{-6}		2×10^{-6}
	(Sol)	Kidney	4	5×10^{-4}	7×10^{-8}	2×10^{-4}	2×10^{-8}
		Spleen	40	6×10^{-3}	8×10^{-7}	2×10^{-3}	3×10^{-7}
		Liver	40	7×10^{-3}	9×10^{-7}	2×10^{-3}	3×10^{-7}
		Total Body	80	0.01	2×10^{-6}	4×10^{-3}	5×10^{-7}
		GI (LLI)		0.01	3×10^{-6}	4×10^{-3}	10^{-6}
	(Insol)	Lung			10^{-7}		4×10^{-8}
		GI (LLI)		3×10^{-3}	6×10^{-7}	10^{-3}	2×10^{-7}
$_{81}Tl^{200}$ (ϵ, γ)	(Sol)	GI (LLI)		0.01	3×10^{-6}	4×10^{-3}	9×10^{-7}
		Kidney	40	0.08	8×10^{-6}	0.03	3×10^{-6}
		Total Body	50	0.1	10^{-5}	0.04	4×10^{-6}
		Muscle	100	0.3	3×10^{-5}	0.09	9×10^{-6}
		Liver	200	0.4	5×10^{-5}	0.2	2×10^{-5}
		Lung	800	2	2×10^{-4}	0.6	6×10^{-5}
		Bone	10^3	2	2×10^{-4}	0.8	9×10^{-5}
	(Insol)	GI (LLI)		7×10^{-3}	10^{-6}	2×10^{-3}	4×10^{-7}
		Lung			4×10^{-6}		10^{-6}
$_{81}Tl^{201}$ (ϵ, γ, e^-)	(Sol)	GI (LLI)		9×10^{-3}	2×10^{-6}	3×10^{-3}	7×10^{-7}
		Kidney	40	0.04	5×10^{-6}	0.02	2×10^{-6}
		Total Body	100	0.1	10^{-5}	0.04	4×10^{-6}
		Muscle	300	0.3	3×10^{-5}	0.1	10^{-5}
		Liver	300	0.3	3×10^{-5}	0.1	10^{-5}
		Bone	400	0.4	5×10^{-5}	0.2	2×10^{-5}
		Lung	10^3	1	10^{-4}	0.4	4×10^{-5}
	(Insol)	GI (LLI)		5×10^{-3}	9×10^{-7}	2×10^{-3}	3×10^{-7}
		Lung			2×10^{-6}		7×10^{-7}

Radionuclide and type of decay	Organ of reference (critical organ in boldface)	Maximum permissible burden in total body $q(\mu c)$	Maximum permissible concentrations			
			For 40 hour week		For 168 hour week**	
			$(MPC)_w$ μc/cc	$(MPC)_a$ μc/cc	$(MPC)_w$ μc/cc	$(MPC)_a$ μc/cc
$_{81}Tl^{202}$ (ε, γ, e⁻)	**GI (LLI)** (Sol)	—	4×10^{-3}	8×10^{-7}	10^{-3}	3×10^{-7}
	Kidney	20	0.01	10^{-6}	3×10^{-3}	4×10^{-7}
	Total Body	50	0.03	3×10^{-6}	0.01	10^{-6}
	Muscle	100	0.07	7×10^{-6}	0.02	2×10^{-6}
	Liver	100	0.08	8×10^{-6}	0.03	3×10^{-6}
	Bone	200	0.1	10^{-5}	0.03	4×10^{-6}
	Lung	400	0.3	3×10^{-5}	0.09	9×10^{-6}
	Lung (Insol)	—	—	2×10^{-7}	—	8×10^{-8}
	GI (LLI)	—	2×10^{-3}	4×10^{-7}	7×10^{-4}	10^{-7}
$_{81}Tl^{204}$ (β⁻)	**GI (LLI)** (Sol)	—	3×10^{-3}	7×10^{-7}	10^{-3}	2×10^{-7}
	Kidney	10	6×10^{-3}	6×10^{-7}	2×10^{-3}	2×10^{-7}
	Total Body	80	0.03	3×10^{-6}	0.01	10^{-6}
	Bone	100	0.04	5×10^{-6}	0.02	2×10^{-6}
	Liver	100	0.06	6×10^{-6}	0.02	2×10^{-6}
	Muscle	200	0.07	7×10^{-6}	0.02	3×10^{-6}
	Lung	500	0.2	2×10^{-5}	0.07	7×10^{-6}
	Lung (Insol)	—	—	3×10^{-8}	—	9×10^{-9}
	GI (LLI)	—	2×10^{-3}	3×10^{-7}	6×10^{-4}	10^{-7}

Nuclide		Organ	q (μCi)	$(MPC)_w$	$(MPC)_a$	$(MPC)_w$	$(MPC)_a$
$_{82}Pb^{203}(\epsilon, \gamma)$	(Sol)	GI (LLI)	—	0.01	3×10^{-6}	4×10^{-3}	9×10^{-7}
		Kidney	30	0.1	4×10^{-6}	0.05	10^{-6}
		Total Body	90	0.5	10^{-5}	0.2	5×10^{-6}
		Liver	200	1	3×10^{-5}	0.3	10^{-5}
		Bone	400	2	7×10^{-5}	0.8	2×10^{-5}
	(Insol)	GI (LLI)	—	0.01	2×10^{-6}	4×10^{-3}	6×10^{-7}
		Lung	—	—	4×10^{-6}	—	10^{-6}
$_{82}Pb^{210}(\alpha, \beta^-, \gamma)$	(Sol)	Kidney	0.4	4×10^{-6}	10^{-10}	10^{-6}	4×10^{-11}
		Total Body	4	4×10^{-6}	10^{-9}	10^{-6}	4×10^{-10}
		Bone	0.7	6×10^{-6}	2×10^{-10}	2×10^{-6}	7×10^{-11}
		Liver	1	10^{-5}	4×10^{-10}	5×10^{-6}	10^{-10}
		GI (LLI)	—	6×10^{-3}	10^{-6}	2×10^{-3}	4×10^{-7}
	(Insol)	Lung	—	—	2×10^{-10}	—	8×10^{-11}
		GI (LLI)	—	5×10^{-3}	9×10^{-7}	2×10^{-3}	3×10^{-7}
$_{82}Pb^{212}(\alpha, \beta^-, \gamma, e^-)$	(Sol)	Kidney	0.02	6×10^{-4}	2×10^{-8}	2×10^{-4}	6×10^{-9}
		GI (LLI)	—	6×10^{-4}	10^{-7}	2×10^{-4}	4×10^{-8}
		Bone	0.1	2×10^{-3}	7×10^{-8}	8×10^{-4}	3×10^{-8}
		Liver	0.2	6×10^{-3}	2×10^{-7}	2×10^{-3}	6×10^{-8}
		Total body	0.2	6×10^{-3}	2×10^{-7}	2×10^{-3}	6×10^{-8}
	(Insol)	Lung	—	—	2×10^{-8}	—	7×10^{-9}
		GI (LLI)	—	5×10^{-4}	9×10^{-8}	2×10^{-4}	3×10^{-8}
$_{83}Bi^{206}(\epsilon, \gamma)$	(Sol)	GI (LLI)	—	10^{-3}	2×10^{-7}	4×10^{-4}	8×10^{-8}
		Kidney	1	0.04	2×10^{-7}	0.02	6×10^{-8}
		Liver	7	0.2	10^{-6}	0.08	4×10^{-7}
		Total Body	10	0.4	2×10^{-6}	0.1	5×10^{-7}
		Spleen	20	0.5	2×10^{-6}	0.2	8×10^{-7}
		Bone	300	10	4×10^{-5}	3	10^{-5}
	(Inso.)	Lung	—	—	10^{-7}	—	5×10^{-8}
		GI (LLI)	—	10^{-3}	2×10^{-7}	4×10^{-4}	7×10^{-8}

Radionuclide and type of decay	Organ of reference (critical organ in boldface)	Maximum permissible burden in total body $q(\mu c)$	Maximum permissible concentrations			
			For 40 hour week		For 168 hour week**	
			$(MPC)_w$ $\mu c/cc$	$(MPC)_a$ $\mu c/cc$	$(MPC)_w$ $\mu c/cc$	$(PMPC)_a$ $\mu c/cc$
$_{83}\text{Bi}^{207}$ (ϵ, γ)						
(Sol)	GI (LLI)	—	2×10^{-3}	4×10^{-7}	6×10^{-4}	10^{-7}
	Kidney	2	0.04	2×10^{-7}	0.02	6×10^{-8}
	Liver	7	0.1	6×10^{-7}	0.05	2×10^{-7}
	Spleen	20	0.4	2×10^{-6}	0.1	5×10^{-7}
	Total Body	20	0.4	2×10^{-6}	0.1	5×10^{-7}
	Bone	300	6	2×10^{-5}	2	8×10^{-6}
(Insol)	**Lung**	—	—	10^{-8}	—	5×10^{-9}
	GI (LLI)	—	2×10^{-3}	3×10^{-7}	6×10^{-4}	10^{-7}
$_{83}\text{Bi}^{210}$ (α, β^-)						
(Sol)	GI (LLI)	—	10^{-3}	3×10^{-7}	4×10^{-4}	9×10^{-8}
	Kidney	0.04	2×10^{-3}	6×10^{-9}	5×10^{-3}	2×10^{-9}
	Liver	0.5	0.02	8×10^{-8}	6×10^{-3}	3×10^{-8}
	Spleen	0.6	0.02	10^{-7}	8×10^{-3}	3×10^{-8}
	Total Body	20	0.07	3×10^{-7}	0.03	10^{-7}
	Bone	6	0.2	10^{-6}	0.08	3×10^{-7}
(Insol)	**Lung**	—	—	6×10^{-9}	—	2×10^{-9}
	GI (LLI)	—	10^{-3}	2×10^{-7}	4×10^{-4}	7×10^{-8}

$_{83}Bi^{212}$ (α, β^-, γ)	(Sol)	**GI (S)**		**0.01**	2×10^{-6}	**4×10^{-3}**	8×10^{-7}
		Kidney	0.01	0.02	**10^{-7}**	8×10^{-3}	**3×10^{-8}**
		Liver	0.1	0.3	10^{-6}	0.09	4×10^{-7}
		Spleen	0.2	0.4	10^{-6}	0.1	5×10^{-7}
		Total Body	0.2	0.5	2×10^{-6}	0.2	8×10^{-7}
		Bone	0.9	2	8×10^{-6}	0.7	3×10^{-6}
	(Insol)	**Lung**			**2×10^{-7}**		**7×10^{-8}**
		GI (S)		**0.01**	2×10^{-6}	**4×10^{-3}**	6×10^{-7}
$_{84}Po^{210}$ (α)	(Sol)	**Spleen**	0.03	**2×10^{-5}**	**5×10^{-10}**	7×10^{-6}	**2×10^{-10}**
		Kidney	0.04	**2×10^{-5}**	**5×10^{-10}**	8×10^{-6}	**2×10^{-10}**
		Liver	0.1	7×10^{-5}	2×10^{-9}	3×10^{-5}	6×10^{-10}
		Total Body	0.4	2×10^{-4}	5×10^{-9}	8×10^{-5}	2×10^{-9}
		Bone	0.5	3×10^{-4}	7×10^{-9}	10^{-4}	2×10^{-9}
		GI (LLI)		9×10^{-4}	2×10^{-7}	3×10^{-4}	7×10^{-8}
	(Insol)	**Lung**			**2×10^{-10}**		**7×10^{-11}**
		GI (LLI)		**8×10^{-4}**	2×10^{-7}	**3×10^{-4}**	5×10^{-8}
$_{85}At^{211}$ (α, ϵ, γ)	(Sol)	**Thyroid**	0.02	**5×10^{-5}**	**7×10^{-9}**	**2×10^{-5}**	**2×10^{-9}**
		Ovary	0.02	**5×10^{-5}**	**7×10^{-9}**	**2×10^{-5}**	3×10^{-9}
		Spleen	0.06	2×10^{-4}	3×10^{-8}	6×10^{-5}	9×10^{-9}
		Total Body	0.3	8×10^{-4}	10^{-7}	3×10^{-4}	4×10^{-8}
		GI (S)		0.02	4×10^{-6}	7×10^{-3}	2×10^{-6}
	(Insol)	**Lung**			**3×10^{-8}**		**10^{-8}**
		GI (ULI)		**2×10^{-3}**	4×10^{-7}	**7×10^{-4}**	10^{-7}
$_{86}Rn^{220}$† (α, β^-, γ, e^-)		**Lung**			**3×10^{-7}**		**10^{-7}**
$_{86}Rn^{222}$† (α, β, γ)		**Lung**			**3×10^{-8}**		**10^{-8}**

†The daughter isotopes of Rn^{220} and Rn^{222} are assumed present to the extent they occur in unfiltered air. For all other isotopes the daughter elements are not considered as part of the intake and if present must be considered on the basis of the rules for mixtures.

Radionuclide and type of decay		Organ of reference (critical organ in boldface)	Maximum permissible burden in total body $q(\mu c)$	Maximum permissible concentrations			
				For 40 hour week		For 168 hour week**	
				$(MPC)_w$ $\mu c/cc$	$(MPC)_a$ $\mu c/cc$	$(MPC)_w$ $\mu c/cc$	$(MPC)_a$ $\mu c/cc$
$_{88}Ra^{223}$ $(\alpha, \beta^-, \gamma)$	(Sol)	**Bone**	0.05	2×10^{-5}	2×10^{-9}	7×10^{-6}	6×10^{-10}
		Total Body	0.07	4×10^{-5}	3×10^{-9}	10^{-5}	10^{-9}
		GI (LLI)	---	2×10^{-4}	4×10^{-8}	6×10^{-5}	10^{-8}
	(Insol)	**Lung**	---	---	2×10^{-10}	---	8×10^{-11}
		GI (LLI)	---	10^{-4}	2×10^{-8}	4×10^{-5}	7×10^{-9}
$_{88}Ra^{224}$ $(\alpha, \beta^-, \gamma, e^-)$	(Sol)	**Bone**	0.06	7×10^{-5}	5×10^{-9}	2×10^{-5}	2×10^{-9}
		Total Body	0.07	9×10^{-5}	8×10^{-9}	3×10^{-5}	3×10^{-9}
		GI (LLI)	---	2×10^{-4}	5×10^{-8}	7×10^{-5}	2×10^{-8}
	(Insol)	**Lung**	---	---	7×10^{-10}	---	2×10^{-10}
		GI (LLI)	---	2×10^{-4}	3×10^{-8}	5×10^{-5}	9×10^{-9}
$_{88}Ra^{226}$ $(\alpha, \beta^-, \gamma)$	(Sol)	**Bone**	0.1	4×10^{-7}	3×10^{-11}	10^{-7}	10^{-11}
		Total Body	0.2	6×10^{-7}	5×10^{-11}	2×10^{-7}	2×10^{-11}
		GI (LLI)	---	10^{-3}	3×10^{-7}	5×10^{-4}	10^{-7}
	(Insol)	**GI (LLI)**	---	9×10^{-4}	2×10^{-7}	3×10^{-4}	6×10^{-8}
$_{88}Ra^{228}$ $(\alpha, \beta^-, \gamma, e^-)$	(Sol)	**Bone**	0.06	8×10^{-7}	7×10^{-11}	3×10^{-7}	2×10^{-11}
		Total Body	0.09	10^{-6}	9×10^{-11}	4×10^{-7}	3×10^{-11}
		GI (LLI)	---	10^{-3}	2×10^{-7}	4×10^{-4}	8×10^{-8}

Isotope	Solubility	Organ of reference	q				
$_{89}\text{Ac}^{227}$ (α, β^-, γ)	(Insol)	Lung		—	4×10^{-11}	—	10^{-11}
		GI (LLI)		7×10^{-4}	10^{-7}	3×10^{-4}	4×10^{-8}
	(Sol)	Bone	0.03	6×10^{-5}	2×10^{-12}	2×10^{-5}	8×10^{-13}
		Total Body	0.1	2×10^{-4}	7×10^{-12}	6×10^{-5}	3×10^{-12}
		Liver	0.2	2×10^{-4}	10^{-11}	8×10^{-5}	3×10^{-12}
		Kidney	0.4	7×10^{-4}	3×10^{-11}	2×10^{-4}	9×10^{-12}
		GI (LLI)		9×10^{-3}	2×10^{-6}	3×10^{-3}	7×10^{-7}
$_{89}\text{Ac}^{228}$ ($\alpha, \beta^-, \gamma, e^-$)	(Insol)	Lung		—	3×10^{-6}	—	9×10^{-12}
		GI (LLI)		9×10^{-3}	2×10^{-6}	3×10^{-3}	5×10^{-7}
	(Sol)	GI (ULI)		3×10^{-3}	6×10^{-7}	9×10^{-4}	2×10^{-7}
		Bone	0.04	2	9×10^{-8}	0.5	3×10^{-8}
		Liver	0.05	2	8×10^{-8}	0.6	3×10^{-8}
		Total Body	0.09	3	10^{-7}	1	5×10^{-8}
		Kidney	0.5	20	6×10^{-7}	6	2×10^{-9}
$_{90}\text{Th}^{227}$ (α, β^-, γ)	(Insol)	Lung		—	2×10^{-8}	—	6×10^{-9}
		GI (ULI)		3×10^{-3}	4×10^{-7}	9×10^{-4}	2×10^{-7}
	(Sol)	GI (LLI)		5×10^{-4}	10^{-7}	2×10^{-4}	4×10^{-8}
		Bone	0.02	8×10^{-3}	3×10^{-10}	3×10^{-3}	10^{-10}
		Kidney	0.08	0.04	2×10^{-9}	0.01	6×10^{-10}
		Total Body	0.1	0.05	2×10^{-9}	0.02	7×10^{-10}
		Liver	0.5	0.2	10^{-8}	0.08	4×10^{-9}
$_{90}\text{Th}^{228}$ ($\alpha, \beta^-, \gamma, e^-$)	(Insol)	Lung		—	2×10^{-10}	—	6×10^{-11}
		GI (LLI)		5×10^{-4}	9×10^{-8}	2×10^{-4}	3×10^{-8}
	(Sol)	Bone	0.02	2×10^{-4}	9×10^{-12}	7×10^{-5}	3×10^{-12}
		GI (LLI)		4×10^{-4}	8×10^{-8}	10^{-4}	3×10^{-8}
		Kidney	0.09	10^{-3}	5×10^{-11}	4×10^{-4}	2×10^{-11}
		Total Body	0.09	10^{-3}	5×10^{-11}	4×10^{-4}	2×10^{-11}
		Liver	0.5	7×10^{-3}	3×10^{-10}	2×10^{-2}	10^{-10}
	(Insol)	Lung		—	6×10^{-12}	—	2×10^{-12}
		GI (LLI)		4×10^{-4}	7×10^{-8}	10^{-4}	2×10^{-8}

Radionuclide and type of decay	Organ of reference (critical organ in boldface)	Maximum permissible burden in total body $q(\mu c)$	Maximum permissible concentrations			
			For 40 hour week		For 168 hour week**	
			$(MPC)_w$ $\mu c/cc$	$(MPC)_a$ $\mu c/cc$	$(MPC)_w$ $\mu c/cc$	$(MPC)_a$ $\mu c/cc$
$_{90}Th^{230}$ (α, γ)						
(Sol)	**Bone**	0.05	5×10^{-5}	2×10^{-12}	2×10^{-5}	8×10^{-13}
	Kidney	0.3	10^{-4}	4×10^{-12}	3×10^{-5}	2×10^{-12}
	Total Body	0.4	3×10^{-4}	2×10^{-11}	10^{-4}	5×10^{-12}
	Liver	0.6	5×10^{-4}	2×10^{-11}	2×10^{-4}	7×10^{-12}
	GI (LLI)	---	9×10^{-4}	2×10^{-7}	3×10^{-4}	7×10^{-8}
(Insol)	**Lung**	---	---	10^{-11}	---	3×10^{-12}
	GI (LLI)	---	9×10^{-4}	2×10^{-7}	3×10^{-4}	6×10^{-8}
$_{90}Th^{231}$ $(\alpha, \beta^-, \gamma)$						
(Sol)	**GI (LLI)**	---	7×10^{-3}	10^{-6}	2×10^{-3}	5×10^{-7}
	Bone	30	200	10^{-5}	80	4×10^{-6}
	Kidney	40	300	10^{-5}	100	5×10^{-6}
	Total Body	100	900	4×10^{-5}	300	10^{-5}
	Liver	300	2×10^{3}	10^{-4}	800	3×10^{-5}
(Insol)	**GI (LLI)**	---	7×10^{-3}	10^{-6}	2×10^{-3}	4×10^{-7}
	Lung	---	---	6×10^{-6}	---	2×10^{-6}
$_{90}Th^{232}$ $(\alpha, \beta^-, \gamma, e^-)$						
(Sol)	**Bone**	0.04	5×10^{-5}	$\left\{\begin{array}{l}2 \times 10^{-12} \\ 5 \times 10^{-12}\end{array}\right.$	2×10^{-5}	$\left\{\begin{array}{l}7 \times 10^{-12} \\ 2 \times 10^{-12}\end{array}\right.$ ‡
	Kidney	0.3	10^{-4}	10^{-11}	4×10^{-5}	4×10^{-12}
	Total Body	0.3	3×10^{-4}	3×10^{-11}	9×10^{-5}	9×10^{-12}
	Liver	0.7	6×10^{-4}	2×10^{-7}	2×10^{-4}	8×10^{-8}
	GI (LLI)	---	10^{-3}		4×10^{-4}	

Nuclide		Organ	(MPC)$_w$ (40 hr)	(MPC)$_w$ (168 hr)	(MPC)$_a$ (40 hr)	(MPC)$_a$ (168 hr)
$_{90}$Th^{234} (β^-, γ)	(Insol)	Lung			10^{-11}	4×10^{-12}
		GI (LLI)	10^{-3}	5×10^{-4}	2×10^{-7}	7×10^{-8}
	(Sol)	Bone	4	1	6×10^{-8}	2×10^{-8}
		Kidney	6	2	9×10^{-8}	3×10^{-8}
		Total Body	20	8	4×10^{-7}	10^{-7}
		Liver	30	10	5×10^{-7}	2×10^{-7}
		GI (LLI)		5×10^{-4}	10^{-7}	4×10^{-8}
$_{90}$Th–Nat (α, β^-, γ, e$^-$)	(Insol)	Lung			3×10^{-8}	10^{-8}
		GI (LLI)		5×10^{-4}	9×10^{-8}	3×10^{-8}
	(Sol)	Bone	0.01	3×10^{-5}	(2×10^{-12})	(6×10^{-12})‡
		Kidney	0.07	10^{-4}	(4×10^{-12})	(2×10^{-12})
		Total Body	0.07	2×10^{-4}	(9×10^{-12})	(3×10^{-12})
		GI (LLI)		3×10^{-4}	6×10^{-8}	2×10^{-8}
		Liver	0.3	5×10^{-4}	(2×10^{-11})	(8×10^{-12})
					(4×10^{-12})	(10^{-12})
$_{91}$Pa^{230} (α, β^-, ϵ, γ)	(Insol)	Lung			2×10^{-6}	5×10^{-7}
		GI (LLI)		7×10^{-3}	10^{-6}	6×10^{-10}
	(Sol)	Bone	0.07	0.04	2×10^{-9}	2×10^{-9}
		Kidney	0.2	0.1	5×10^{-9}	5×10^{-9}
		Total Body	0.3	0.2	8×10^{-9}	8×10^{-10}
		GI (LLI)		7×10^{-3}	10^{-6}	3×10^{-9} / 3×10^{-10}
						4×10^{-7}
$_{91}$Pa^{231} (α, β^-, γ)	(Sol)	Bone	0.02	3×10^{-5}	10^{-12}	4×10^{-13}
		Kidney	0.06	7×10^{-5}	3×10^{-12}	10^{-12}
		Total Body	0.1	10^{-4}	5×10^{-12}	2×10^{-12}
		Liver	0.3	4×10^{-4}	2×10^{-11}	5×10^{-12}
		GI (LLI)		8×10^{-4}	2×10^{-7}	6×10^{-8}
					2×10^{-10}	4×10^{-11}
	(Insol)	Lung			10^{-7}	4×10^{-11}
		GI (LLI)		8×10^{-4}	10^{-7}	5×10^{-8}

‡Provisional values for Th^{232} and Th-nat. Although calculations and animal experiments suggest that Th-nat is perhaps as hazardous as Pu and indicate the values listed above, industrial experience to date has suggested that the hazard of Th-nat is not much greater than that of U-nat. The NCRP has recognized that a certain period of time may be required for adjustment of operations to comply with new recommendations. Therefore, pending further investigation the values (MPC)$_a$=3×10^{-11} μc/cc for the 40-hour week and (MPC)$_a$=10^{-11} μc/cc for continuous occupational exposure (168 hr/wk) are recommended as permissible levels. These values are essentially those that have been generally used in this country (Federal Register 1957). However, the values given in Table I are listed to indicate the possibility that further evidence may require lower values and to urge especially that exposure levels of Th-nat be kept as low as is operationally possible. The exception indicated here applies only to the (MPC), values for Th-nat and Th^{232}.

Radionuclide and type of decay	Organ of reference (critical organ in boldface)	Maximum permissible burden in total body $q(\mu c)$	Maximum permissible concentrations			
			For 40 hour week		For 168 hour week	
			$(MPC)_w$ $\mu c/cc$	$(MPC)_a$ $\mu c/cc$	$(MPC)_w$ $\mu c/cc$	$(MPC)_a$ $\mu c/cc$
$_{91}Pa^{233}$ (β^-, γ) (Sol)	**GI (LLI)**	---	4×10^{-3}	8×10^{-7}	10^{-3}	3×10^{-7}
	Kidney	40	10	6×10^{-7}	5	2×10^{-7}
	Bone	60	20	9×10^{-7}	7	3×10^{-7}
	Total Body	60	20	9×10^{-7}	7	3×10^{-7}
	Liver	200	50	2×10^{-6}	20	8×10^{-7}
(Insol)	Lung	---	---	2×10^{-7}	---	6×10^{-8}
	GI (LLI)	---	3×10^{-3}	6×10^{-7}	10^{-3}	2×10^{-7}
$\dagger_{92}U^{230}$ $(\alpha, \beta^-, \gamma)$ (Sol)	**Kidney**	0.01	7×10^{-5}	3×10^{-10}	2×10^{-5}	10^{-10}
	GI (LLI)	---	10^{-4}	3×10^{-8}	5×10^{-5}	10^{-8}
	Total Body	0.06	3×10^{-4}	10^{-9}	10^{-4}	5×10^{-10}
	Bone	7×10^{-3}	4×10^{-4}	2×10^{-9}	2×10^{-4}	6×10^{-10}
(Insol)	Lung	---	---	10^{-10}	---	4×10^{-11}
	GI (LLI)	---	10^{-4}	2×10^{-8}	5×10^{-5}	8×10^{-9}
$\dagger_{92}U^{232}$ $(\alpha, \beta^-, \gamma, e^-)$ (Sol)	**Bone**	0.01	2×10^{-5}	10^{-10}	8×10^{-6}	3×10^{-11}
	Total Body	0.07	6×10^{-5}	3×10^{-10}	2×10^{-5}	10^{-10}
	Kidney	0.04	10^{-4}	6×10^{-10}	4×10^{-5}	2×10^{-10}
	GI (LLI)	---	8×10^{-4}	2×10^{-7}	3×10^{-4}	6×10^{-8}
(Insol)	Lung	---	---	3×10^{-11}	---	9×10^{-12}
	GI (LLI)	---	8×10^{-4}	10^{-7}	3×10^{-4}	5×10^{-8}

Isotope	Form	Organ	q				
$_{92}U^{233}$ (α, γ)	(Sol)	**Bone**	0.05	10^{-4}	5×10^{-10}	4×10^{-5}	2×10^{-10}
		Kidney	0.08	3×10^{-4}	10^{-9}	10^{-4}	4×10^{-10}
		Total Body	0.4	4×10^{-4}	2×10^{-9}	10^{-4}	5×10^{-10}
		GI (LLI)	—	9×10^{-4}	2×10^{-7}	3×10^{-4}	7×10^{-8}
	(Insol)	**Lung**	—	—	10^{-10}	—	4×10^{-11}
		GI (LLI)	—	9×10^{-4}	2×10^{-7}	3×10^{-4}	6×10^{-8}
$_{92}U^{234}$ (α, γ)	(Sol)	**Bone**	0.05	10^{-4}	6×10^{-10}	4×10^{-5}	2×10^{-10}
		Kidney	0.08	3×10^{-4}	10^{-9}	10^{-4}	4×10^{-10}
		Total Body	0.4	4×10^{-4}	2×10^{-9}	10^{-4}	6×10^{-10}
		GI (LLI)	—	9×10^{-4}	2×10^{-7}	3×10^{-4}	7×10^{-8}
	(Insol)	**Lung**	—	—	10^{-10}	—	4×10^{-11}
		GI (LLI)	—	9×10^{-4}	2×10^{-7}	3×10^{-4}	6×10^{-8}
$_{92}U^{235}$ $(\alpha, \beta^-, \gamma)$	(Sol)	**Kidney**	0.03	10^{-4}	5×10^{-10}	4×10^{-5}	2×10^{-10}
		Bone	0.06	10^{-4}	6×10^{-10}	5×10^{-5}	2×10^{-10}
		Total Body	0.4	4×10^{-4}	2×10^{-9}	10^{-4}	6×10^{-10}
		GI (LLI)	—	8×10^{-4}	2×10^{-7}	3×10^{-4}	6×10^{-8}
	(Insol)	**Lung**	—	—	10^{-10}	—	4×10^{-11}
		GI (LLI)	—	8×10^{-4}	10^{-7}	3×10^{-4}	5×10^{-8}
$_{92}U^{236}$ (α, γ)	(Sol)	**Bone**	0.06	10^{-4}	6×10^{-10}	5×10^{-5}	2×10^{-10}
		Kidney	0.08	3×10^{-4}	10^{-9}	10^{-4}	4×10^{-10}
		Total Body	0.4	4×10^{-4}	2×10^{-9}	10^{-4}	6×10^{-10}
		GI (LLI)	—	10^{-3}	2×10^{-7}	3×10^{-4}	7×10^{-8}
	(Insol)	**Lung**	—	—	10^{-10}	—	4×10^{-11}
		GI (LLI)	—	10^{-3}	2×10^{-7}	3×10^{-4}	6×10^{-8}
$_{92}U^{238}$ (α, γ, e^-)	(Sol)	**Kidney**	5×10^{-3}	2×10^{-5}	7×10^{-11}	6×10^{-6}	3×10^{-11}
		Bone	0.06	10^{-4}	6×10^{-10}	5×10^{-5}	2×10^{-10}
		Total Body	0.5	4×10^{-4}	2×10^{-9}	10^{-4}	6×10^{-10}
		GI (LLI)	—	10^{-3}	2×10^{-7}	4×10^{-4}	8×10^{-8}
	(Insol)	**Lung**	—	—	10^{-10}	—	5×10^{-11}
		GI (LLI)	—	10^{-3}	2×10^{-7}	4×10^{-4}	6×10^{-8}

Radionuclide and type of decay	Organ of reference (critical organ in boldface)	Maximum permissible burden in total body $q\,(\mu c)$	Maximum permissible concentrations			
			For 40 hour week		For 168 hour week	
			$(MPC)_w$ $\mu c/cc$	$(MPC)_a$ $\mu c/cc$	$(MPC)_w$ $\mu c/cc$	$(PMPC)_a$ $\mu c/cc$
$^{†}_{92}$U-nat ($\alpha, \beta^-, \gamma, e^-$) (Sol)	**Kidney**	5×10^{-3}	2×10^{-5}	7×10^{-11}	6×10^{-6}	3×10^{-11}
	Bone	0.03	6×10^{-5}	3×10^{-10}	2×10^{-5}	10^{-10}
	Total Body	0.2	2×10^{-4}	8×10^{-10}	7×10^{-5}	3×10^{-10}
	GI (LLI)	--	5×10^{-4}	10^{-7}	2×10^{-4}	4×10^{-8}
(Insol)	**Lung**	--	--	6×10^{-11}	--	2×10^{-11}
	GI (LLI)	--	5×10^{-4}	8×10^{-8}	2×10^{-4}	3×10^{-8}
$^{†}_{92}$U^{240}+$_{93}$Np^{240} ($\alpha, \beta^-, \gamma, e^-$) (Sol)	**GI (LLI)**	--	10^{-3}	2×10^{-7}	3×10^{-4}	8×10^{-8}
	Kidney	4.0	0.4	2×10^{-6}	0.1	5×10^{-7}
	Total Body	20.0	2.0	10^{-5}	0.8	4×10^{-6}
	Bone	2.0	3.0	10^{-5}	1.0	4×10^{-6}
(Insol)	**GI (LLI)**	--	10^{-3}	2×10^{-7}	3×10^{-4}	6×10^{-8}
	Lung	--	--	10^{-6}	--	5×10^{-7}
$_{93}$Np^{237} (α, β^-, γ) (Sol)	**Bone**	0.06	9×10^{-5}	4×10^{-12}	3×10^{-5}	10^{-12}
	Kidney	0.1	2×10^{-4}	7×10^{-12}	6×10^{-5}	2×10^{-12}
	Total Body	0.5	4×10^{-4}	2×10^{-11}	10^{-4}	6×10^{-12}
	Liver	0.5	6×10^{-4}	2×10^{-11}	2×10^{-4}	8×10^{-12}
	GI (LLI)	--	9×10^{-4}	2×10^{-7}	3×10^{-4}	7×10^{-8}
(Insol)	**Lung**	--	--	10^{-10}	--	4×10^{-11}
	GI (LLI)	--	9×10^{-4}	2×10^{-7}	3×10^{-4}	5×10^{-8}
$_{93}$Np^{239} (α, β^-, γ) (Sol)	**GI (LLI)**	--	4×10^{-3}	8×10^{-7}	10^{-3}	3×10^{-7}
	Bone	30	100	4×10^{-6}	30	2×10^{-6}
	Kidney	40	200	7×10^{-6}	50	2×10^{-6}
	Total Body	70	300	10^{-5}	90	4×10^{-6}
	Liver	100	500	2×10^{-5}	200	8×10^{-6}

Nuclide	Solubility	Organ	q (μc)	$(MPC)_w$	$(MPC)_a$	$(MPC)_w$	$(MPC)_a$
$_{94}Pu^{238}$ (α, γ)	(Insol)	GI (LLI)	—	4×10^{-3}	7×10^{-7}	10^{-3}	2×10^{-7}
		Lung	—	—	2×10^{-6}	—	7×10^{-7}
	(Sol)	Bone	0.04	10^{-4}	2×10^{-12}	5×10^{-5}	7×10^{-13}
		Liver	0.2	6×10^{-4}	8×10^{-12}	2×10^{-4}	3×10^{-12}
		Kidney	0.3	8×10^{-4}	10^{-11}	3×10^{-4}	4×10^{-12}
		GI (LLI)	—	8×10^{-4}	2×10^{-7}	3×10^{-4}	6×10^{-8}
		Total Body	0.3	10^{-3}	10^{-11}	4×10^{-4}	5×10^{-12}
	(Insol)	Lung	—	—	3×10^{-11}	—	10^{-11}
		GI (LLI)	—	8×10^{-4}	10^{-7}	3×10^{-4}	5×10^{-8}
$_{94}Pu^{239}$ (α, γ)	(Sol)	Bone	0.04	10^{-4}	2×10^{-12}	5×10^{-5}	6×10^{-13}
		Liver	0.4	5×10^{-4}	7×10^{-12}	2×10^{-4}	2×10^{-12}
		Kidney	0.5	7×10^{-4}	9×10^{-12}	2×10^{-4}	3×10^{-12}
		GI (LLI)	—	8×10^{-4}	2×10^{-7}	3×10^{-4}	6×10^{-8}
		Total Body	0.4	10^{-3}	10^{-11}	3×10^{-4}	5×10^{-12}
	(Insol)	Lung	—	—	4×10^{-11}	—	10^{-11}
		GI (LLI)	—	8×10^{-4}	2×10^{-7}	3×10^{-4}	5×10^{-8}
$_{94}Pu^{240}$ (α, γ)	(Sol)	Bone	0.04	10^{-4}	2×10^{-12}	5×10^{-5}	6×10^{-13}
		Liver	0.4	5×10^{-4}	7×10^{-12}	2×10^{-4}	2×10^{-12}
		Kidney	0.5	7×10^{-4}	9×10^{-12}	2×10^{-4}	3×10^{-12}
		GI (LLI)	—	8×10^{-4}	2×10^{-7}	3×10^{-4}	6×10^{-8}
		Total Body	0.4	10^{-3}	10^{-11}	3×10^{-4}	5×10^{-12}
	(Insol)	Lung	—	—	4×10^{-11}	—	10^{-11}
		GI (LLI)	—	8×10^{-4}	2×10^{-7}	3×10^{-4}	5×10^{-8}
$_{94}Pu^{241}$ (α, β⁻, γ)	(Sol)	Bone	0.9	7×10^{-3}	9×10^{-11}	2×10^{-3}	3×10^{-11}
		Kidney	5	0.04	5×10^{-10}	0.01	2×10^{-10}
		GI (LLI)	—	0.04	8×10^{-6}	0.01	3×10^{-6}
		Total Body	9	0.06	8×10^{-10}	0.02	3×10^{-10}
		Liver	10	0.07	10^{-9}	0.03	3×10^{-10}
	(Insol)	Lung	—	—	4×10^{-8}	—	10^{-8}
		GI (LLI)	—	0.04	7×10^{-6}	0.01	2×10^{-6}

Radionuclide and type of decay	Organ of reference (critical organ in boldface)	Maximum permissible burden in total body $q(\mu c)$	Maximum permissible concentrations			
			For 40 hour week		For 168 hour week	
			$(MPC)_w$ $\mu c/cc$	$(MPC)_a$ $\mu c/cc$	$(MPC)_w$ $\mu c/cc$	$(MPC)_a$ $\mu c/cc$
$_{94}Pu^{242}(\alpha)$ (Sol)	**Bone**	0.05	10^{-4}	2×10^{-12}	5×10^{-5}	6×10^{-13}
	Liver	0.4	6×10^{-4}	7×10^{-12}	2×10^{-4}	3×10^{-12}
	Kidney	0.5	7×10^{-4}	10^{-11}	3×10^{-4}	3×10^{-12}
	GI (LLI)	---	9×10^{-4}	2×10^{-7}	3×10^{-4}	7×10^{-8}
	Total Body	0.4	10^{-3}	10^{-11}	4×10^{-4}	5×10^{-12}
(Insol)	**Lung**	---	---	4×10^{-11}	---	10^{-11}
	GI (LLI)	---	9×10^{-4}	2×10^{-7}	3×10^{-4}	5×10^{-8}
$^{\dagger}_{94}Pu^{243}$ $(\alpha,\beta^-,\gamma,e^-)$ (Sol)	**GI (ULI)**	---	0.01	2×10^{-6}	3×10^{-3}	6×10^{-7}
	Bone	7.0	10^3	10^{-5}	400	5×10^{-6}
	Kidney	30.0	5×10^3	7×10^{-5}	2×10^3	2×10^{-5}
	Liver	40.0	6×10^3	7×10^{-5}	2×10^3	3×10^{-5}
	Total Body	50.0	8×10^3	10^{-4}	3×10^3	4×10^{-5}
(Insol)	**GI (ULI)**	---	0.01	2×10^{-6}	3×10^{-3}	8×10^{-7}
	Lung	---	---	2×10^{-5}	---	6×10^{-6}
$^{\dagger}_{94}Pu^{244}$ $(\alpha,\beta^-,\gamma,e^-)$ (99.7%) Spontaneous fission (0.3%) (Sol)	**Bone**	0.04	10^{-4}	2×10^{-12}	4×10^{-5}	6×10^{-13}
	GI (LLI)	---	3×10^{-4}	7×10^{-8}	10^{-4}	2×10^{-8}
	Kidney	0.4	6×10^{-4}	8×10^{-12}	2×10^{-4}	3×10^{-12}
	Total Body	0.3	9×10^{-4}	10^{-11}	3×10^{-4}	4×10^{-12}
(Insol)	**Lung**	---	---	3×10^{-11}	---	10^{-11}
	GI (LLI)	---	3×10^{-4}	6×10^{-8}	10^{-4}	2×10^{-8}

Element		Organ of reference	q (μc)	$(MPC)_w$ (40 hr)	$(MPC)_a$ (40 hr)	$(MPC)_w$ (168 hr)	$(MPC)_a$ (168 hr)
$_{95}\mathrm{Am}^{241}$ (α, γ)	(Sol)	Kidney	0.1	10^{-4}	6×10^{-12}	4×10^{-5}	2×10^{-12}
		Bone	0.05	10^{-4}	6×10^{-12}	5×10^{-5}	2×10^{-12}
		Liver	0.4	2×10^{-4}	9×10^{-12}	7×10^{-5}	3×10^{-12}
		Total Body	0.3	4×10^{-4}	2×10^{-11}	10^{-4}	5×10^{-12}
		GI (LLI)	—	8×10^{-4}	2×10^{-7}	3×10^{-4}	6×10^{-8}
	(Insol)	Lung	—	—	10^{-10}	—	4×10^{-11}
		GI (LLI)	—	8×10^{-4}	10^{-7}	2×10^{-4}	5×10^{-8}
†$_{95}\mathrm{Am}^{242m}$ $(\alpha, \beta^-, \gamma, \varepsilon, e^-)$	(Sol)	Bone	0.07	10^{-4}	6×10^{-12}	4×10^{-5}	2×10^{-12}
		Kidney	0.1	10^{-4}	6×10^{-12}	5×10^{-5}	2×10^{-12}
		Liver	0.3	2×10^{-4}	9×10^{-12}	7×10^{-5}	3×10^{-12}
		Total Body	0.3	3×10^{-4}	2×10^{-11}	10^{-4}	5×10^{-12}
		GI (LLI)	—	3×10^{-3}	6×10^{-7}	9×10^{-4}	2×10^{-7}
	(Insol)	Lung	—	—	3×10^{-10}	—	9×10^{-11}
		GI (LLI)	—	3×10^{-3}	5×10^{-7}	9×10^{-4}	2×10^{-7}
†$_{95}\mathrm{Am}^{242}$ $(\alpha, \beta^-, \gamma, \varepsilon, e^-)$	(Sol)	GI (LLI)	—	4×10^{-3}	8×10^{-7}	10^{-3}	3×10^{-7}
		Liver	0.06	0.9	4×10^{-8}	0.3	10^{-8}
		Kidney	0.1	2.0	8×10^{-8}	0.6	3×10^{-8}
		Bone	0.1	2.0	8×10^{-8}	0.6	3×10^{-8}
		Total Body	0.3	4.0	2×10^{-7}	1.0	6×10^{-8}
	(Insol)	Lung	—	—	5×10^{-8}	—	2×10^{-8}
		GI (LLI)	—	4×10^{-3}	7×10^{-7}	10^{-3}	2×10^{-7}
$_{95}\mathrm{Am}^{243}$ $(\alpha, \beta^-, \gamma)$	(Sol)	Bone	0.05	10^{-4}	6×10^{-12}	4×10^{-5}	2×10^{-12}
		Kidney	0.1	10^{-4}	6×10^{-12}	5×10^{-5}	2×10^{-12}
		Liver	0.4	2×10^{-4}	9×10^{-12}	7×10^{-5}	3×10^{-12}
		Total Body	0.4	4×10^{-4}	2×10^{-11}	10^{-4}	5×10^{-12}
		GI (LLI)	—	8×10^{-4}	2×10^{-7}	3×10^{-4}	6×10^{-8}
†$_{95}\mathrm{Am}^{244}$ $(\alpha, \beta^-, \gamma, e^-)$	(Sol)	GI (SI)	—	0.1	3×10^{-5}	0.05	10^{-5}
		Bone	0.2	90.0	4×10^{-6}	30.0	10^{-6}
		Kidney	0.2	100.0	4×10^{-6}	30.0	10^{-6}
		Liver	0.2	100.0	5×10^{-6}	40.0	2×10^{-6}
		Total Body	0.4	200.0	10^{-5}	80.0	3×10^{-6}
	(Insol)	Lung	—	—	2×10^{-5}	—	8×10^{-6}
		GI (SI)	—	0.1	2×10^{-5}	0.05	8×10^{-6}

Radionuclide and type of decay		Organ of reference (critical organ in boldface)	Maximum permissible burden in total body $q(\mu c)$	Maximum permissible concentrations			
				For 40 hour week		For 168 hour week	
				$(MPC)_w$ $\mu c/cc$	$(MPC)_a$ $\mu c/cc$	$(MPC)_w$ $\mu c/cc$	$(MPC)_a$ $\mu c/cc$
$_{96}Cm^{242}$ (α, γ)	(Insol)	**Lung**	—	—	10^{-10}	—	4×10^{-11}
		GI (LLI)	—	8×10^{-4}	10^{-7}	3×10^{-4}	5×10^{-8}
	(Sol)	**GI (LLI)**	—	7×10^{-4}	2×10^{-7}	2×10^{-4}	5×10^{-8}
		Liver	0.05	3×10^{-3}	10^{-10}	9×10^{-4}	4×10^{-11}
		Bone	0.09	5×10^{-3}	2×10^{-10}	2×10^{-3}	8×10^{-11}
		Kidney	0.2	9×10^{-3}	4×10^{-10}	3×10^{-3}	10^{-10}
		Total Body	0.2	0.01	6×10^{-10}	5×10^{-3}	2×10^{-10}
$_{96}Cm^{243}$ (α, γ)	(Insol)	**Lung**	—	—	2×10^{-10}	—	6×10^{-11}
		GI (LLI)	—	7×10^{-4}	10^{-7}	3×10^{-4}	4×10^{-8}
	(Sol)	**Bone**	0.09	10^{-4}	6×10^{-12}	5×10^{-5}	2×10^{-12}
		Liver	0.2	2×10^{-4}	10^{-11}	8×10^{-5}	3×10^{-12}
		Kidney	0.2	3×10^{-4}	10^{-11}	10^{-4}	4×10^{-12}
		Total Body	0.3	5×10^{-4}	2×10^{-11}	2×10^{-4}	7×10^{-12}
		GI (LLI)	—	7×10^{-4}	2×10^{-7}	2×10^{-4}	5×10^{-8}
$_{96}Cm^{244}$ (α, γ)	(Insol)	**Lung**	—	—	10^{-10}	—	3×10^{-11}
		GI (LLI)	—	7×10^{-4}	10^{-7}	2×10^{-4}	4×10^{-8}
	(Sol)	**Bone**	0.1	2×10^{-4}	9×10^{-12}	7×10^{-5}	3×10^{-12}
		Liver	0.2	3×10^{-4}	10^{-11}	9×10^{-5}	4×10^{-12}
		Kidney	0.2	4×10^{-4}	2×10^{-11}	10^{-4}	6×10^{-12}
		Total Body	0.3	6×10^{-4}	3×10^{-11}	2×10^{-4}	9×10^{-12}
		GI (LLI)	—	8×10^{-4}	2×10^{-7}	3×10^{-4}	6×10^{-8}
	(Insol)	**Lung**	—	—	10^{-10}	—	3×10^{-11}
		GI (LLI)	—	8×10^{-4}	10^{-7}	3×10^{-4}	5×10^{-8}

$_{96}Cm^{245}$ (α, β^-, γ)	(Sol)	Bone	0.04	10^{-4}	5×10^{-12}	4×10^{-5}	2×10^{-12}
		Liver	0.5	2×10^{-4}	8×10^{-12}	7×10^{-5}	3×10^{-12}
		Kidney	0.2	2×10^{-4}	9×10^{-12}	7×10^{-5}	3×10^{-12}
		Total Body	0.4	3×10^{-4}	10^{-11}	10^{-4}	5×10^{-12}
		GI (LLI)	—	8×10^{-4}	2×10^{-10}	3×10^{-4}	6×10^{-8}
	(Insol)	Lung	—	—	10^{-10}	—	4×10^{-11}
		GI (LLI)	—	8×10^{-4}	10^{-7}	3×10^{-4}	5×10^{-8}
$_{96}Cm^{246}$ (α)	(Sol)	Bone	0.05	10^{-4}	5×10^{-12}	4×10^{-5}	2×10^{-12}
		Liver	0.5	2×10^{-4}	8×10^{-12}	7×10^{-5}	3×10^{-12}
		Kidney	0.2	2×10^{-4}	9×10^{-12}	7×10^{-5}	3×10^{-12}
		Total Body	0.4	3×10^{-4}	10^{-11}	10^{-4}	5×10^{-12}
		GI (LLI)	—	8×10^{-4}	2×10^{-10}	3×10^{-4}	6×10^{-8}
	(Insol)	Lung	—	—	10^{-10}	—	4×10^{-11}
		GI (LLI)	—	8×10^{-4}	10^{-7}	3×10^{-4}	5×10^{-8}
$\ddagger_{96}Cm^{247}$ ($\alpha, \beta^-, \gamma, e^-$)	(Sol)	Bone	0.04	10^{-4}	5×10^{-12}	4×10^{-5}	2×10^{-12}
		Liver	0.5	2×10^{-4}	9×10^{-12}	7×10^{-5}	3×10^{-12}
		Kidney	0.2	2×10^{-4}	9×10^{-12}	7×10^{-5}	3×10^{-12}
		Total Body	0.4	3×10^{-4}	10^{-11}	10^{-4}	5×10^{-12}
		GI (LLI)	—	6×10^{-4}	10^{-7}	2×10^{-4}	4×10^{-8}
	(Insol)	Lung	—	—	10^{-10}	—	4×10^{-11}
		GI (LLI)	—	6×10^{-4}	10^{-7}	2×10^{-4}	4×10^{-8}
$\ddagger_{96}Cm^{248}$ α (89%) Spontaneous fission (11%)	(Sol)	Bone	5×10^{-3}	10^{-5}	6×10^{-13}	4×10^{-6}	2×10^{-13}
		Liver	0.06	2×10^{-5}	10^{-12}	8×10^{-6}	4×10^{-13}
		Kidney	0.03	3×10^{-5}	10^{-12}	9×10^{-6}	4×10^{-13}
		Total Body	0.04	4×10^{-5}	2×10^{-12}	10^{-5}	6×10^{-13}
		GI (LLI)	—	4×10^{-5}	8×10^{-9}	10^{-5}	3×10^{-9}
	(Insol)	Lung	—	—	10^{-11}	—	4×10^{-12}
		GI (LLI)	—	4×10^{-5}	7×10^{-9}	10^{-5}	2×10^{-9}
$\ddagger_{96}Cm^{249}$ ($\alpha, \beta, \gamma, e^-$)	(Sol)	GI (S)	—	0.06	10^{-5}	0.02	5×10^{-6}
		Bone	1.0	300.0	10^{-5}	100.0	4×10^{-6}
		Total Body	4.0	800.0	3×10^{-5}	300.0	10^{-5}
	(Insol)	GI (S)	—	0.06	10^{-5}	0.02	4×10^{-6}
		Lung	—	—	5×10^{-5}	—	2×10^{-5}

Radionuclide and type of decay		Organ of reference (critical organ in boldface)	Maximum permissible burden in total body $q(\mu c)$	Maximum permissible concentrations			
				For 40 hour week		For 168 hour week	
				$(MPC)_w$ $\mu c/cc$	$(MPC)_a$ $\mu c/cc$	$(MPC)_w$ $\mu c/cc$	$(PMPC)_a$ $\mu c/cc$
$_{97}Bk^{249}$ (α, β⁻, γ)	(Sol)	**GI (LLI)**	—	0.02	4×10^{-6}	6×10^{-3}	10^{-6}
		Bone	0.7	0.07	9×10^{-10}	0.02	3×10^{-10}
		Total Body	5	0.5	7×10^{-9}	0.2	2×10^{-9}
		Lung	—	—	10^{-7}	—	4×10^{-8}
	(Insol)	**GI (LLI)**	—	0.02	3×10^{-6}	6×10^{-3}	10^{-6}
†$_{97}Bk^{250}$ (α, β⁻, γ, e⁻)	(Sol)	**GI (ULI)**	—	6×10^{-3}	10^{-6}	2×10^{-3}	5×10^{-7}
		Bone	0.05	10.0	10^{-7}	4.0	5×10^{-8}
		Total Body	0.3	80.0	10^{-6}	30.0	4×10^{-7}
	(Insol)	**GI (ULI)**	—	6×10^{-3}	10^{-6}	2×10^{-3}	4×10^{-7}
		Lung	—	—	2×10^{-6}	—	8×10^{-7}
$_{98}Cf^{249}$ (α, γ)	(Sol)	**Bone**	0.04	10^{-4}	2×10^{-12}	4×10^{-5}	5×10^{-13}
		GI (LLI)	—	7×10^{-4}	2×10^{-7}	2×10^{-4}	5×10^{-8}
		Total Body	0.3	9×10^{-4}	10^{-11}	3×10^{-4}	4×10^{-12}
	(Insol)	**Lung**	—	—	10^{-10}	—	3×10^{-11}
		GI (LLI)	—	7×10^{-4}	10^{-7}	2×10^{-4}	4×10^{-8}
$_{98}Cf^{250}$ (α)	(Sol)	**Bone**	0.04	4×10^{-4}	5×10^{-12}	10^{-4}	2×10^{-12}
		GI (LLI)	—	7×10^{-4}	2×10^{-7}	3×10^{-4}	6×10^{-8}
		Total Body	0.3	3×10^{-3}	4×10^{-11}	10^{-3}	10^{-11}
	(Insol)	**Lung**	—	—	10^{-10}	—	3×10^{-11}
		GI (LLI)	—	7×10^{-4}	10^{-7}	3×10^{-4}	4×10^{-8}
†$_{98}Cf^{251}$ (α, γ)	(Sol)	**Bone**	0.04	10^{-4}	2×10^{-12}	4×10^{-5}	6×10^{-13}
		GI (LLI)	—	8×10^{-4}	2×10^{-7}	3×10^{-4}	6×10^{-8}
		Total Body	0.3	9×10^{-4}	10^{-11}	3×10^{-4}	4×10^{-12}
	(Insol)	**Lung**	—	—	10^{-10}	—	3×10^{-11}
		GI (LLI)	—	8×10^{-4}	10^{-7}	3×10^{-4}	5×10^{-8}

$_{98}$Cf^{252} (α,γ)	(Sol)	GI (LLI)	—	7×10^{-4}	2×10^{-7}	2×10^{-4}	5×10^{-8}
		Bone	0.04	2×10^{-3}	2×10^{-7}	6×10^{-4}	7×10^{-12}
		Total Body	0.3	0.01	2×10^{-10}	4×10^{-3}	5×10^{-11}
	(Insol)	Lung	—	—	10^{-10}	—	4×10^{-11}
		GI (LLI)	—	7×10^{-4}	10^{-7}	2×10^{-4}	4×10^{-8}
$^{\dagger}_{98}$Cf^{253} ($\alpha,\beta^-,\gamma,e^-$)	(Sol)	GI (LLI)	—	4×10^{-3}	9×10^{-7}	10^{-3}	3×10^{-7}
		Bone	0.04	0.06	8×10^{-10}	0.02	3×10^{-10}
		Total Body	0.3	0.5	6×10^{-9}	0.2	2×10^{-9}
	(Insol)	Lung	—	—	8×10^{-10}	—	3×10^{-10}
		GI (LLI)	—	4×10^{-3}	7×10^{-7}	10^{-3}	3×10^{-7}
$^{\dagger}_{98}$Cf^{254} Spontaneous fission	(Sol)	GI (LLI)	—	4×10^{-6}	8×10^{-10}	10^{-6}	3×10^{-10}
		Bone	7×10^{-4}	4×10^{-4}	5×10^{-12}	10^{-4}	2×10^{-12}
		Total Body	5×10^{-3}	3×10^{-3}	4×10^{-11}	10^{-3}	10^{-11}
	(Insol)	Lung	—	—	5×10^{-12}	—	2×10^{-12}
		GI (LLI)	—	4×10^{-6}	6×10^{-10}	10^{-6}	2×10^{-10}
$^{\dagger}_{100}$Es^{253} ($\alpha,\beta^-,\gamma,e^-$)	(Sol)	GI (LLI)	—	7×10^{-4}	10^{-7}	2×10^{-4}	5×10^{-8}
		Bone	0.04	0.06	8×10^{-10}	0.02	3×10^{-9}
		Total Body	0.3	0.4	5×10^{-9}	0.1	2×10^{-9}
	(Insol)	Lung	—	—	6×10^{-10}	—	2×10^{-10}
		GI (LLI)	—	7×10^{-4}	10^{-7}	2×10^{-4}	4×10^{-8}
$^{\dagger}_{100}$Es^{254}m ($\alpha,\beta^-,\gamma,e^-$)	(Sol)	GI (LLI)	—	5×10^{-4}	10^{-7}	2×10^{-4}	4×10^{-8}
		Bone	0.02	0.4	5×10^{-9}	0.1	2×10^{-9}
		Total Body	0.1	3.0	4×10^{-8}	1.0	10^{-8}
	(Insol)	Lung	—	—	6×10^{-9}	—	2×10^{-9}
		GI (LLI)	—	5×10^{-4}	10^{-7}	2×10^{-4}	3×10^{-8}
$^{\dagger}_{100}$Es^{254} ($\alpha,\beta^-,\gamma,e^-$)	(Sol)	GI (LLI)	—	4×10^{-4}	9×10^{-8}	10^{-4}	3×10^{-8}
		Bone	0.02	10^{-3}	2×10^{-11}	5×10^{-4}	6×10^{-12}
		Total Body	0.2	10^{-2}	10^{-10}	3×10^{-3}	5×10^{-11}
	(Insol)	Lung	—	—	10^{-10}	—	4×10^{-11}
		GI (LLI)	—	4×10^{-4}	7×10^{-8}	10^{-4}	3×10^{-8}

Radionuclide and type of decay		Organ of reference (critical organ in boldface)	Maximum permissible burden in total body $q(\mu c)$	Maximum permissible concentrations			
				For 40 hour week		For 168 hour week	
				$(MPC)_w$ $\mu c/cc$	$(MPC)_a$ $\mu c/cc$	$(MPC)_w$ $\mu c/cc$	$(MPC)_a$ $\mu c/cc$
$\dagger_{99}Es^{255}$ (α, β^-, γ)	(Sol)	GI (LLI)	---	8×10^{-4}	2×10^{-7}	3×10^{-4}	6×10^{-8}
		Bone	0.04	0.04	5×10^{-10}	0.01	2×10^{-10}
		Total Body	0.3	0.3	4×10^{-9}	0.09	10^{-9}
		Lung	---	---	4×10^{-10}	---	10^{-10}
	(Insol)	GI (LLI)	---	8×10^{-4}	10^{-7}	3×10^{-4}	5×10^{-8}
$\dagger_{100}Fm^{254}$ α, γ, e^- (99.9448%) Spontaneous fission (5.52×10^{-2}%)	(Sol)	GI (ULI)	---	4×10^{-3}	8×10^{-7}	10^{-3}	3×10^{-7}
		Bone	0.02	5.0	6×10^{-8}	2.0	2×10^{-8}
		Total Body	0.1	30.0	4×10^{-7}	10.0	10^{-7}
		Lung	---	---	7×10^{-8}	---	2×10^{-8}
	(Insol)	GI (ULI)	---	4×10^{-3}	6×10^{-7}	10^{-3}	2×10^{-7}
$\dagger_{100}Fm^{255}$ (α, γ)	(Sol)	GI (LLI)	---	10^{-3}	2×10^{-7}	3×10^{-4}	7×10^{-8}
		Bone	0.04	1.0	2×10^{-8}	0.4	6×10^{-9}
		Total Body	0.3	9.0	10^{-7}	3.0	4×10^{-8}
		Lung	---	---	10^{-8}	---	4×10^{-9}
	(Insol)	GI (LLI)	---	10^{-3}	2×10^{-7}	3×10^{-4}	6×10^{-8}
$\dagger_{100}Fm^{256}$ Spontaneous fission	(Sol)	GI (ULI)	---	3×10^{-5}	6×10^{-9}	9×10^{-6}	2×10^{-9}
		Bone	8×10^{-4}	0.2	3×10^{-9}	0.07	10^{-9}
		Total Body	5×10^{-3}	1.0	2×10^{-8}	0.5	7×10^{-9}
		Lung	---	---	2×10^{-9}	---	6×10^{-10}
	(Insol)	GI (ULI)	---	3×10^{-5}	5×10^{-9}	9×10^{-6}	2×10^{-9}

When recommending changes in MPD levels, the NCRP suggests that a 5-year transition period be allowed during which the new values may be put into effect.

APPENDIX III

The Standard Man

TABLE 1

MASS AND EFFECTIVE RADIUS OF ORGANS OF THE ADULT HUMAN BODY[a]

	Mass (grams)	Percent of Total Body[b]	Effective Radius (centimeters)
Total body[b]	70,000	100	30
Muscle	30,000	43	30
Skin and subcutaneous tissue[c]	6,100	8.7	0.1
Fat	10,000	14	20
Skeleton:			
Without bone marrow	7,000	10	5
Red marrow[d]	1,500	2.1	—
Yellow marrow	1,500	2.1	—
Blood	5,400	7.7	—
Gastrointestinal tract[b]	2,000	2.9	30
Contents of GI tract:			
Lower large intestine	150	—	5
Stomach	250	—	—
Small intestine	1,100	—	—
Upper large intestine	135	—	—
Liver	1,700	2.4	10
Brain	1,500	2.1	—
Lungs (2)	1,000	1.4	30
Lymphoid tissue	700	1.0	—
Kidneys (2)	300	0.43	7
Heart	300	0.43	—
Spleen	150	0.21	7
Urinary bladder	150	0.21	—

TABLE 1

MASS AND EFFECTIVE RADIUS OF ORGANS OF THE ADULT HUMAN BODY[a]
(Continued)

	Mass (grams)	Percent of Total Body[b]	Effective Radius (centimeters)
Pancreas	70	0.10	—
Salivary glands (6)	50	0.071	—
Testes (2)	40	0.057	—
Spinal cord	30	0.043	—
Eyes (2)	30	0.043	—
Thyroid gland	20	0.029	3
Teeth	20	0.029	—
Prostate gland	20	0.029	—
Adrenal glands or Suprarenal (2)	20	0.029	—
Thymus	10	0.014	—
Miscellaneous (blood vessels, cartilage, nerves, etc.)	390	0.56	—

[a] The reports, "Standard Man", by Hermann Lisco, ANL-4253, Nov. 1948-Feb. 1949, p. 96 and "A Survey Report of the Characteristics of the Standard Man", Sept. 1948, by M. J. Cook were used as the principal sources of reference in the original selection of values given in this table. (Data taken from report of the International Sub-Committee II on Permissible Dose for External Radiations, K. Z. Morgan, Chairman, ICRP/54/4)

[b] Does not include contents of gastrointestinal tract.

[c] The mass of the skin alone is taken as 2000 grams. The minimum thickness of the epidermis is 0.07 mm.

[d] The average depth of the blood forming organs is assumed to be at 5 cm.

TABLE 2

CHEMICAL COMPOSITION

Element	Proportion (percent)	Approximate mass in the body (grams)
Oxygen	65.0	45,500
Carbon	18.0	12,600
Hydrogen	10.0	7,000
Nitrogen	3.0	2,100
Calcium	1.5	1,050
Phosphorus	1.0	700
Sulfur	0.25	175
Potassium	0.2	140
Sodium	0.15	105
Chlorine	0.15	105
Magnesium	0.05	35
Iron	0.006	4
Manganese	0.00003	0.02
Copper	0.0002	0.1
Iodine	0.00004	0.03

Note. The figures for a given organ may differ considerably from these averages for the whole body. For example, the nitrogen content of the dividing cells of the basal layer of skin is probably nearer 6 percent than 3 percent.

TABLE 3

APPLIED PHYSIOLOGY

(Average data for normal activity in a temperate zone)

Water Balance

Daily Water Intake

Water of Oxidation	0.3 liters
In food	0.7 liters
As fluids	1.5 liters
Total	2.5 liters

Calculations of maximum permissible levels for radioactive isotopes in water have been based on the total intake figure of 2.5 liters a day.

TABLE 3
Applied Physiology
(Continued)

Daily Water Output

Sweat	0.5 liters
From Lungs	0.4 liters
In feces	0.1 liters
Urine	1.5 liters
Total	2.5 liters

(The total water content of the body is 50 liters)

Respiration[a]

Area of Respiratory Tract

Respiratory interchange area	50m²
Nonrespiratory area (upper tract and trachea to bronchioles)	20m²
Total	70m²

Respiratory Exchange

Physical activity	Hours per day	Tidal air (liters)	Respiration per minute	Volume per 8 hours (m³)m³(m³)	Volume per day, (m³)m³(m³)
At work	8	1.0	20	10	20
Not at work	16	0.5	20	5	20

Carbon Dioxide Content (by Volume) of Air

Inhaled air (dry, at sea level)	0.03%
Alveolar air	5.5%
Exhaled air	4.0%

Retention of Partculate Matter in the Respiratory Tract

Retention of particulate matter in the lungs depends on many factors, such as the size, shape and density of the particles, the chemical form and whether or not the person is a mouth breather; however, when specific data are lacking it is assumed the distribution is as follows:

Distribution	Readily Soluble Compounds	Other Compounds
Exhaled	25%	25%
Deposited in upper respiratory passages and subsequently swallowed	50%	50%
Deposited in the lungs (lower respiratory passages)	25% (this is taken up into the body)	25%[b]

[a] As stated in U.S. Department of Commerce, Bureau of Standards Handbook 47, 1950, Appendix I.

[b] Of this, half is eliminated from the lungs and swallowed in the first 24 hours making a total of 62½% swallowed. The remaining 12½% is retained in the lungs with a half-life of 120 days, it being assumed that this portion is taken up into body fluids.

TABLE 4

DURATION OF EXPOSURE

Duration of occupational exposure
 The following figures have been adopted in calculations pertaining to occupational exposure:

8 hours per day
40 hours per week
50 weeks per year
50 years continuous work period

Duration of "lifetime" for nonoccupational exposure
 A conventional figure of 70 years has been adopted.

Relative and Absolute Values For Leukocyte Counts in Normal Adults per Cubic Millimeter of Blood[a]

Type of Cell	Percent	Absolute Number Average	Minimum	Maximum
Total leukocytes	—	7,000	5,000	10,000
Myelocytes	0	0	0	0
Juvenile neutrophils	3–5	300	150	400
Segmented neutrophils	54–62	4,000	3,000	5,800
Eosinophils	1–3	200	50	250
Basophils	0–0.75	25	15	50
Lymphocytes	25–33	2,100	1,500	3,000
Monocytes	3–7	375	285	500

Blood Count
Normal Range of Values—Human Adults[a]

	Male	Female
Red Cell Count (millions per cu. mm)	5.4 ± 0.8	4.8 ± 0.6
Hemoglobin (gm. per 100 cc.)	16.0 ± 2.0	14.0 ± 2.0
Hematocrite or Vol. packed R.B.C. (cc. per 100 cc.)	47.0 ± 7.0	42.0 ± 5.0

[a] M. M. Wintrobe, Clinical Hematology, Lea & Febiger, Philadelphia, Third Edition.

Index

484